Air Pollution: Sources, Impacts and Controls

———————————————

Air Pollution: Sources, Impacts and Controls

Edited by

Pallavi Saxena

Department of Environmental Sciences, Hindu College,
University of Delhi, Delhi, India

and

Vaishali Naik

Atmospheric Chemistry & Climate Group, National Oceanic and Atmospheric
Administration's Geophysical Fluid Dynamics Laboratory (NOAA GFDL),
Princeton, NJ, USA

CABI is a trading name of CAB International

CABI	CABI
Nosworthy Way	745 Atlantic Avenue
Wallingford	8th Floor
Oxfordshire OX10 8DE	Boston, MA 02111
UK	USA
Tel: +44 (0)1491 832111	Tel: +1 (617)682-9015
Fax: +44 (0)1491 833508	E-mail: cabi-nao@cabi.org
E-mail: info@cabi.org	
Website: www.cabi.org	

A catalogue record for this book is available from the British Library, London, UK.

Library of Congress Cataloging-in-Publication Data

Names: Saxena, Pallavi, editor. | Naik, Vaishali, editor.
Title: Air pollution : sources, impacts and controls / editors, Pallavi
    Saxena and Vaishali Naik.
Description: Boston : CAB International, [2019] | Includes bibliographical
    references and index.
Identifiers: LCCN 2018027231 (print) | LCCN 2018029242 (ebook) | ISBN
    9781786393906 (ePDF) | ISBN 9781786393913 (ePub) | ISBN
    9781786393890 (pbk: alk. paper)
Subjects: LCSH: Air--Pollution. | Air quality management. |
    Air--Pollution--Health aspects.
Classification: LCC TD883 (ebook) | LCC TD883 .A4783 2019 (print) |
    DDC 628.5/3--dc23
LC record available at https://lccn.loc.gov/2018027231

ISBN-13: 978 1 78639 389 0 (pbk)
          978 1 78639 390 6 (ePDF)
          978 1 78639 391 3 (ePub)

Commissioning editor: Ward Cooper
Editorial assistant: Emma McCann
Production editor: Tim Kapp

Typeset by SPi, Pondicherry, India
Printed and bound in the UK by Severn, Gloucester

Contents

List of Contributors

Amit Awasthi, Assistant Professor, Department of Physics, University of Petroleum and Energy Studies, Bidholi, Dehradun, India. E-mail: awasthitiet@gmail.com

Sangeeta Bansal, Research Scholar, University School of Environment Management, Guru Gobind Singh Indraprastha University, Delhi, India. E-mail: sangi.725@gmail.com

Monojit Chakraborty, Postdoctoral Fellow, Department of Environmental System Science, ETH Zurich, Tannenstrasse 1, Switzerland. E-mail: monojit.chakraborty@gmail.com

Naveen Chandra, Postdoctoral Fellow, Research and Development Center for Global Change, JAMSTEC, Yokohama 2360001, Japan. E-mail: nav.phy09@gmail.com

Arti Choudhary, Postdoctoral Fellow, Division of Environmental Sciences, Central Road Research Institute, New Delhi, India. E-mail: choudharyarti12@gmail.com

Manisha Gaur, Research scholar, Division of Environmental Sciences, Central Road Research Institute, New Delhi, India. E-mail: manisha.gaur@gmail.com

Vineet Goswami, Postdoctoral Fellow, AIRIE Program, Department of Geosciences, Warner College of Natural Resources, Colorado State University, Fort Collins, CO 80523-1482, USA. E-mail: vineet.goswami@colostate.edu

Tarun Gupta, Professor, Atmospheric Particle Technology Lab (APTL) at CESE and Department of Civil Engineering at Indian Institute of Technology Kanpur, Kanpur – 208016, India. E-mail: tarun@iitk.ac.in

Harpreet Kaur, Research Scholar, School of Environmental Sciences, Jawaharlal Nehru University, New Delhi, India. E-mail: hkaur.ecology@gmail.com

Priyanka Kulshreshtha, Assistant Professor, Department of Resource Management and Design Application, Lady Irwin College, University of Delhi, Delhi, India. E-mail: priya.kulsh@gmail.com

Amit Kumar, Research Scholar, Department of Agricultural Soil Science, Georg-August University of Göttingen, Büsgenweg 2, Göttingen, Germany. E-mail: aksoni089@gmail.com

Ruchi Kumari, Research Scholar, Institute of Neurology, University College London, United Kingdom. E-mail: ruchikaush@gmail.com

Chinmay Mallik, Postdoctoral Fellow, Max Planck Institute for Chemistry, Hahn Meitner Weg 1, 55128 Mainz, Germany. E-mail: mallik.chinmay@gmail.com

Renu Masiwal, Junior Research Fellow, Environment and Biomedical Metrology Department, CSIR-National Physical Laboratory, New Delhi, India. E-mail: tuliprenu@gmail.com

Vandana Maurya, Research scholar, School of Social Sciences, Jawaharlal Nehru University, New Delhi, India. E-mail: maurya.vandana09@gmail.com

Neha Mishra, Assistant Professor, Department of Environmental Sciences, Indraprastha College, University of Delhi, Delhi, India. E-mail: nehamishra2706@gmail.com

Vaishali Naik, Physical Scientist, DOC/NOAA/OAR/Geophysical Fluid Dynamics Laboratory, Atmospheric Chemistry and Climate Group, 201 Forrestal Road, Princeton, NJ 08540, USA. E-mail: vaishali.naik@noaa.gov

Prashant Rajput, CSIR-Senior Research Associate, Department of Civil Engineering at Indian Institute of Technology Kanpur, Kanpur – 208016, India, and Research Fellow, Global Centre for Clean Air Research (GCARE), Department of Civil and Environmental Engineering, University of Surrey, Guildford, GU2 7XH, UK. E-mail: prajput.prl@gmail.com

Pallavi Saxena, Assistant Professor, Department of Environmental Sciences, Hindu College, University of Delhi, Delhi, India. E-mail: pallavienvironment@gmail.com.

Anuradha Shukla, Chief Scientist, Division of Environmental Sciences, Central Road Research Institute, New Delhi, India. E-mail: anuradha.crri@gmail.com

Gyanesh Kumar Singh, Junior Research Fellow, Atmospheric Particle Technology Lab (APTL) at CESE and Department of Civil Engineering at Indian Institute of Technology Kanpur, Kanpur – 208016, India. E-mail: gyanesh@iitk.ac.in

Ravi Prakash Singh, Research Scientist, National Remote Sensing Centre, Hyderabad, Telangana, India. E-mail: singhravi004@gmail.com

Ruchi Singh, Assistant Professor, Department of Environmental Sciences, College of Vocational Studies, University of Delhi, Delhi, India. E-mail: ruchisingh1907@gmail.com

Saumya Singh, Postdoctoral Fellow, Department of Chemistry, University of Toronto, 80 St George Street, Toronto, Ontario M5S 3H6, Canada. E-mail: saumya.8singh@gmail.com.

Saurabh Sonwani, Research scholar, School of Environmental Sciences, Jawaharlal Nehru University, New Delhi, India. E-mail: sonwani.s19@gmail.com

Shani Tiwari, Postdoctoral Fellow, Graduate School of Environmental Studies, Nagoya University, Nagoya, Japan. E-mail: pshanitiwari@gmail.com

Introduction

Pallavi Saxena[1]* and Vaishali Naik[2]
*[1]University of Delhi, Delhi, India; [2]Geophysical Fluid
Dynamics Laboratory, Princeton, NJ, USA*

Air pollution is a major problem all over the
world in both developed and developing countries.
Rapid increase in population and demand for energy have resulted in emission of toxic air pollutants that affect the surrounding environment
as well as human health. According to the World
Health Organization (WHO), about 4 million
deaths, along with numerous cases of respiratory illness, result annually from air pollution in
developing countries (WHO, 2015, 2016). The
major air pollutants that are responsible for
deteriorating air quality are oxides of nitrogen
(NO_x), sulfur dioxide (SO_2), carbon monoxide
(CO), particulate matter (PM) and volatile organic compounds (VOCs) and ozone (O_3). The
major source of air pollution, especially in urban
cities, is the transport sector. It has been reported
that in developing countries most of the air pollution (approx. 70–80%) is caused by vehicular
emissions particularly from larger numbers of
older vehicles with low vehicle maintenance,
low fuel quality and improper road infrastructure (Wang *et al.*, 2010; Bigazzi and Figliozzi,
2014). Air pollution was considered to be a local
problem with large numbers of point sources but
due to the application of tall stacks, biomass
burning and long range transport of pollutants,
it has become a regional to global problem. Most
of the developing countries, including India,

have experienced a drastic decline in air quality
due to rapid economic growth over the last three
decades. There has been an increasing trend towards urbanization in developing countries. In
fact, it is a significant factor for the increase in
air pollution in many cities in Asia, Africa, the
Near East and Latin America (Ashmore, 2005).
In the year 1960, less than 22% of the developing
world's population was urban but by 1990 it had
increased to 34% (Grant, 2012), and by 2020,
global urban population is projected to increase
by 50% (World Bank, 2009; UN, 2017).

The effects of air pollution on living systems
like plants, animals and human beings and on other
materials is worse. It may affect the biochemical
and physiological processes of plants and ultimately lead to yield loss (Heck *et al.*, 1988).
A number of scientific studies have linked air
pollution to a variety of health problems including: (i) aggravation of respiratory and cardiovascular disease; (ii) decreased lung function;
(iii) increased frequency and severity of respiratory symptoms such as difficulty in breathing and
coughing; (iv) increased susceptibility to respiratory infections; (v) effects on the nervous system,
including the brain, such as IQ loss and impacts
on learning, memory and behaviour; (vi) cancer; and (vii) premature death. Some sensitive
individuals appear to be at greater risk for air

---

* Corresponding author: pallavienvironment@gmail.com

pollution-related health effects, for example, those with pre-existing heart and lung diseases (e.g. heart failure/ischaemic heart disease, asthma, emphysema and chronic bronchitis), diabetics, older adults and children (Sadanaga *et al.*, 2003). Air pollution also damages our environment. Ground-level ozone (O_3) can damage vegetation, adversely impacting the growth of plants and trees. These impacts can reduce the ability of plants to uptake CO_2 from the atmosphere and indirectly affect entire ecosystems. Visibility is reduced by particles in the air that scatter and absorb light. Pollution in the form of acids and acid-forming compounds (such as sulfur dioxide (SO_2) and oxides of nitrogen (NO_x)) can deposit from the atmosphere to the Earth's surface. This acid deposition can be either dry or wet (Pant and Harrison, 2013). Wet deposition is more commonly known as acid rain. Acid rain can occur anywhere and, in some areas, rain can be 100 times more acidic than natural precipitation. Acid deposition can be a very serious regional problem, particularly in areas downwind from high SO_2 and NO_x-emitting sources (e.g. coal burning power plants, smelters and factories) (Davis, 2008). Acid deposition can have many harmful ecological effects in both land and water systems. While acid deposition can damage tree foliage directly, it more commonly stresses trees by changing the chemical and physical characteristics of the soil (Liu and Diamond, 2005). In lakes, acid deposition can kill fish and other aquatic life. Air pollution can also impact the Earth's climate. Different types of pollutants affect the climate in different ways, depending on their specific properties and the amount of time they stay in the atmosphere (Karanasiou *et al.*, 2014; Fiore *et al.*, 2015). Any pollutant that affects the Earth's energy balance is known as a 'climate forcer' (IPCC, 2013). Climate forcers, such as greenhouse gases (GHGs), perturb the Earth's radiation balance by absorbing or emitting longwave terrestrial radiation and warm the climate. Major GHGs include carbon dioxide (CO_2), nitrous oxide (N_2O), methane (CH_4) and tropospheric ozone (O_3). Some climate forcers either absorb or reflect the incoming or reflected shortwave solar radiation leading to warming or cooling, respectively. Aerosol particles (solid or liquid droplets suspended in the air) modify the incoming or reflected shortwave radiation.

Black carbon (BC), a component of particle pollution, directly absorbs incoming and reflected solar radiation and reduces the reflection of sunlight off snow and ice (SAFAR, 2013), contributing to increased absorption of energy at the Earth's surface and warming of the atmosphere. Recent studies suggest that BC may be having a significant impact on the Earth's climate (Novakov and Rosen, 2013). Other types of particles – particularly sulfates (SO_4^{2-}), nitrates (NO_3^-) and organic carbon (OC) – are largely reflective and therefore have a net cooling impact on the atmosphere. Particles can also have important indirect effects on climate through impacts on clouds and precipitation. The longer a pollutant stays in the atmosphere, the longer the effect associated with that pollutant will persist. Some climate forcing pollutants stay in the atmosphere for decades or centuries after they are emitted, meaning today's emissions will affect the climate far into the future (Tao *et al.*, 2012). These pollutants, like CO_2, tend to accumulate in the atmosphere so their net warming impact continues over time. Other climate forcers, such as ozone and BC, remain in the atmosphere for shorter periods of time, so reducing emissions of these pollutants may have beneficial impacts on climate in the near term. These short-lived climate forcers originate from a variety of sources, including the burning of fossil fuels and biomass, wildfires and industrial processes. Short-lived climate-forcing pollutants (SLCPs) and their chemical precursors can be transported long distances and may produce particularly harmful warming effects in sensitive regions such as the Arctic (Von Schneidemesser and Monks, 2013). In the United States, the Environmental Protection Agency (USEPA) has established air quality standards to protect public health, including the health of 'sensitive' populations such as children, older adults and people with asthma. USEPA also sets limits to protect public welfare. This includes protecting ecosystems, such as plants and animals, from harm, as well as protecting against decreased visibility and damage to crops, vegetation and buildings (Johnson and Graham, 2005). USEPA has set National Ambient Air Quality Standards (NAAQS) for six principal air pollutants: nitrogen oxides (expressed as NO_2), ozone, sulfur dioxide, PM, carbon monoxide (CO) and lead (Pb). Four of these pollutants (CO, Pb, NO and SO_2) are emitted directly from a variety of sources. Tropospheric ozone is not directly emitted, but is formed when nitrogen oxides (NO_x) react with carbon monoxide (CO) and VOCs in the presence of sunlight. PM is both emitted directly and formed in the atmosphere through gas- and aqueous phase chemical reactions. Directly emitted components of PM include dust, sea salt, BC and OC from incomplete combustion, while secondary components include sulfates, nitrates and organics.

Summary of Chapters

On the basis of discussion of a number of issues, it has been noticed that the problem of air pollution especially in Asian cities still persists because of the increasing number of vehicles. This book covers the important topics related to air pollution sources, their impacts and control strategies. Chapter 1 provides a comprehensive overview and classification of the origin of the most important air pollutants, including criteria, toxic and other non-criteria pollutants. This chapter particularly highlights the anthropogenic sources of major air pollutants like NO_x, SO_x, CO, PM and tropospheric ozone along with their trends and source apportionment profiles. Chapter 2 deals with the role of important biogenic volatile organic compounds (BVOCs) in atmospheric chemistry, such as ozone formation, secondary organic aerosol formation, etc. and also focuses on soil as a contributor of these volatiles. This chapter serves as an important part of this book because information on BVOC is still limited, particularly in highlighting the overall impact of an increase in BVOC emissions which will affect their ecological, physiological and environmental roles. Chapter 3 summarizes the transport process of air pollutants from different regions, and highlights some of the important dynamical processes in the atmosphere and their impacts on the burden of pollutants in the troposphere. Chapter 4 describes an overall view of different measurement techniques/methods involved in various air pollutants, particularly criteria air pollutants. This chapter also highlights the feasibility and cost-effectiveness of techniques and, importantly, covers most of the pollutant measurement methodologies, with the support of case studies. Chapter 5 explains

the philosophical and practical discussion and its implications vis-à-vis mathematical air quality modelling. An evaluation of pertinent mathematical modelling techniques – Box, Gaussian, Eulerian, Lagrangian and Particle modelling approaches – is provided. Chapter 6 summarizes the tools used for assessment of air quality, popularly known as air quality indices. This chapter gives an overview of various case studies of air quality indices and their implementation in assessment of levels of different air pollutants which act as indicators for human health. These indices can identify the impact of air pollutants on plant as well as human health. Chapter 7 focuses on the effects of air pollution on the environment and on the economy. This chapter provides a basic understanding of how air pollution affects the global economic growth and environment. It also puts forth ideas to stabilize ecological and economic conditions and global equilibrium that can be designed to fulfil the 'needs' of each person so that all have an equal opportunity to realize their individual human potential. Chapter 8 emphasizes the adverse health effects of major air pollutants (both ambient and indoors) on the people living in urban areas, especially young children, pregnant women, the elderly and people with respiratory and cardiovascular conditions. This chapter particularly highlights $PM_{2.5}$ and PM_{10} pollution as a number of cases have been registered of patients suffering from respiratory illness and nowadays it is a serious agenda to take into account. Chapter 9 provides a brief introduction to megacity air pollution. It also brings together recent comprehensive reviews from particular megacities of the developing world. Megacities in developing countries are suffering from the highest PM loads that are associated with increased mortality rates. This chapter highlights the use of cleaner energies such as natural gas, efficient fuel burning and increasing reliance on renewable sources of energy (solar, hydro, wind and geothermal) for controlling air pollution without limiting economic growth. Chapter 10 emphasizes cost-effective technologies to curb air pollution. It highlights that the cost-effectiveness of technologies varies with the level of operation, nature of pollutants and achievable regulatory compliance. It explains some emerging cost-effective and economically viable technologies that are in progress and some in the developmental stage, such as biofiltration and carbon capture technologies, which have shown the potential to curb atmospheric air pollutants. This chapter ends with a discussion on technologies still in the infancy stage and with the limitation of efficiency, requiring further research and development to commercially exploit them. Chapter 11 highlights the importance of laws and policies and their role in reducing air pollution. It critically analyses various regulatory measures and policies in the most polluted countries of the world. It also explains what has been achieved through treaties and conventions organized by the United Nations such as the Earth summit, Kyoto Protocol, Montreal Protocol, etc. This is also a significant chapter as it focuses on the developmental agenda for the future, and reflects the links between socioeconomic and environmental sustainability, protecting and reinforcing the environmental pillar. This book ends with Chapter 12, which highlights the various sources of air contaminants and their chemistry in the atmosphere. Further, various health and environmental hazards of air pollution have also been highlighted and discussed. As a final note, the chapter suggests that adequate policies should be made and efficiently applied to alleviate and constrain air pollution emissions in order to mitigate their direct and indirect effects. Reduction and eventual elimination of the usage of fossil fuels to meet various energy requirements and a drive towards more efficient and cleaner energy resources are the primary steps.

Conclusion

This book, based on several studies, concludes that air pollutant concentrations are still increasing in developing countries even after the adoption of several new policies and control technologies. The source apportionment studies of many air pollutants, most likely particulate matter, are still facing the challenge to minimize their levels so that they can come under permissible limits. Very much less attention is given to biogenic sources than anthropogenic ones, though in the case of volatile organic compounds, ozone and secondary aerosol production, they contribute largely to the air quality budget. Impact studies related to air pollution and human health also indicate a serious problem, which needs to be addressed because the death rate due to respiratory illness is increasing day by day. At this sensitive stage, there is an urgent call for stringent control policies or methods to curb air pollution to help save the environment. Thus, the different chapters in this book explain sources of air pollution, both anthropogenic and biogenic; impact studies on human health; the environment and economy; and control methods and policies to curb air pollution.

References

Ashmore, M.R. (2005) Assessing the future global impacts of ozone on vegetation. *Plant, Cell and Environment* 28(8), 949–964.

Bigazzi, A.Y. and Figliozzi, M.A. (2014) Review of urban bicyclists intake and uptake of traffic-related air pollution. *Transport Reviews* 34(2), 221–245.

Davis, L.W. (2008) The effect of driving restrictions on air quality in Mexico City. *Journal of Political Economy* 116(1), 38–81.

Fiore, A.M., Naik, V. and Leibensperger, E.M. (2015) Air quality and climate connections. *Journal of the Air & Waste Management Association* 65(6), 645–685.

Grant, U. (2012) Urbanization and the Employment Opportunities of Youth in Developing Countries. UNESCO Report. Available at: http://unesdoc.unesco.org/images/0021/002178/217879e.pdf (accessed 2 July 2018).

Heck, W.W., Taylor, O.C. and Tingey, D.T. (eds) (1988) *Assessment of Crop Loss From Air Pollutants.* Springer, Dordrecht, The Netherlands.

IPCC (2013) *Climate Change 2013: The Physical Science Basis. Working Group I Contribution to the Fifth Assessment Report of the Intergovernmental Panel on Climate Change*, ed. Stocker, T.F., Qin, D., Plattner, G.-K., Tignor, M.M.B., Allen, S.K. *et al.* Cambridge University Press, Cambridge and New York. DOI: 10.1017/CBO9781107415324.

Johnson, P.R.S. and Graham, J.J. (2005) Fine particulate matter national ambient air quality standards: public health impact on populations in the Northeastern United States. *Environmental Health Perspectives* 113(9), 1140–1147.

Karanasiou, A., Viana, M., Querol, X., Moreno, T. and de Leeuw, F. (2014) Assessment of personal exposure to particulate air pollution during commuting in European cities – Recommendations and policy implications. *Science of the Total Environment* 490, 785–797.

Liu, J.G. and Diamond, J. (2005) China's environment in a globalizing world. *Nature* 435, 1179–1186.

Novakov, T. and Rosen, H. (2013) The black carbon story: early history and new perspectives. *Ambio* 42(7), 840–851.

Pant, P. and Harrison, R.M. (2013) Estimation of the contribution of road traffic emissions to particulate matter concentrations from field measurements: a review. *Atmospheric Environment* 77, 78–97.

Sadanaga, Y., Matsumoto, J. and Kajii, Y. (2003) Photochemical reactions in the urban air: recent understandings of radical chemistry. *Journal of Photochemistry and Photobiology C: Photochemistry Reviews* 4(1), 85–104.

SAFAR (2013) System of Air Pollution Forecasting and Research (SAFAR). Indian Institute of Tropical Meteorology, Pune, India. Available at: http://safar.tropmet.res.in (accessed 2 July 2018).

Tao, W.-K., Chen, J.-P., Li, Z., Wang, C. and Zhang, C. (2012) Impact of aerosols on convective clouds and precipitation. *Reviews of Geophysics* 50(2), 1–62.

UN (2017) World Population Prospects: Key Findings & Advance Tables, 2017 Revision. Working Paper No. ESA/P/WP/248. United Nations, Department of Economic and Social Affairs, Population Division. Available at: https://esa.un.org/unpd/wpp/Publications/Files/WPP2017_KeyFindings.pdf (accessed 2 July 2018).

Von Schneidemesser, E. and Monks, P.S. (2013) Air quality and climate – synergies and trade-offs. *Environmental Science: Processes & Impacts* 15(7), 1315–1325.

Wang, S., Zhao, M., Xing, J., Wu, Y., Zhou, Y. *et al.* (2010) Quantifying the air pollutants emission reduction during the 2008 Olympic Games in Beijing. *Environmental Science & Technology* 44(7), 2490–2496.

WHO (2015) World Health Statistics 2015. World Health Organization, Geneva, Switzerland. Available at: http://apps.who.int/iris/bitstream/10665/170250/1/9789240694439_eng.pdf (accessed 2 July 2018).

WHO (2016) World Health Statistics 2016. World Health Organization, Geneva, Switzerland. Available at: http://apps.who.int/iris/bitstream/10665/206498/1/9789241565264_eng.pdf (accessed 2 July 2018).

World Bank (2009) The World Bank Annual Report 2009: Year in Review. Available at: http://siteresources.worldbank.org/EXTAR2009/Resources/6223977-1252950831873/AR09_Complete.pdf (accessed 2 July 2018).

1 Anthropogenic Sources of Air Pollution

Chinmay Mallik*
Max Planck Institute for Chemistry, Mainz, Germany

Abstract

While 'Air Pollution' is an alarming term for environmentalists, policy makers, governments and common people, its 'anthropogenic sources' make it a formidable hazard to deal with. The universal dependence of humanity on fossil fuels, which has changed the way human beings live and breathe on this planet, has been a catastrophe for human health and the Earth's environment. Although there exists a myriad of anthropogenic air pollutants, their human-made sources, dominated by the combustion of fossil fuels, can be conveniently grouped into a few major sectors (energy, industry, agriculture and waste) for the purpose of comparison across various temporal and spatial domains, and the formulation of strategies to monitor and control their emissions. While hundreds of anthropogenic air pollutants are toxic, there exist six ubiquitous air pollutants which are regulated by the governments in most countries due to their significant harmful impacts on human health and the environment. The association of anthropogenic air pollutants and their emission sources is documented in the form of emission inventories, spanning local, regional and global domains. This chapter provides an overview of the types of air pollutants, their primary sources, and the estimate of their global emission strengths as represented in emission inventories.

1.1 Introduction

The word 'anthropogenic' refers to anything produced due to human activities. Thus, any biological, chemical, radioactive or physical substance that is emitted into the air as a result of human activities, and results in concentrations higher than that which would be present in the natural atmosphere, leading to adverse impacts on human, animals, vegetation and other biotic as well as abiotic components of the Earth and its atmosphere, would classify as an 'anthropogenic air pollutant'. Specifically, anthropogenic emissions are those that are produced as a result of human activities but not necessarily those produced biologically from humans, e.g. emissions of ammonia (NH_3) from human breath/sweat is a natural source of NH_3. The clearest indication of anthropogenic influence is visible in the growing difference in carbon dioxide (CO_2) emissions and uptake, leading to a dramatic increase in measured atmospheric CO_2 levels, now crossing 400 ppm. The impacts of anthropogenic air pollution are clearly manifested in the form of global warming, premature mortality in the millions due to ozone (O_3) and particulate matter (PM), extreme precipitation and droughts leading to crop loss, intensification of cyclones, fog and haze jeopardizing daily lives, impacts on ecosystems, loss of flora and fauna, increased

---

* E-mail: mallik.chinmay@gmail.com

© CAB International 2019. *Air Pollution: Sources, Impacts and Controls*
(eds P. Saxena and V. Naik)

ecosystem carbon storage, and changes in soil nitrogen and phosphate, to name but a few (IPCC, 2013).

Air pollution is not only an emission problem but also a chemical problem. Hence, the sources of air pollution are both primary and secondary. 'Primary' pollutants are those that are directly emitted into the atmosphere from various emission sources, e.g. sulfur dioxide (SO_2) emissions from coal burning in power plants, and nitrogen oxides (NO_x: $NO+NO_2$) from the transport sector. The concentration of primary pollutants is likely to be greater near their emission sources, but depending on their chemical lifetime and meteorological conditions, they can be transported over long distances (thousands of kilometres) in a short time (days). 'Secondary' pollutants are produced in the atmosphere due to various physical and chemical processes involving atmospheric constituents

(gases and particles) including primary pollutants, e.g. formation of ozone (O_3) from NO_x and hydrocarbons. High concentrations of secondary pollutants can occur even in places far removed from large emission sources, e.g. O_3 formation can take place in rural and even pristine remote areas due to photochemical reactions among O_3 precursors, as they are transported away from their emission sources (Mallik *et al.*, 2013). Contemporary sources and impacts of anthropogenic air pollutants span all environment regimes, spanning urban, semi-urban and rural areas, farmlands, forests, lakes, mountains and even oceans (Fig. 1.1).

1.2 Anthropogenic Air Pollutants

Although a variety of chemical substances emitted into the atmosphere due to human activities

Fig. 1.1. The illustration shows the sources of various criteria and toxic air pollutants, as well as sources of greenhouse gases, encountered by a healthy farmer during his migration to a city. The various emission sources depicted are livestock, agriculture, industries, road dust, vehicular emissions and waste. As he experiences high levels of air pollution, his health deteriorates, cutting his journey short, and he is in a dilemma as to which path to tread next.

exert significant harmful effects on human health and the environment, the United States Environment Protection Agency (USEPA) identifies six ubiquitous air pollutants that need to be regularly monitored and regulated, hence referred to as 'criteria pollutants' (USEPA, 1990). These criteria pollutants are ground level O_3, carbon monoxide (CO), nitrogen dioxide (NO_2), lead (Pb), particulate matter (PM) and SO_2. Regulation of criteria pollutants mandates development of air quality standards, which vary from country to country. The 'non-criteria pollutants' do not have an air quality standard assigned to them, and include the entire range of air contaminants including toxic and hazardous substances, most of which are volatile organic compounds (VOCs). Gases such as carbon dioxide (CO_2) and methane (CH_4), co-emitted with various criteria pollutants, are studied under the realm of 'greenhouse gases' because of their significant control on the radiative balance of the Earth's atmosphere and, hence, its climate. Many short-lived air pollutants also influence the radiative balance of the Earth by absorbing terrestrial infrared radiation (e.g. tropospheric O_3), absorbing (e.g. black carbon) or scattering (sulfate aerosols) solar radiation or by interacting with clouds. These influences on radiation induce changes in the Earth's climate. Therefore, such radiatively active pollutants are also known as short-lived climate forcers (IPCC, 2013).

1.2.1 Toxic pollutants

Toxic air pollutants, popularly known as 'air toxics' or as 'hazardous air pollutants (HAPs)', are known or suspected to cause serious health effects including cancer, reproductive and birth defects, or to cause adverse environmental effects (EPA, n.d.a). The USEPA identifies about 187 air toxics, emission sources for some of which are given in Table 1.1. The sources of air toxics can be classified as major and area sources. 'Major source' refers to a singular or a group of stationary sources juxtaposed in an area of common control, and can potentially emit 10 tons of a HAP or 25 tons of a combination of HAPs annually (USEPA, 1992). A 'stationary source' implies any standing structure e.g. building, stack, or set-up which emits or may emit an air pollutant. An 'area source' refers to a stationary source or an aggregate of stationary sources of HAP that

do not constitute a major source but are still a threat to human health and the environment.

1.2.2 Criteria air pollutants

1.2.2.1 Sulfur dioxide (SO_2)

Sulfur is emitted into the atmosphere in various states of oxidation. Despite its ubiquity in all spheres of the globe, the most recognizable form of sulfur in the atmosphere is SO_2 as it is a precursor for sulfate aerosol – a key component of PM. The atmospheric sources of SO_2 are natural as well as anthropogenic but, over the years, the anthropogenic component has increased overwhelmingly. The primary source of SO_2 is combustion of coal and oil (which contain 1–2 % sulfur by weight) with smaller contributions from other industrial activities such as metal smelting and manufacture of H_2SO_4. The global SO_2 emissions were of the order of 115 Gg-SO_2 during 2005 with China contributing 32 Gg-SO_2 (~28%; Smith et al., 2011). Due to its profound impacts on human health, and aquatic and terrestrial ecosystems including acid rain, SO_2 has been regulated in power plants and transport sectors in various developed countries employing desulfurization and end-of-pipe abatement techniques. However, over the Asian region, anthropogenic SO_2 emissions are not well controlled and are projected to increase under current regulations (Wang et al., 2014). SO_2 can be toxic at high levels causing reduced respiration, inflammation of the airways, and lung damage (ATSDR, 1998). Plants exposed to high levels of SO_2 incur acute foliar injury, where it can be oxidized to sulfite, which is very toxic and can interfere with photosynthesis and energy metabolism. The SO_2 emissions over India were estimated at 8.8 Tg for 2010 with sector-wise contribution of 66% and 32% from power and industries, and fuel-wise contribution of 76% and 19% from coal and oil, respectively, to the national SO_2 emissions (Lu et al., 2011). High SO_2 levels have been detected in ambient air of megacities like Beijing (60 ppbv in winter; Sun et al., 2004) and Kolkata (6.4 ppbv in winter; Mallik et al., 2014). For India, the national ambient air-quality standard requires annual average SO_2 to be less than 19 and 7.6 ppbv for industrial/residential and sensitive areas, respectively (CPCB, 2013; EPA, n.d.c).

Table 1.1. Major sources and health effects of hazardous/toxic air pollutants. (Adapted from EPA, n.d.b)

Air toxic	Major source	Health effect	References
Acetaldehyde	Intermediate in synthesis of chemicals, preservatives, solvent in rubber, tanning and paper industries, silvering of mirrors, incomplete wood combustion	Irritation (eyes, skin, throat), probable human carcinogen	USEPA (1987); HSDB (1993)
Acrolein	Manufacture of acrylic acid, modacrylic fibers	Mucous membrane irritation, probable human carcinogen	IRIS (2003)
Acrylonitrile	Automobile exhausts, manufacturing facilities, solvents for spinning fibers, lithium batteries	Irritation of mucous membranes, headaches, numbness, tremors	USEPA (1985)
Arsenic compounds (arsine)	Volcanoes, weathering of arsenic mineral ores, diet, fish, soil, water, air, metal smelters, industrial processes (semiconductor industries)	Nausea, diarrhoea, abdominal pain, nervous system disorders, inorganic arsenic is human carcinogen, (arsine is toxic)	USEPA (1984)
Benzene	Gasoline, oil and coal burning, vehicle exhaust, industrial solvents, tobacco smoke	Dizziness, drowsiness, headaches, irritation (skin, eye, respiratory tract), unconsciousness at high levels	ATSDR (2007)
Beryllium compounds	Mining and processing areas, oil and coal burning, tobacco smoke	Inflammation of lungs, chronic beryllium disease, probable human carcinogen	IRIS (1998a)
1-3-butadiene	Motor vehicle exhaust	Irritation (eyes, throat, nasal passages, lungs)	IRIS (2002)
Cadmium compounds	Oil and coal burning, incineration of waste	Pulmonary irritation, kidney disease, probable human carcinogen	IRIS (1989)
Carbon tetrachloride	Accidental releases in production, use and releases during disposal, e.g. in landfills	Headache, nausea, vomiting, liver and kidney damage, probable human carcinogen	IRIS (2010)
Carbonyl sulfide	Combustion, commercial processes, natural sources	Narcotic effects, irritation of skin and eyes	IRIS (1991); Mallik et al. (2016)
Chloroform	Formation during the chlorination of drinking water, treatment of wastewater and swimming pools, landfills, waste sites, pulp and paper mills	Nervous system depression, hepatitis and jaundice, probable human carcinogen	IRIS (2001)
Chromium compounds	Ferrochrome production, ore refining, refractory, cement, chemical and automobile industries	Respiratory effects, e.g. shortness of breath, coughing, bronchitis, lung cancer	IRIS (1998b)
Coke oven emissions	Steel, graphite, aluminium, power and construction industries	Conjunctivitis, dermatitis, respiratory and digestive system lesions, cancer	IRIS (1998b)
1,3-dichloropropene	Soil fumigants	Mucous membrane irritation, chest pain, breathing problems, probable human carcinogen	IRIS (2000)

Continued

Table 1.1. Continued.

Air toxic	Major source	Health effect	References
Ethylene compounds:	Additive to leaded gasoline	Probable human carcinogens, impact on reproductive functions	IRIS (1993, 2004); ATSDR (1990)
dibromide	fumigation		
dichloride	chemical industry, vinyl chloride production, rubber and plastic industry	damage to nervous and respiratory systems, liver, kidney	
oxide	textile, detergent, ethane foam, solvents, antifreeze, adhesives, sterilants	irritation to eye and skin, nervous system effects	
Formaldehyde	Power plants, chemical industry, incinerators, automobile exhaust, wood products, e.g. building materials and home furnishings, concrete and plaster additives	Eye, nose throat irritation, respiratory effects, probable human carcinogen	IRIS (1990)
Hydrazine	Chemical blowing agents/ pneumatogens, pesticides, boiler water treatment, textile dyes, photography and pharmaceutical intermediates, tobacco smoke	Eye, nose throat irritation, headache, dizziness, seizures, nausea, damage to nervous system, liver, kidney, probable human carcinogen	IRIS (1988)
Lead compounds	Manufacture of batteries, lead alloys, lead products like sheets, pipes, ammunition and paint industries	Impacts circulatory, nervous, immune, renal and cardiovascular systems	USEPA (2006)
Manganese compounds	Natural sources, diet, iron and steel production, power plants, coke ovens, mining	Nervous and respiratory system effects, manganism	ATSDR (2012)
Mercury compounds	Thermometers, barometers, batteries, lamps, industrial processes, lubricating oils, dental amalgams, inhalation in occupational settings, diet, e.g. fish	Toxic, central nervous system effects, e.g. tremors, slowed sensory and nerve functions, nausea, kidney pain, methyl mercury (blindness, deafness, impaired consciousness)	IRIS (1995)
Methylene chloride	Solvent and paint stripper by a number of industries, e.g. drugs, pharmaceuticals, metal cleaning, electronics, insect sprays	Nervous system effects, e.g. decreased visual, auditory and motor functions	ATSDR (2000)
Nickel compounds	Diet, contact with nickel-containing jewellery and stainless utensils, nickel metal refining, oil and coal combustion, tobacco smoking	Gastrointestinal distress, neurological effects, pulmonary fibrosis, renal edema, dermatitis; nickel refinery dust and nickel sub-sulfide are human carcinogens	ATSDR (2005)
Polycyclic organic matter (POM)*	Cigarette smoke, vehicle exhaust, agricultural burning, residential wood burning, asphalt roads, laying tar, meat grilling	Skin disorders, cancer	IRIS (1999)
Toluene	Automobile emissions, solvent for paint, adhesives, nail polish, print, etc., cigarette smoke, benzene production	Nervous system disorders, irritation of eyes, throat, respiratory tract, headaches, dizziness	IRIS (2005)
Vinyl chloride	Manufacture of PVC plastic, vinyl products, e.g. pipes, cable coatings	Nervous system disorders, liver damage, human carcinogen	ATSDR (2006)

*Various polycyclic aromatic hydrocarbon compounds (PAHs), including benzopyrene.

Several precursors of SO_2 in a lower oxidized state (reduced sulfur compounds, RSCs), including dimethyl sulfide (DMS), hydrogen sulfide (H_2S), carbon disulfide (CS_2) and carbonyl sulfide (COS), contribute significantly to the global sulfur budget. Once released into the atmosphere, these are oxidized to produce SO_2. Landfills are a major source of anthropogenic RSCs. Due to its comparatively long lifetime, COS is able to penetrate into the stratosphere where its photolysis and subsequent oxidation contributes to the stratospheric sulfate layer. Being the major precursors of sulfate aerosols which exert a negative radiative forcing on the atmosphere, sulfur gases indirectly play a crucial role in the Earth's radiative balance and are of great interest to geoengineering (climate engineering) experts.

1.2.2.2 Nitrogen oxides (NO_x)

NO_x is composed of both NO and NO_2. NO has both natural (e.g. soils, lighting) and anthropogenic sources (e.g. vehicle exhaust). NO_2 is formed from the oxidation of NO. NO_x is the major precursor of tropospheric O_3. NO converts atmospheric HO_2 into OH, the most important oxidizing agent in the Earth's atmosphere. Thus, NO_x exerts pivotal control in the chemical cycling of atmospheric oxidants. In regions of high NO_x, ozone is generally photochemically produced as NO oxidizes to NO_2 via peroxy radicals (formed during oxidation of CO, CH_4 and VOCs) and NO_2 is photolyzed at λ<424 nm. In regions of low NO_x, ozone is catalytically destroyed. Further, under very high NO_x conditions, NO_2 acts as a sink of atmospheric OH, resulting in the formation of acids (HNO_3) and simultaneously terminating the cycling between atmospheric OH and HO_2. The formation of HNO_3 brings back nitrogen into the biosphere. While NO_x is a useful catalyst in atmospheric chemistry, it has harmful effects on the health of human beings and the environment. High NO_2 levels can aggravate asthma and other respiratory diseases, and long-term exposure can actually lead to them. NO_x also leads to acid rain, haze, photochemical smog and algal bloom and has been designated as a criteria pollutant by the USEPA, making it mandatory to monitor and control NO_x emissions. The National Ambient Air Quality Standard (NAAQS) for NO_x over USA is 53 ppbv over a year, as determined by the USEPA. For India, the

national ambient air-quality standard requires annual average NO_2 to be less than 21 and 16 ppbv for industrial/residential and sensitive areas, respectively (CPCB, 2013).

The major source of NO_x is fossil fuel combustion. For India, the road and power sectors constituted 34% and 31% of the national NO_x emissions, respectively, while industry and biomass burning accounted for 17% and 13% of NO_x emissions during 2005 (Garg et al., 2006). The Emissions Database for Global Atmospheric Research (EDGAR) inventory reveals that the top five NO_x emitters globally are China, USA, international shipping, India and Russia with contributions of 20.7, 14.2, 13.8, 7.1 and 4.3 Tg, respectively, in 2008 (JRC, n.d.). For the EDGAR inventory, NO_x emissions in China showed about 40% increase during 2000–2005 and a further 26% increase in 2008 compared to 2005. However, over Europe NO_x levels have declined from 14 Tg in 2005 to 8 Tg in 2014 (EEA, n.d.).

1.2.2.3 Tropospheric O_3

Tropospheric O_3 (secondary air pollutant) is not emitted directly into the atmosphere but chemically formed due to photochemical reactions involving NO_x (primary pollutants) and volatile organic compounds (VOCs). While O_3 itself is an oxidant, it is the primary source of atmospheric OH (R1–R3), the 'detergent' of the atmosphere.

$$O_3 + h\nu \rightarrow O\left(^1D\right) + O_2 \quad \text{(Eq. 1.1)}, \lambda < 330 \text{ nm}$$

$$O\left(^1D\right) + M \rightarrow O\left(^3P\right) + M\left(M = N_2, O_2\right)$$
$$\text{(Eq. 1.2)}$$

$$O\left(^1D\right) + H_2O \rightarrow 2OH \quad \text{(Eq. 1.3)}$$

While O_3 plays a crucial role in atmospheric chemistry, acting both as a regional and a global pollutant because of its lifetime and formation during transport of its precursors, it also plays a harmful role with respect to the environment, climate, crops and human health, and hence is designated as a criteria pollutant by the USEPA. In humans, O_3 can cause shortness of breath, inflamed respiratory tract, aggravated lung disease and chronic obstructive pulmonary disease (COPD). In plants, O_3 impacts sensitive vegetation and ecosystems, reduces photosynthesis and stunts growth, damaging crops. Sicard et al.

(2017) have found that key biodiversity areas in South and North Asia are at risk from high O_3 concentrations. The USEPA NAAQS requires annual fourth-highest daily maximum 8-hour O_3 concentration, averaged over 3 years, to be below 70 ppbv.

Apart from its pivotal role in atmospheric chemistry, O_3 is a greenhouse gas (because of its strong absorption band centred at 9.6 µm). The radiative forcing (RF) of tropospheric O_3 was estimated (with a medium level of scientific understanding) at 0.35 Wm^{-2} with an uncertainty range of 0.25–0.65 Wm^{-2} (Forster et al., 2007), making it the third most important anthropogenic greenhouse gas next to carbon dioxide (CO_2) and methane (CH_4). Long-term measurements show increasing O_3 levels over past decades mostly due to increases in anthropogenic emissions of ozone precursors (Oltmans et al., 2006; Mallik et al., 2015). Paoletti et al. (2014) estimated that O_3 has increased much faster over urban areas compared to rural areas in the USA and Europe. Studies estimate a positive increase in surface O_3 up to 5 ppbv for the Asian region by 2050 (Wild et al., 2012).

1.2.2.4   Carbon monoxide (CO)

Carbon monoxide (CO) is not only a formidable primary atmospheric pollutant but also a crucial player in regional and global atmospheric chemistry. As a primary pollutant, it reduces the oxygen-carrying capacity of the blood, which can lead to unconsciousness and even death at high CO levels (WHO, 1999). Being a precursor to important greenhouse gases, O_3 and CO_2, CO emissions may be attributed an indirect radiative forcing of around 0.2 Wm^{-2} (Forster et al., 2007). CO is a major sink for atmospheric OH, the main cleansing agent of the atmosphere (R4). The oxidation of CO by OH (R4) has manifold implications to atmospheric chemistry, including the HO_x cycling, production of tropospheric ozone, and the abundance of greenhouse gases like methane and carbon dioxide. Thus, this single reaction (R1) exerts immense control over the chemical and radiation budget of the atmosphere regionally and globally.

$$CO + OH + O_2 \rightarrow CO_2 + HO_2 \qquad \text{(Eq. 1.4)}$$

The sources of CO are ubiquitous, as it is formed whenever combustion and burning of carbon-based products occur; this includes incomplete combustion of fuels in industry, transport and domestic sectors, burning of crops and forest fires. It is also formed by oxidation of CH_4 and non-methane volatile organic compounds (NMVOCs). Due to its fairly long lifetime (<few months), CO emissions over a region can influence its zonal-mean concentration levels. It can be conceived that with increasing needs for food and fuel, the emission of CO will increase, particularly over developing regions of the world. However, using Atmospheric Infrared Sounder CO measurements, Warner et al. (2013) found that the mixing ratio of CO in the Northern Hemisphere has decreased by 1.28 ppb yr^{-1} during 2003–2012, but there is still some uncertainty regarding regional CO trends (Jiang et al., 2017). The EDGAR inventory shows that global CO emissions decreased by 1.5% in 2008 compared to 2005. Globally during 2008, the top five emitters of CO were China, Africa, Sudan, India and the Congo, contributing 106, 105, 61, 58 and 53 Tg, respectively (EDGAR v4.2). Global CO emissions are dominated by anthropogenic emissions (500–600 Tg yr^{-1}) followed by biomass burning (Granier et al., 2011) with significant inter-annual variability (300–600 Tg yr^{-1}). The National Ambient Air Quality Standard (NAAQS) set by the USEPA for CO is 9 ppmv for an 8-hour average and this value should not be exceeded once over a year. For India, the national ambient air-quality standard requires an 8-hour average CO to be less than 1.8 ppmv for industrial, residential and sensitive areas (CPCB, 2013).

1.2.2.5   Particulate matter (PM)

Perhaps, the most notorious of all air pollutants is PM. There are two main categories of PM: particles with a diameter less than 10 µm are called PM_{10} and those with a diameter less than 2.5 µm are known as $PM_{2.5}$. Particles less than 1 µm ($PM_{1.0}$) include black carbon (BC) and organic carbon (OC). The major chemical constituents of PM are generally inorganic ions, e.g. ammonium, sulfate and nitrate. Sea salt, mineral dust, organic and elemental carbon form minor constituents of PM. Despite thousands of publications every year on particulate matter, a large fraction of PM remains unapportioned (Fuzzi et al., 2015). The severest impacts of PM are encountered in the form of human mortality and morbidity (up to several millions globally due to respiratory and cardiovascular impacts), loss in visibility, and haze and fog, leading to slower traffic, greater vehicular emissions and economic loss (due to delays). The climate impacts of PM include atmospheric warming caused by black carbon ($+0.4$ to $+0.8$ W m^{-2}; IPCC, 2013), and change in precipitation patterns mediated by organic aerosols, etc. The major anthropogenic sources of PM are grassland and agricultural waste burning, construction and demolition activities, road dust, combustion of fuels (fossil fuels, wood, etc.), industrial emissions, mining and quarrying operations with more details provided later. The NAAQS set by the USEPA for PM$_{10}$ and PM$_{2.5}$ are 150 and 35 µgm^{-3}, respectively, for 24-hour averages over a 3-year period. For India, the national ambient air-quality standard requires 24-hour PM$_{10}$ (RSPM) to be less than 100 and 60 µgm^{-3} for industrial/residential and sensitive areas, respectively (CPCB, 2013; https://www.epa.gov/pm-pollution/particulate-matter-pm-basics#PM).

1.2.3 Greenhouse gases

1.2.3.1 Carbon dioxide (CO_2)

CO_2 is the most important and most discussed anthropogenically emitted greenhouse gas and the poster child for global warming. CO_2 levels have steadily increased from 278 ppmv during the beginning of the industrial era (1750) and recently reached the 400 ppmv mark, showing more than a 40% increase (NOAA, n.d.). The major causes of CO_2 enhancement are increasing anthropogenic emissions, especially fossil fuel combustion, cement production and land use change. Between 1750 and 2011, CO_2 emissions from fossil fuel combustion and cement production are estimated to have released 375 GtC, while land use changes and deforestation have released 180 GtC, totalling 555 GtC for net anthropogenic emissions (IPCC, 2013). The annual CO_2 emissions from fossil fuel combustion and cement production was 9.8 GtC for 2011 alone (IPCC, 2013). In 2011, fossil fuel emissions were 9.5 PgC (IPCC, 2013). Of the 555 Gt CO_2 emitted during 1750–2011, only 155 was taken up by the ocean, yet there has been 26% increase in hydrogen ion concentration in

surface ocean water during the same period (IPCC, 2013). While the total anthropogenic radiative forcing (RF) for 2011 relative to 1750 is 2.29, CO_2 alone is responsible for a RF of 1.68, i.e. 73% of the RF. Total CO_2 emissions are linearly related to global surface temperatures, hence substantial reduction in CO_2 emissions is required to arrest rising temperatures (IPCC, 2013). In fact, 65% of our carbon budget compatible with a 2°C goal is already used, and to limit global surface temperatures to below 2°C, we are allowed to further release a net 1000 GtCO$_2$ only into the atmosphere (IPCC, 2013). Fossil fuel-derived CO_2 emissions in China have increased by more than a factor of two over the recent decade and China surpassed the United States as the world's largest fossil CO_2 emitter in 2006 (Gregg et al., 2008; WRI, n.d.). The impact of CO_2 emissions on humanity has been so profound that it has given rise to 'carbon trading', a process of buying and selling permits to emit CO_2. However, despite intense efforts from the global scientific community, there still exist large discrepancies between estimates of CO_2 emissions based on *in situ* measurements and those from emission inventories (Chandra et al., 2016).

1.3 Sources of Anthropogenic Pollutants

While there are hundreds of primary anthropogenic air pollutants in the Earth's atmosphere, their emission sources can be broadly classified into four sectors, as shown in Fig. 1.2 (Vallack and Rypdal, 2012). The energy sector, accounting for combustion and handling of fuels, is by far the most notorious contributor to anthropogenic air pollution. Emissions from various industries, barring the combustion of fuels, constitute the sector for industrial processes, solvent and other product use. The sector for agriculture, forestry and vegetation fires comprises all agricultural activities including crop burning, savanna burning, etc., but again excluding fuel combustion. The sector for waste comprises emissions related to waste storage, collection and disposal. The approach for nomenclature of the sectors varies from inventory to inventory (described under 'Emission Inventories' below), for example, the

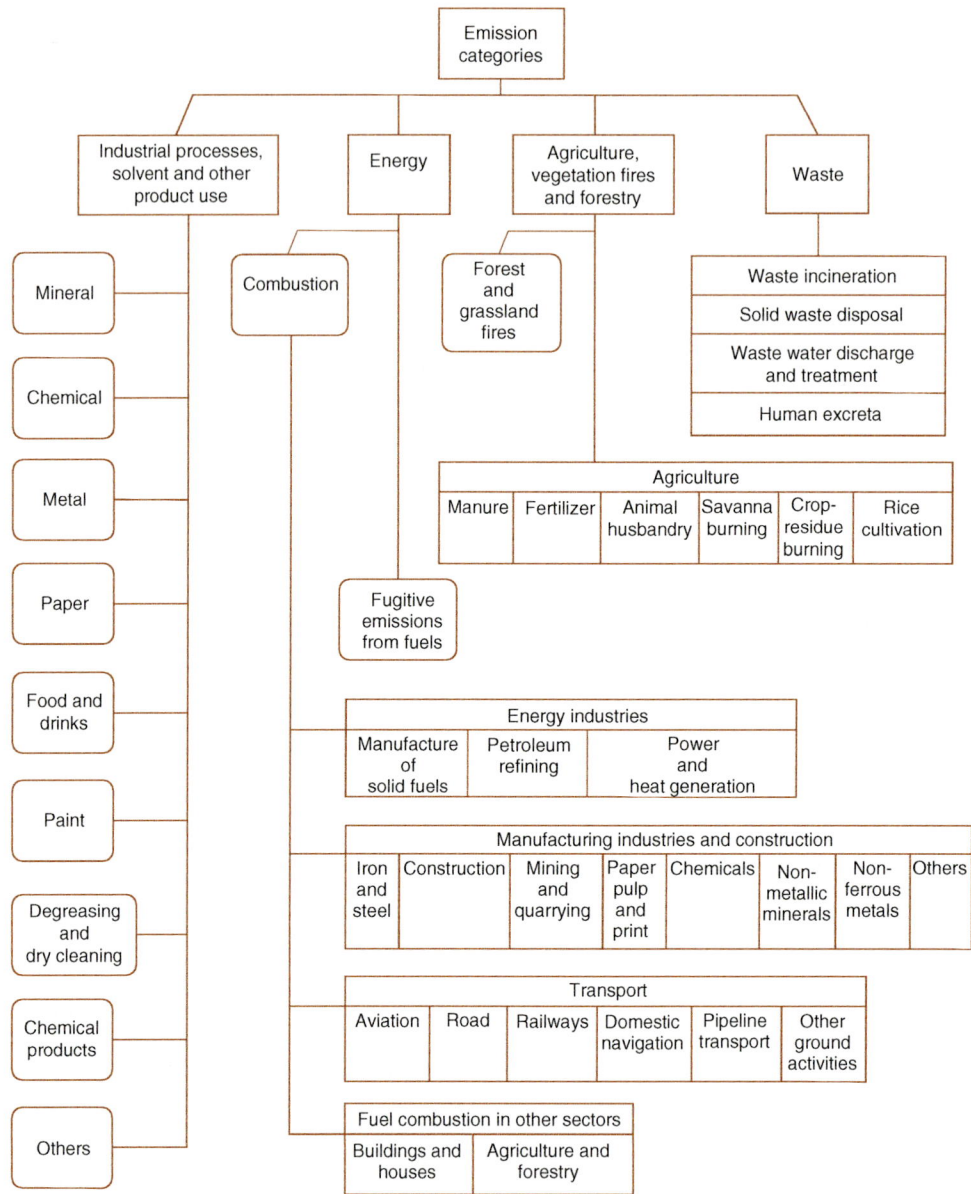

Fig. 1.2. Sector-wise distribution of anthropogenic emission sources. While the processes shown are self-explanatory, 'fugitive emissions' refers to non-combustion activities pertaining to exploration, processing, production, storage, distribution and use of fuels such as natural gas, gasoline, etc. (Adapted from Vallack and Rypdal, 2012.)

Intergovernmental Panel on Climate Change (IPCC) uses the nomenclature 'agriculture, land use change and forestry' while the Global Atmospheric Pollution Forum uses the nomenclature 'agriculture, vegetation fires and forestry' (IPCC, 1996; Vallack and Rypdal, 2012). The discussion in the present section is based mainly on the manual of the Global Atmospheric Pollution Forum (Vallack and Rypdal, 2012) but tries to offer an easy-to-understand simplistic overview.

1.3.1 Emissions from the 'energy' sector

Emissions from the energy sector can be sub-divided into two main components: direct combustion from fuel; and fugitive emissions from transport and handling of fuels. Combustion of fossil fuels is by far the largest source of most anthropogenic air pollutants and greenhouse gases. During 2002–2011, CO_2 emissions from fossil fuel combustion and cement manufacturing alone have grown at 3.2% yr^{-1} with an average emission rate of 8.3 petagram carbon per year (PgC yr^{-1}; IPCC, 2013).[1] 'Fossil fuel' is a common term for representing buried combustible geologic deposits of organic materials, formed from animal and plant remains that have been metamorphosed to oil, gas or coal by the action of heat and pressure in the Earth's crust over millions of years. The global energy demand has been steadily increasing from 356 quadrillion British thermal units (quad Btu) in 1990 and 410 quad Btu in 2000 (decadal increase of 15%) to 523 quad Btu in 2010 (decadal increase of 27%), and projected at 600 quad Btu in 2020 (AEO, 2017; IEO, 2017). The major contributors to the reference projection of 580.7 quad Btu in 2016 are 136.3 quad Btu from China (23.5%), 97 from USA (16.7%), 80.7 from Europe (OECD countries only, 13.9%), 34 from Middle-East Asia (5.8%), 29.8 from India (5.1%) and 29.7 quad Btu from Russia (5.1%), all other countries together accounting for less than 30% of the world's energy consumption (IEO, 2017). For the 550 quad Btu energy consumed around the globe in 2012, liquid fuels, e.g. crude oil, petroleum, etc., comprised 183.5 quad Btu (33.3%), coal comprised 153 (28%), natural gas 124 (22.7%), renewables 64 (11.5%) and nuclear energy about 24.5 (4.4%) Btu, respectively (AEO, 2017; IEO, 2017). However, the percentage contribution of fuels varies greatly from country to country (Fig. 1.3; Table 1.2).

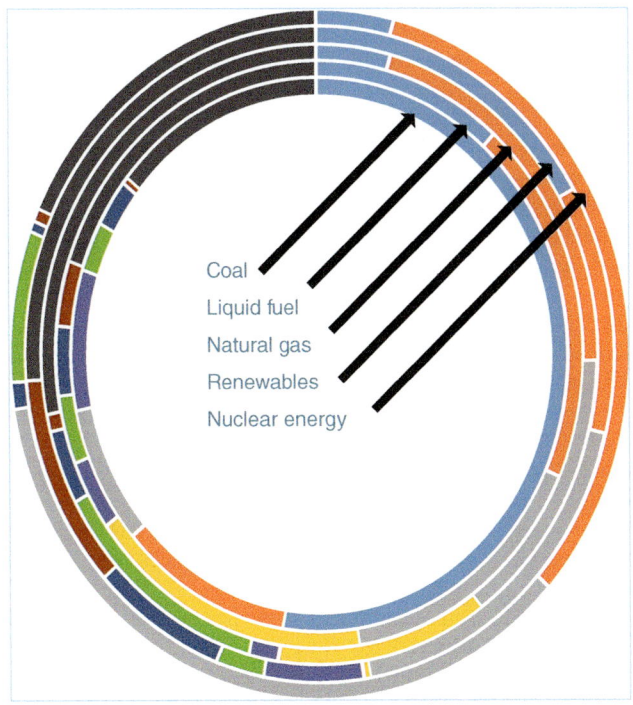

Coal
Liquid fuel
Natural gas
Renewables
Nuclear energy

■ China   ■ USA   ■ Europe OECD
■ Middle East   ■ India   ■ Russia
■ Africa   ■ Brazil   ■ Others

Fig. 1.3. Region-wise distribution of energy demand across the globe based on fuel type. (Based on data from IEO, 2017.)

Table 1.2. Country-wise division of global energy consumption during 2012. The values for Europe are for OECD (Organisation for Economic Co-operation and Development) countries only. (From EIA, n.d.b)

Country	Liquid fuel (million barrels/day)	Coal (British thermal units)	Natural gas (trillion cubic feet)	Renewables (British thermal units)	Nuclear (billion kilowatt hour)
China	10.6	80.6	5.1	10.6	93
USA	18.5	17.3	25.5	7.8	769
Europe	14.4	12.5	17.9	11.7	837
Middle East	16	0.1	14.8	0.2	1
India	3.6	12.6	2.1	3.5	30
Russia	3.4	4.5	15.7	1.7	166
Africa	3.5	4.3	2.7	4.7	12
Brazil	3.3	0.5	1.1	6.8	15
World	91.4	153.9	120.8	63.7	2345

Figure 1.3, based on IEO (2017), reveals the huge dependence of China on coal for its energy demands and accounts for over 50% of the energy from coal worldwide in 2012. On the other hand, USA and Europe consumed nearly 70% of the world's nuclear energy during 2012. The ratio of energy consumption from coal to renewables was 7.6 for China while it was only 1.1 for Europe (OECD), showing that Europe depended nearly equally on coal and renewables, while China depended overwhelmingly on coal in 2012. This value was 3.6 for India and 2.2 for USA. Moreover, the ratio of coal to natural gas consumption was 15.8 for China, while this value was 6.0 for India, 0.7 for USA and OECD Europe and only 0.3 for Russia, again showing natural gas as a more favourable energy source in the latter countries. Further, it has been projected that the use of coal would slowly stabilize over the years with decreases in China offset by increases in India. The use of liquid fuels and natural gas continue to increase unabated (IEO, 2017). Global oil production grew by only 0.4 million barrels/day in 2016, while oil consumption growth averaged 1.6 million barrels/day (EIA, n.d.a). However, the dependence on renewable energies is projected to increase at a higher rate (2.3% per year between 2015 and 2040) compared to other energy sources (IEO, 2017).

Product-wise, oil products can be classified as distillates: light (motor and aviation gasolines and light distillate feedstock) and middle (jet, gas, kerosene and diesel oils), fuel oil (marine bunkers and crude oil used without refining)

and others (liquefied petroleum gas or LPG, refinery gas, solvents, petroleum coke, bitumen, wax and lubricants). Crude oil is a yellow to black mineral oil of varying density comprising several naturally originating hydrocarbons. Natural gas is composed of gases in gaseous or liquefied forms in underground deposits, the main constituents being overwhelmingly methane and some ethane. Natural gas liquids (NGLs) are the liquefied or liquid hydrocarbons produced in the purification/manufacture of natural gas, the major constituents being propane, ethane, pentane, butane, natural gasoline, etc. Refinery gas is a non-condensable gas recovered during treatment of oil products in refineries and petrochemical industries or during distillation of crude oil, and consists mainly of olefins, hydrogen, ethane and methane. Liquefied petroleum gases (LPG) are the light hydrocarbons consisting mainly of propane (C_3H_8) or butane (C_4H_{10}) or both, extracted from crude oil plants, refinery processes and natural gas processing plants and are normally liquefied under pressure for transportation and storage with a net calorific value of about 47 terajoules per kiloton (TJ/kt; Vallack and Rypdal, 2012). Gasoline consists of various C_5–C_{12} hydrocarbons; motor gasoline is a light hydrocarbon oil distilled between 35°C and 215°C with additives, oxygenates and octane enhancers and is used as a fuel for land-based spark ignition engines (net calorific value 44.8TJ/kt; Vallack and Rypdal, 2012). Kerosene is a refined petroleum distillate comprising mainly C_{13}–C_{15} hydrocarbons obtained from distilling medium oil between 150°C and 300°C and its volatility is intermediate between gasoline and diesel oil with a net calorific value of about 43.75 TJ/kt (Vallack and Rypdal, 2012). Diesel oil (mostly C_{14}–C_{18} hydrocarbons) is obtained by distilling crude oil between 180°C and 380°C (net calorific value 43.34 TJ/kt; Vallack and Rypdal, 2012). The pollutants released from various emission sources can be identified from the measurements of NMVOCs in ambient air and robust studies have been carried out in various parts of the world using correlations of different NMVOCs to identify and characterize such emissions (Derwent *et al.*, 2000; Xie and Berkowitz, 2006; Mallik *et al.*, 2014). In general, diesel cars, although more energy efficient compared to petrol cars, emit more NO_x and particulate matter, while petrol cars (with or without catalysts) emit more hydrocarbons, CO and CO_2 (Air Pollution, n.d.). The NO_x emission factors for gasoline in passenger cars (uncontrolled) is estimated at 1.8 g/km, while it is 2.7 for diesel and about 2.1 for CNG and LPG (Vallack and Rypdal, 2012). For heavy-duty vehicles (uncontrolled), NO_x emission factors are 14.3 and 5.7 g/km for diesel and LPG, respectively (Vallack and Rypdal, 2012).

1.3.2 Emissions from the sector 'Industrial processes, solvent and other product use'

This sector includes industries and activities-related to the production and usage of minerals, chemicals, metals, paper, food and drink, paint, grease and laundry, except those related to energy, e.g. emissions from coke during iron manufacture in a blast furnace due to their use as a reducing agent, or evaporative emissions of sulfur and hydrocarbon compounds in activities related to paints. Emissions from combustion processes for the generation of heat and electricity within the industries is accounted for under the energy sector. The emissions under the mineral subsector include production and manufacture of bricks, cement, lime, paving of roads with asphalt releasing a variety of pollutants such as SO_2, NO_x, CO, non-methane hydrocarbons (NMHCs), CO_2 and PM. Emissions due to the production of various chemicals like NH_3, HNO_3, urea, etc. are included under the chemical industry subsector. Metal production includes industries related to the

production of iron, aluminium, zinc, copper, etc. 'Solvent and other product use' includes the application of paints in domestic, commercial and industrial sectors (including painting of buildings, homes, ships, vehicles, etc.), manufacture (polyester resin, PVC, ink, glue, rubber, etc.), print industry, extraction of non-edible oils, etc., releasing a variety of NMHCs into the atmosphere. Food production includes all processes that occur after the harvesting of crops or slaughtering of animals.

1.3.3 Emissions from the sector 'Agriculture, vegetation fires and forestry'

This sector includes emissions related to the cultivation of various crops including preparing land, manure management, fertilizer application, animal husbandry including enteric fermentation, field burning of agricultural residues as well as grassland burning for crop plantation. This sector, along with fossil fuel combustion and food production, is a major channel for release of reactive nitrogen into the atmosphere. The use of manure and fertilizers leads to increased nitrous oxide (N_2O) in the atmosphere. N_2O is very effective as a greenhouse gas (310 times more potent than CO_2 in global warming) and O_3 scavenger. Rice cultivation alone occupies more than one-tenth of the total arable land area of the Earth (1.5 billion hectares which is about 10% of the global total land area) and is a major source of atmospheric methane (FAOSTAT, 2017). Animal grazing uses over 3 billion hectares of meadows and pastures globally (Klein Goldewijk *et al.*, 2011). Livestock is a major contributor to global greenhouse gas emissions (FAO, 2006) and a meat-based diet may have twice the carbon footprint over a vegetarian diet (Center for Sustainable Systems, University of Michigan, 2017). Animal husbandry or rearing is a major source of atmospheric methane. The agriculture and waste sector accounted for one-third, i.e. 188 of the 558 Tg CH_4 yr^{-1} emissions during 2003–12 (Saunois *et al.*, 2016; GCP, n.d.). Between 1750 and 2011, land-use change (mainly deforestation) is estimated to have released 180 (100 to 260) PgC (IPCC, 2013). The major factor responsible for land-use change emissions between 2002 and 2011 was found to be tropical deforestation, releasing about 0.9 PgC yr^{-1}, with gross deforestation emissions alone accounting for about 3 PgC yr^{-1} (IPCC, 2013). The CO emission factors for combustion of different woods and waste vary in the range of 1000–5000 kg TJ^{-1}, which is more than 50 times higher than that for the burning of coal or oil (Vallack and Rypdal, 2012). The emission factors for NMVOCs for wood and waste burning are about 600 kg TJ^{-1}, OC from wood burning is about 1.77 kg ton^{-1} fuel, much higher compared to only 0.9 from coke ovens emissions from anthracite and bituminous coal (Vallack and Rypdal, 2012).

1.3.4 Emissions from the sector 'Waste'

This sector includes emissions due to the disposal of waste by incineration/burning or chemical treatment in landfills and sewages. Activities comprise combustion of solid wastes, NH$_3$ emissions from latrines, waste-water treatment and disposal, etc. This sector is a source of a large number of hydrocarbon, nitrogen and sulfur compounds in the atmosphere (Kim et al., 2006). The SO$_2$ emission factors vary between 1.25 and 1.75 kg ton^{-1} for different kinds of municipal waste burnings except in the open where this factor is about 0.5 only (Vallack and Rypdal, 2012). Similarly, the NO$_x$ emission factors for municipal waste burning vary between 1.2 and 3 kg ton^{-1}, the highest being for open burning (Vallack and Rypdal, 2012). For open burning, the emission factors for CO, NMHCs, PM$_{10}$ and PM$_{2.5}$ are 42, 3.3, 8 and 6.4 kg ton^{-1}, respectively (Vallack and Rypdal, 2012).

1.4 Emission Inventories

The emission sources of anthropogenic air pollutants and greenhouse gases are reported under the purview of various emission inventories. As the name suggests, an 'emission inventory' is a catalogue or database of various emission sources for air pollutants and greenhouse gases, and is supposed to account for the emissions of the pollutants over several geographic grids (usually in squares of degrees/kilometres) during a specific time (usually years). To build up an emission inventory, information is required regarding the potential of the source to release a particular pollutant (emission factor, e.g. kilograms of SO$_2$ emitted due to combustion of each

ton of coal) and the activity/amount of the source (e.g. amount of coal consumed by a power plant per day). An emission factor represents the relationship between the amounts of a pollutant released and an activity leading to the release of that pollutant. Emission factor is generally expressed as the weight of the pollutant released per unit volume, weight or distance of the activity emitting the pollutant. The net emission is the product of the emission factor and the activity rate (EPA, 1995).

The emission factor of a pollutant for the same source can vary significantly across the globe, e.g. the sulfur content of coal varies from region to region, hence the SO$_2$ emission factors for coal will vary by region. Similarly, NO$_x$ and hydrocarbon emissions from cars will vary from country to country depending on technology, fuel, regulations, implementation and usage. The information regarding emission factors is derived from laboratory experiments, chamber measurements as well as in situ measurements of atmospheric concentration of gases and particles. The activity rate is mostly based on data from government agencies, e.g. number of power plants, cement factories, vehicles operating in a given region and their produce/usage. The ground measurements are also supplemented by remote sensing measurements, e.g. from satellites. Several global, regional and local emission inventories have been developed (Granier et al., 2011; ECCAD, n.d.). These inventories form the basis of several atmospheric chemistry and climate models to study the interrelationships of anthropogenic emissions with the past, present and future of the Earth's atmospheric composition. Further, there exist assimilation models like MACC (Monitoring Atmospheric Composition and Climate) that assimilate data from various sources (in situ, satellite) and provide gridded datasets for various atmospheric constituents (ECMWF, n.d.). There are also inter-community initiatives, such as Global Emissions InitiAtive (GEIA), that aim to be a key forum for emissions knowledge (GEIA, n.d.). Information regarding a few emission inventories are given in Table 1.3.

1.5 Concluding Remarks

As the influence of humans has now reached every nook and cranny of planet Earth, anthropogenic

Table 1.3. Summary of common emission inventories. (Adapted from ECCAD, n.d.)

Inventory	Region	Grid size	Temporal coverage	Gas	Reference
EMEP: Co-operative programme for monitoring and evaluation of long-range transmission of air pollutants in Europe	Europe	$0.5° \times 0.5°$	1980–2020	CO, NO_x, NMVOCs SO_2, NH_3, $PM_{2.5}$, PM_{10}, heavy metals, persistent organic pollutants	EEA, 2009; Muntean et al., 2014
REAS (Regional Emission inventory in Asia)	Asia	$0.25° \times 0.25°$	2000–2008	SO_2, NO_x, CO, NMVOC, PM_{10}, $PM_{2.5}$, BC, OC, NH_3, CH_4, N_2O, CO_2	Kurokawa et al., 2013
MACC City (Monitoring Atmospheric Composition and Climate)	Global	$0.5° \times 0.5°$	1960–present	CO, NO_x, SO_2, BC, OC, NH_3, C_2H_6, C_3H_8, C_4H_{10} and higher alkanes, C_2H_4, C_3H_6, C_4H_8 and higher alkenes, CH_3OH and other alcohols HCHO and other aldehydes, acetone and other ketones, aromatics	Granier et al., 2011
HYDE (Hundred Year Database for Integrated Environmental Assessment)	Global	$1° \times 1°$	1890–1990	CO_2, CO, CH4, NMVOC, SO_2, NO_x, N_2O, NH_3	van Aardenne et al., 2001
EDGAR (Emissions Database for Global Atmospheric Research)	Global	$0.1° \times 0.1°$	1970–2012	Greenhouse gases: CH_4, CO_2, N_2O, PFCs, HFCs, CF_6, SF_6, NF_3 air pollutants: CO, NO_x, NMVOC, NH3 and SO_2 Aerosol:PM_{10}	EC-JRC/PBL, 2011
HTAP (Hemispheric Transport of Air Pollution)	Global	$0.1° \times 0.1°$	2008–2010	CO, CH_4, NMVOC, NO_x, SO_2, NH_3, PM_{10}, $PM_{2.5}$, OC and BC	Maenhout et al., 2015

air pollution has become ubiquitous. While 187 atmospheric compounds are now identified as toxic, leading to health hazards such as cancer and birth defects, six ubiquitous air contaminants have been declared as criteria pollutants.

The criteria pollutants: O_3, CO, NO_x, Pb, PM and SO_2 have air quality standards associated with them to enable their monitoring and regulation. O_3 and PM pollution have been shown to account for 2–4 million deaths globally every year

(Silva *et al.*, 2013; Lelieveld *et al.*, 2015). Fossil fuel combustion is a major anthropogenic source of NO_x and over 60% of the global energy demand is met by coal and oil, followed by natural gas, renewables and nuclear energy. The concentration of O_3, CO, NO_x and hydrocarbons greatly influences the radical balance and the self-cleaning capacity of our atmosphere. While there are hundreds of air pollutants in our atmosphere, their anthropogenic sources can be conveniently grouped into four primary sectors related to energy, industrial processing, agriculture and waste. The geographical distribution of these source sectors largely determines the concentration and trends of primary pollutants, such as CO, SO_2 and NO_x. While over 70% of the global energy consumption is shared by China (23.5%), USA (16.7%), Europe (OECD countries only, 13.9%), Middle-East Asia (5.8%), India (5.1%) and Russia (5.1%), the type of emissions vary for different regions, e.g. unlike Europe, China depends more heavily on coal compared to renewables. Further, studies show that India is overtaking China in terms of anthropogenic

SO_2 emissions, showing the importance of implementation of regulations in curbing air pollution (Li *et al.*, 2017). A prerequisite for implementing regulations is proper knowledge of the scale of the problem, which requires reliable estimation of concentrations and emissions based on measurements and emission inventories.

The well-established link between human emissions/activity and climate and its repercussions is a matter of serious concern (IPCC, 2013). The impact of human activities on the Earth's ecosystem has been so profound that scientists are pressing for declaration of 'Anthropocene' as an epoch (Crutzen, 2002). While the damage done cannot be completely undone, people, societies and governments must seriously start making concerted efforts to limit and abate polluting the atmosphere, as each of us is a stakeholder in the present and future of the Earth's atmosphere and climate.

Ours is a generation that has produced and caused air pollution to evolve to dangerous levels. Will our legacy be a polluted Earth for our children to suffer and suffocate?

Note

[1] 1 PgC is equivalent to 3.667 Gigatonne CO_2. A Btu represents the amount of heat that is needed to raise the temperature of one pound of water by one degree Fahrenheit.

References

AEO (2017) National Energy Modeling System. run ref2017.d120816a. *Annual Energy Outlook*. US Energy Information Administration. Available at: https://www.eia.gov/outlooks/aeo/pdf/0383(2017).pdf (accessed 24 December 2017).

ATSDR (1990) *Toxicological Profile for Ethylene Oxide*. Agency for Toxic Substances and Disease Registry. US Public Health Service, US Department of Health and Human Services, Atlanta, GA.

ATSDR (1998) *Toxicological profile for Sulfur Dioxide*. Agency for Toxic Substances and Disease Registry. US Public Health Service, US Department of Health and Human Services, Atlanta, GA.

ATSDR (2000) *Toxicological Profile for Methylene Chloride (Update), Draft for Public Comment*. Agency for Toxic Substances and Disease Registry. US Public Health Service, US Department of Health and Human Services, Atlanta, GA.

ATSDR (2005) *Toxicological Profile for Nickel (Update)*. Agency for Toxic Substances and Disease Registry. US Public Health Service, US Department of Health and Human Services, Atlanta, GA.

ATSDR (2006) *Toxicological Profile for Vinyl Chloride (Update)*. Agency for Toxic Substances and Disease Registry. US Public Health Service, US Department of Health and Human Services, Atlanta, GA.

ATSDR (2007) *Toxicological Profile for Benzene*. Agency for Toxic Substances and Disease Registry. US Public Health Service, US Department of Health and Human Services, Atlanta, GA.

ATSDR (2012) *Toxicological Profile for Manganese*. Agency for Toxic Substances and Disease Registry. US Public Health Service, US Department of Health and Human Services, Atlanta, GA.

Air Pollution (n.d.) Motor Vehicle Emission Controls: Fuel Types. Air Pollution. Available at: http://www.air-quality.org.uk/26.php (accessed 28 August 2017).

Center for Sustainable Systems, University of Michigan (2017) Carbon Footprint Factsheet. Pub. No. CSS09-05. Center for Sustainable Systems, University of Michigan. Available at: http://css.umich. edu/factsheets/carbon-footprint-factsheet (accessed 28 August 2017).

Chandra, N., Lal, S., Venkataramani, S., Patra, P.K. and Sheel, V. (2016) Temporal variations of atmospheric CO_2 and CO at Ahmedabad in western India. *Atmospheric Chemistry Physics* 16, 6153–6173. DOI: 10.5194/acp-16-6153-2016.

CPCB (2013) *Guidelines for the Measurement of Ambient Air Pollutants*, vol. 1. Central Pollution Control Board, Government of India. National Ambient Air Quality Monitoring Series: NAAQMS/36/2012-13. Available at: http://www.indiaenvironmentportal.org.in/files/file/NAAQMS_Volume-I.pdf (accessed 1 August 2018).

Crutzen, P.J. (2002) The 'anthropocene'. *Journal de Physique IV France* 12(10), 1–5. DOI: 10.1051/ jp4:20020447.

Derwent, R.G., Davies, T.J., Delaney, M., Dollard, G.J., Field, R.A. *et al.* (2000) Analysis and interpretation of the continuous hourly monitoring data for 26 C_2–C_8 hydrocarbons at 12 United Kingdom sites during 1996. *Atmospheric Environment* 34, 297–312. DOI: 10.1016/S1352-2310(99)00203-4.

EC-JRC/PBL (2011) Emission Database for Global Atmospheric Research (EDGAR), release version 4.2. European Commission, Joint Research Center/ Netherlands Environmental Assessment Agency. Available at: http://edgar.jrc.ec.europa.eu (accessed 24 December 2017).

ECCAD (n.d.) About ECCAD. Emissions of Atmospheric Compounds and Compilation of Ancillary Data. Available at: http://eccad.aeris-data.fr/#DatasetPlace (accessed 28 August 2017).

ECMWF (n.d.) Public Datasets. European Centre for Medium-Range Weather Forecasts. Available at: http://apps.ecmwf.int/datasets/ (accessed 28 August 2017).

EEA (n.d.) Air pollutant emissions data viewer (Gothenburg Protocol, LRTAP Convention) 1990–2016. European Environment Agency (EEA). Available at: https://www.eea.europa.eu/data-and-maps/ dashboards/air-pollutant-emissions-data-viewer-1 (accessed 15 July 2018).

EEA (2009) EMEP/EEA air pollutant emission inventory guidebook. EEA Technical Report 9. European Environment Agency. Available at: http://www.eea.europa.eu/publications/emep-eea-emission-inventory-guidebook-2009 (accessed 15 July 2018).

EIA (n.d.a) Short-Term Energy Outlook – Global Liquid Fuels. U.S. Energy Information Administration, Washington, D.C. Available at: https://www.eia.gov/outlooks/steo/report/global_oil.php (accessed 28 August 2017).

EIA (n.d.b) Annual Energy Outlook. U.S. Energy Information Administration, Washington, D.C. Available at: https://www.eia.gov/outlooks/aeo/data/browser/ (accessed 28 August 2017).

EPA (n.d.a) Hazardous Air Pollutants. U.S. Environmental Protection Agency, Environmental Criteria and Assessment Office, Office of Health and Environmental Assessment, Office of Research and Development, Washington, D.C. Available at: https://www.epa.gov/haps (accessed 28 August 2017).

EPA (n.d.b) Health Effects Notebook for Hazardous Air Pollutants. U.S. Environmental Protection Agency, Environmental Criteria and Assessment Office, Office of Health and Environmental Assessment, Office of Research and Development, Washington, D.C. Available at: https://www.epa.gov/haps/ health-effects-notebook-hazardous-air-pollutants (accessed 28 August 2017).

EPA (n.d.c) Particulate Matter (PM) Basics. U.S. Environmental Protection Agency, Environmental Criteria and Assessment Office, Office of Health and Environmental Assessment, Office of Research and Development, Washington, D.C. Available at: https://www.epa.gov/pm-pollution/particulate-matter-pm-basics (accessed 28 August 2017).

EPA (1995) Introduction to AP-42, *Compilation of Air Pollutant Emissions Factors*, 5th edn, vol. 1. Available at: https://www3.epa.gov/ttn/chief/ap42/c00s00.pdf (accessed 1 August 2018).

FAO (2006) Livestock's long shadow. UN Food and Agriculture Organization, Rome. Available at: http:// www.fao.org/docrep/010/a0701e/a0701e00.HTM (accessed 24 December 2017).

FAOSTAT (2017) Crops/Regions/World list/Production Quantity (pick lists), Rice (paddy), 2014. UN Food and Agriculture Organization, Corporate Statistical Database (FAOSTAT). Available at: http://www. fao.org/faostat/en/#data/QC (accessed 24 December 2017).

Forster, P., Ramaswamy, V., Artaxo, P., Berntsen, T., Betts, R. *et al.* (2007) Changes in atmospheric constituents and in radiative forcing. In: Solomon, S., Qin, D., Manning, M., Chen, Z., Marquis, M. *et al.* (eds) *Climate Change 2007: The Physical Science Basis. Contribution of Working Group I to the Fourth Assessment Report of the Intergovernmental Panel on Climate Change*. Cambridge University Press, Cambridge and New York. Available at: http://www.ipcc.ch/pdf/assessment-report/ar4/wg1/ar4-wg1-chapter2.pdf (accessed 10 October 2018).

Fuzzi, S., Baltensperger, U., Carslaw, K., Decesari, S., Denier van der Gon, H. *et al.* (2015) Particulate matter, air quality and climate: lessons learned and future needs. *Atmospheric Chemistry and Physics* 15, 8217–8299.

GCP (n.d.) Global Methane Budget. Global Carbon Atlas. Available at: http://www.globalcarbonatlas.org/en/CH4-emissions (accessed 28 August 2017).

GEIA (n.d.) About GEIA. GEIA: Global Emissions InitiAtive. Available at: http://www.geiacenter.org/about (accessed 28 August 2017).

Garg, A., Shukla, P. and Kapshe, M. (2006) The sectoral trends of multigas emissions inventory of India. *Atmospheric Environment* 40(24), 4608–4620.

Granier, C., Bessagnet, B., Bond, T., D'Angiola, A., Denier van der Gon, H. *et al.* (2011) Evolution of anthropogenic and biomass burning emissions of air pollutants at global and regional scales during the 1980–2010 period. *Climate Change* 109, 163–190. DOI: 10.1007/s10584-011-0154-1.

Gregg, J.S., Andres, R.J. and Marland, G. (2008) China: emissions pattern of the world leader in CO_2 emissions from fossil fuel consumption and cement production. *Geophysical Research Letters* 35, L08806. DOI: 10.1029/2007GL032887.

HSDB (1993) Hazardous Substances Data Bank (HSDB, online database). US Department of Health and Human Services. National Toxicology Information Program, National Library of Medicine, Bethesda, MD. Available at: https://www.nlm.nih.gov/pubs/factsheets/hsdbfs.html (accessed 10 October 2018).

IEO (2017) World total energy consumption by region and fuel. International Energy Outlook 2017. US Energy Information Administration. Available at: https://www.eia.gov/outlooks/ieo/pdf/0484(2017).pdf (accessed 24 December 2017).

IPCC (1996) *Climate Change 1995: The Science of Climate Change. Contribution of Working Group I to the Second Assessment Report of the Intergovernmental Panel on Climate Change*, ed. Houghton, J.T., Meira Filho, L.G., Callander, B.A., Harris, N., Kattenberg, A. and Maskell, K. Cambridge University Press, Cambridge and New York. Available at: https://www.ipcc.ch/ipccreports/sar/wg_I/ipcc_sar_wg_I_full_report.pdf (accessed 1 August 2018).

IPCC (2013) *Climate Change 2013: The Physical Science Basis. Contribution of Working Group I to the Fifth Assessment Report of the Intergovernmental Panel on Climate Change*, ed. Stocker, T.F., Qin, D., Plattner, G.-K., Tignor, M.M.B., Allen, S.K. *et al.* Cambridge University Press, Cambridge and New York. DOI: 10.1017/CBO9781107415324.

IRIS (1988) *Integrated Risk Information System (IRIS) on Hydrazine/Hydrazine Sulfate*. US Environmental Protection Agency. National Center for Environmental Assessment, Office of Research and Development, Washington, DC.

IRIS (1989) *Integrated Risk Information System (IRIS) on Cadmium*. US Environmental Protection Agency. National Center for Environmental Assessment, Office of Research and Development, Washington, DC.

IRIS (1990) *Integrated Risk Information System (IRIS) on Formaldehyde*. US Environmental Protection Agency. National Center for Environmental Assessment, Office of Research and Development, Washington, DC.

IRIS (1991) *Integrated Risk Information System (IRIS) on Carbonyl Sulfide*. US Environmental Protection Agency. National Center for Environmental Assessment, Office of Research and Development, Washington, DC.

IRIS (1993) *Integrated Risk Information System (IRIS) on 1,2-Dichloroethane*. US Environmental Protection Agency. National Center for Environmental Assessment, Office of Research and Development, Washington, DC.

IRIS (1995) *Integrated Risk Information System (IRIS) on Elemental Mercury*. US Environmental Protection Agency. National Center for Environmental Assessment, Office of Research and Development, Washington, DC.

IRIS (1998a) *Integrated Risk Information System (IRIS) on Beryllium and Compounds*. US Environmental Protection Agency. National Center for Environmental Assessment, Office of Research and Development, Washington, DC.

IRIS (1998b) *Integrated Risk Information System (IRIS) on Chromium VI*. US Environmental Protection Agency. National Center for Environmental Assessment, Office of Research and Development, Washington, DC.

IRIS (1999) *Integrated Risk Information System (IRIS) on Polycyclic Organic Matter*. US Environmental Protection Agency. National Center for Environmental Assessment, Office of Research and Development, Washington, DC.

IRIS (2000) *Integrated Risk Information System (IRIS) on 1,3-Dichloropropene*. US Environmental Protection Agency. National Center for Environmental Assessment, Office of Research and Development, Washington, DC.

IRIS (2001) *Integrated Risk Information System (IRIS) on Chloroform*. US Environmental Protection Agency. National Center for Environmental Assessment, Office of Research and Development, Washington, DC.

IRIS (2002) *Integrated Risk Information System (IRIS) on 1,3-Butadiene*. US Environmental Protection Agency. National Center for Environmental Assessment, Office of Research and Development, Washington, DC.

IRIS (2003) *Integrated Risk Information System (IRIS) on Acrolein*. US Environmental Protection Agency. National Center for Environmental Assessment, Office of Research and Development, Washington, DC.

IRIS (2004) *Integrated Risk Information System (IRIS) on 1,2-Dibromoethane*. US Environmental Protection Agency. National Center for Environmental Assessment, Office of Research and Development, Washington, DC.

IRIS (2005) *Integrated Risk Information System (IRIS) on Toluene*. US Environmental Protection Agency. National Center for Environmental Assessment, Office of Research and Development, Washington, DC.

IRIS (2010) *Integrated Risk Information System (IRIS) on Carbon Tetrachloride*. US Environmental Protection Agency. National Center for Environmental Assessment, Office of Research and Development, Washington, DC.

JRC (n.d.) The Emissions Database for Global Atmospheric Research. European Commission – Joint Research Centre. Available at: http://edgar.jrc.ec.europa.eu/background.php (accessed 28 August 2017).

Jiang, Z., Worden, J.R., Worden, H., Deeter, M., Jones, D.B.A. *et al.* (2017) A 15-year record of CO emissions constrained by MOPITT CO observations. *Atmospheric Chemistry and Physics* 17, 4565–4583. DOI: 10.5194/acp-17-4565-2017.

Kim, K.-H., Jeon, E.-C., Choi, Y.-J. and Koo, Y.-S. (2006) The emission characteristics and the related malodor intensities of gaseous reduced sulfur compounds (RSC) in a large industrial complex. *Atmospheric Environment* 40(24), 4478–4490.

Klein Goldewijk, K., Beusen, A., de Vos, M. and van Drecht, G. (2011) The HYDE 3.1 spatially explicit database of human induced land use change over the past 12,000 years, *Global Ecology and Biogeography* 20(1), 73–86. DOI: 10.1111/J.1466-8238.2010.00587.X.

Kurokawa, J., Ohara, T., Morikawa, T., Hanayama, S., Maenhout, G.-J. *et al.* (2013) Emissions of air pollutants and greenhouse gases over Asian regions during 2000–2008: Regional Emission inventory in ASia (REAS) version 2. *Atmospheric Chemistry and Physics* 13, 11019–11058. DOI: 10.5194/acp-13-11019-2013.

Lelieveld, J., Evans, J.S., Fnais, M., Giannadaki, D. and Pozzer, A. (2015) The contribution of outdoor air pollution sources to premature mortality on a global scale. *Nature* 525, 367. DOI: 10.1038/nature15371.

Li, C., McLinden, C., Fioletov, V., Krotkov, N. and Carn, S. (2017) India is overtaking China as the world's largest emitter of anthropogenic sulfur dioxide. *Scientific Reports* 7, 14304. DOI: 10.1038/s41598-017-14639-8.

Lu, Z., Zhang, Q. and Streets, D.G. (2011) Sulfur dioxide and primary carbonaceous aerosol emissions in China and India, 1996–2010. *Atmospheric Chemistry and Physics* 11, 9839–9864.

Maenhout, G.-J., Crippa, M., Guizzardi, D., Dentener, F., Muntean, M. *et al.* (2015) HTAP_v2.2: a mosaic of regional and global emission grid maps for 2008 and 2010 to study hemispheric transport of air pollution. *Atmospheric Chemistry and Physics* 15, 11411–11432.

Mallik, C., Lal, S., Venkataramani, S., Naja, M. and Ojha, N. (2013) Variability in ozone and its precursors over the Bay of Bengal during postmonsoon: transport and emission effects. *Journal of Geophysical Research* 118(17), 10190–10209.

Mallik, C., Ghosh, D., Ghosh, D., Sarkar, U., Lal, S. and Venkatramani, S. (2014) Variability of SO_2, CO and light hydrocarbons over a megacity in Eastern India: effects of emissions and transport. *Environmental Science Pollution Research* 21(14), 8692–8706. DOI: 10.1007/s11356-014-2795-x.

Mallik C., Lal, S. and Venkataramani, S. (2015) Trace gases at a semi-arid urban site in western India: variability and inter-correlations. *Journal of Atmospheric Chemistry* 72, 143–164.

Mallik, C, Chandra, N., Venkataramani, S. and Lal, S. (2016) Variability of atmospheric carbonyl sulfide at a semi-arid urban site in western India. *Science of the Total Environment* 552, 725–737.

Muntean, M., Maenhout, G.-J., Song, S., Selin, N.E., Olivier, J.G.J. *et al.* (2014) Trend analysis from 1970 to 2008 and model evaluation of EDGARv4 global gridded anthropogenic mercury emissions. *Science of the Total Environment* 494–495, 337–350.

NOAA (n.d.) For first time, Earth's single-day CO_2 tops 400 ppm. NOAA, Scripps Institution of Oceanography, reported at the website of NASA JPL. Available at: https://climate.nasa.gov/news/916/for-first-time-earths-single-day-co2-tops-400-ppm (accessed 28 August 2017).

Oltmans, S.J., Lefohn, A.S., Harris, J.M., Galbally, I., Scheel, H.E. *et al.* (2006) Long-term changes in tropospheric ozone. *Atmospheric Environment* 40, 3156–3173.

Paoletti, E., De Marco, A., Beddows, D.C.S., Harrison, R.M. and Manning, W.J. (2014) Ozone levels in European and USA cities are increasing more than at rural sites, while peak values are decreasing. *Environmental Pollution* 192, 295–299. DOI: 10.1016/j.envpol.2014.04.040.

Saunois, M., Bousquet, P., Poulter, B., Peregon, A., Ciais, P. *et al.* (2016) The global methane budget 2000–2012. *Earth System Science Data* 8, 697–751. DOI: 10.5194/essd-8-697-2016.

Sicard, P., Anav, A., De Marco, A. and Paoletti, E. (2017) Projected global tropospheric ozone impacts on vegetation under different emission and climate scenarios. *Atmospheric Chemistry and Physics* 17, 12177–12196. Available at: https://www.atmos-chem-phys.net/17/12177/2017/acp-17-12177-2017.pdf (accessed 9 July 2018).

Silva, R.A., West, J.J., Zhang, Y.Q., Anenberg, S.C., Lamarque, J.-F. *et al.* (2013) Global premature mortality due to anthropogenic outdoor air pollution and the contribution of past climate change. *Environmental Research Letters* 8, 034005. DOI: 10.1088/1748-9326/8/3/034005.

Smith, S.J., van Aardenne, J., Klimont, Z., Andres, R., Volke, A. and Arias, S.D. (2011) Anthropogenic sulfur dioxide emissions: 1850–2005. *Atmospheric Chemistry and Physics* 11, 1101–1116.

Sun, Y., Zhuang, G., Wang, Y., Han, L., Guo, J. *et al.* (2004) The air-borne particulate pollution in Beijing – concentration, composition, distribution and sources. *Atmospheric Environment* 38, 5991–6004.

USEPA (1984) *Health Assessment Document for Inorganic Arsenic*. EPA/540/1-86/020. US Environmental Protection Agency, Environmental Criteria and Assessment Office, Office of Health and Environmental Assessment, Office of Research and Development, Washington, DC.

USEPA (1985) *Health and Environmental Effects Profile for Acetonitrile*. US Environmental Protection Agency, Environmental Criteria and Assessment Office, Office of Health and Environmental Assessment, Office of Research and Development, Washington, DC. EPA/600/X-85/357.

USEPA (1987) *Health Assessment Document for Acetaldehyde*. EPA/600/8-86-015A. US Environmental Protection Agency, Environmental Criteria and Assessment Office, Office of Health and Environmental Assessment, Office of Research and Development, Research Triangle Park, NC.

USEPA (1990) Amendments to the Clean Air Act. US Environmental Protection Agency. Title 42, Chapter 85, United States code. Available at: https://www.epa.gov/clean-air-act-overview/1990-clean-air-act-amendment-summary.

USEPA (1992) Report on the National Air Toxics Assessment (NATA) version 5. US Environmental Protection Agency. Available at: https://www.epa.gov/national-air-toxics-assessment/2011-national-air-toxics-assessment.

USEPA (2006) *Air Quality Criteria for Lead Final Report*. EPA/600/R-05/144aF-bF. US Environmental Protection Agency, Washington, DC.

Vallack, H. and Rypdal, K. (2012) The Global Atmospheric Pollution Forum Air Pollutant Emission Inventory Manual. Available at: http://ledsgp.org/resource/global-atmospheric-pollution-forum-air-pollutant-emission-inventory/?loclang=en_gb (accessed 28 August 2017).

van Aardenne, J.A., Dentener, F.J., Olivier, J.G.J., Klein Goldewijk, C.G.M. and Lelieveld, J. (2001) A $1°×1°$ resolution data set of historical anthropogenic trace gas emissions for the period 1890–1990. *Global Biogeochemical Cycles* 15, 909–928.

Wang, S.X., Zhao, B., Cai, S.Y., Klimont, Z., Nielsen, C.P. *et al.* (2014) Emission trends and mitigation options for air pollutants in East Asia. *Atmospheric Chemistry and Physics* 14, 6571–6603. DOI: 10.5194/acp-14-6571-2014.

Warner, J., Carminati, F., Wei, Z., Lahoz, W. and Attié, J.L. (2013) Tropospheric carbon monoxide variability from AIRS under clear and cloudy conditions. *Atmospheric Chemistry and Physics* 13, 12469–12479. DOI: 10.5194/acp-13-12469-2013.

WHO (1999) *Environmental Health Criteria 213: Carbon Monoxide (Second Edition)*. World Health Organization. Available at: http://www.who.int/ipcs/publications/ehc/ehc_213/en (accessed 28 August 2017).

WRI (n.d.) Top 10 Emitters in 2012. World Resources Institute, Washington, D.C. Available at: http://www.wri.org/resources/charts-graphs/top-10-emitters-2012 (accessed 28 August 2017).

Wild, O., Fiore, A.M., Shindell, D.T., Doherty, R.M., Collins, W.J. *et al.* (2012) Modelling future changes in surface ozone: a parametrized approach. *Atmospheric Chemistry and Physics* 12, 2037–2054. DOI: 10.5194/acp-12-2037-2012.

Xie, Y.L. and Berkowitz, C.M. (2006) The use of positive matrix factorization with conditional probability functions in air quality studies: an application to hydrocarbon emissions in Houston, Texas. *Atmospheric Environment* 40(17), 3070–3091. DOI: 10.1016/j.atmosenv.2005.12.065.

2 Biogenic Sources of Air Pollution

Harpreet Kaur[1]* and Ruchi Kumari[2]

[1]*Jawaharlal Nehru University, New Delhi, India;* [2]*University College London, London, UK*

Abstract

Volatile organic compounds emitted from a variety of biogenic sources such as plants, soil and microorganisms play an important role in atmospheric chemistry. With the rapidly changing scenarios of land use, land cover, temperature regimes and pollution, significant outcomes can be expected from an increase in such biogenic sources. This chapter is divided into two parts. The first section deals with the role of important biogenic volatile organic compounds in atmospheric chemistry, such as ozone formation and secondary organic aerosol formation, etc. The second section focuses on soil as a contributor of these volatiles. A number of studies have focused on the contribution of various plant species in volatile carbon emission; however, the role of soil has not been discussed in detail. A variety of global change components, such as increased carbon dioxide levels and tropospheric ozone land use, are also discussed.

2.1 Introduction

Volatile organic compounds (VOCs) are organic compounds with vapour pressure high enough to be vaporized into the atmosphere under normal conditions. In addition to anthropogenic sources, these are emitted by a variety of living organisms, including plants and microorganisms. Such VOCs are termed biogenic volatile organic compounds (BVOCs). BVOCs are classified according to their structure and biosynthetic origin (Pichersky *et al.*, 2006). These represent a group of trace gases (except carbon dioxide and carbon monoxide), including isoprenoids (isoprenes and monoterpenes) as well as alkanes, alkenes, carbonyls, esters, ethers, alcohols and acids. The atmospheric lifetime of BVOCs varies from minutes to several days (Kesselmeier and Staudt, 1999).

Various studies report that plants re-emit a substantial fraction of their assimilated carbon as BVOC into the atmosphere, which can further affect the physical and chemical properties of the environment (Peñuelas and Llusià, 2001). It is estimated that 10% of the total carbon stored in plants via photosynthesis is lost through BVOC release (Peñuelas and Llusià, 2003). Isoprene accounts for more than 90% of the total BVOC emission in certain plant species (Blande *et al.*, 2007) and together with monoterpenes, constitutes 65% of the total BVOCs emitted by the terrestrial ecosystems (Guenther *et al.*, 2012).

BVOCs produced by plants and microbes engage in plant reproduction (VOCs attract pollinators), growth, wound healing, development and defence against herbivores. BVOCs also mediate communication between plants, animals

---

* Corresponding author: hkaur.ecology@gmail.com

and insects, and protect plants against high temperature (Peñuelas *et al.*, 1995; Pichersky and Gershenzon, 2002). However, BVOC emissions increase with global warming and may produce both positive and negative feedback loops on temperatures through their indirect influence on the radiative balance of the Earth via aerosol formation (Sonwani *et al.*, 2016). Figure 2.1 gives an overall view of the role of BVOCs in the ecosphere.

The main objective of this chapter is to review the existing knowledge on BVOCs, their effect on atmospheric chemistry and how various components of our environment mediate positive or negative feedback loops in the biosphere–atmosphere–climate system. The first objective is to analyse our understanding of soil as an important BVOC source. Second, the chapter discusses the components of global change that can promote or limit BVOC emissions.

2.2 BVOCs and Atmospheric Chemistry

2.2.1 Production of tropospheric ozone

BVOCs are precursors to secondary air pollutants in the troposphere that also influence the radiative balance of the Earth. Oxidation of BVOCs gives rise to ozone (O_3) and smog in the presence of nitrogen oxides ($NO_x = NO + NO_2$) in the atmosphere (Daum *et al.*, 2000). Hydroxyl (OH) free radicals present in the atmosphere oxidize BVOCs in the presence of sufficient NO_x to form O_3. VOCs participate in oxidation of NO to NO_2 eliminating the role of O_3, which is otherwise eliminated in NO $\leftrightarrow$ NO_2 conversion, thereby increasing its concentration in the troposphere. Important sources of NO_x include fossil-fuel burning, fertilizer application, biomass burning, as well as natural processes, such as lightning. Most of the NO_x is present as NO_2 by dusk; sunlight photolyzes NO_2 during the day. In the presence of VOCs, production of O_3 is one molecule per one molecule of NO_2, which is not the case in the absence of VOCs (Monson and Holland, 2001). Fowler *et al.* (2008) thus highlight the importance of modelling techniques to understand the effects of BVOCs on O_3 concentrations diurnally and seasonally in the troposphere, given the non-linear nature of tropospheric chemistry with respect to emissions of O_3 precursors.

$$RCH_3 + OH + O_2 \rightarrow RCHO_2 + H_2O \quad \text{(Eq. 2.1)}$$

$$RCHO_2 + NO \rightarrow RCHO + NO_2 \quad \text{(Eq. 2.2)}$$

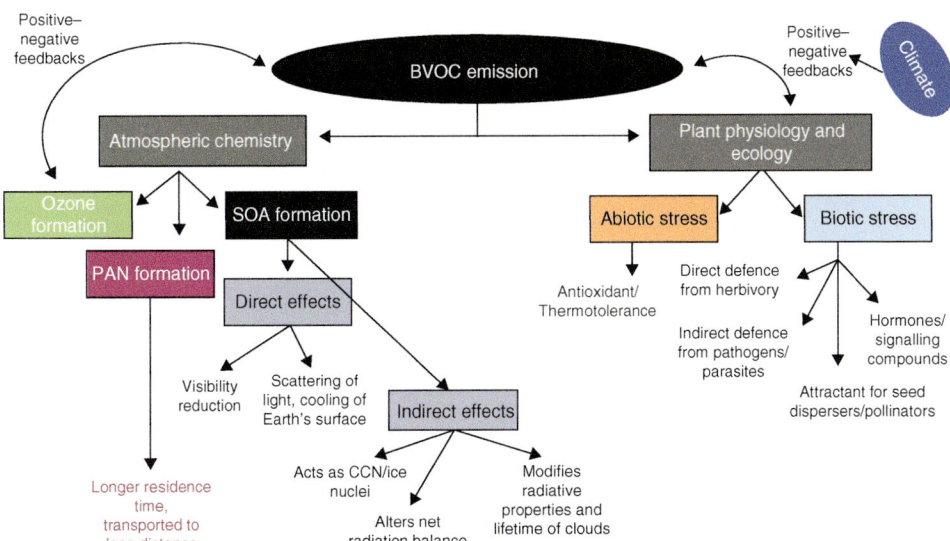

Fig. 2.1. Role of BVOCs in the ecosphere. (Adapted from Sonwani *et al.*, 2016.)

$$HO_2 + NO \rightarrow OH + NO_2 \qquad \text{(Eq. 2.3)}$$

$$2(NO_2 + O_2 \rightarrow NO + O_3) \qquad \text{(Eq. 2.4)}$$

Net result: $\mathbf{RCH_3 + 4O_2} \rightarrow$
$\mathbf{RCHO + 2O_3 + H_2O}$ (Eq. 2.5)

(Adapted from Sonwani et al., 2016.)

Equations 2.1 to 2.4 depict the role that VOCs play in O_3 formation, and equation 2.5 represents the net effect of any C–H bond available in the atmosphere on O_3 photochemistry. Different BVOC molecules have different O_3 formation capacities, depending upon the NO_x concentration. For instance, one molecule of isoprene can exert different outcomes; at low concentrations of NO_x, isoprene emission can reduce O_3 formation following a different set of equations. However, at average to higher NO_x concentrations, a molecule of isoprene can result in many O_3 molecules per molecule of isoprene (Trainer et al., 1987).

Besides affecting O_3 concentrations, it has been reported that in high isoprene emission zones such as extensive tropical forests, oxidation of isoprene reduces OH radical concentration up to 71% compared to CO oxidation (up to 11% reduction) and CH_4 oxidation (reduction up to 5%) (Grosjean, 1995). Peroxyacetyl nitrates, commonly referred to as PAN, are important oxidative products of BVOC oxidation. Owing to their longer residence times in the atmosphere, PAN travel longer distances and act as a source of NO_x on thermal decomposition (Fehsenfeld et al., 1992; Poisson et al., 2000). This could result in high levels of NO_x pollution in areas with even no NO_x emissions and contribute to O_3 in such areas (Sonwani et al., 2016).

2.2.2   Secondary organic aerosol formation

Though little is understood about the mechanism of secondary organic aerosol (SOA) formation via BVOC oxidation, it is known that the latter results in the formation of lower vapour pressure compounds that trigger condensation on pre-existing atmospheric molecules (Kulmala, 2003; Joutsensaari et al., 2005). Isoprene plays an important role in aerosol formation in the atmosphere. The

number of SOAs produced per molecule of isoprene is lower than the number of SOAs produced per molecule of monoterpene (Ng et al., 2006). Isoprene is an important source because of its higher atmospheric concentration (Kroll et al., 2005; Ng et al., 2006). However, terpenes and sesquiterpenes contribute significant amounts of SOA as compared to isoprene (Ng et al., 2006).

2.3   Soil: A BVOC Reservoir

Terrestrial biosphere plays a significant role in regulating atmospheric chemistry and climate. Vegetation cover, as well as interactions between atmosphere and terrestrial biosphere, has changed in the past few decades due to climate change, and a similar responsiveness can be observed for terrestrial biogeochemistry. The positive radiative forcing is estimated to reach up to 1.5 W/m²/K towards the end of 21st century resulting from the feedbacks between terrestrial biosphere and atmosphere (Arneth et al., 2010). The impact of global warming, drought, and high concentration of CO_2 on exchange of CO_2 on carbon budgets have been assessed in many studies; less attention has been paid to understanding how BVOC strengths are changing in response to these alterations (Monson et al., 2007). Several studies have shown that soil acts as a huge reservoir of BVOCs (Spielmann et al., 2017). These BVOCs come from decomposing litter and dead organic material, or may even be synthesized by microorganisms present in the soil. However, little is known regarding the types and quantities of BVOCs produced, their sources and sinks, and the factors that control their emission and diffusion. Nonetheless, it can be assumed that they play important roles in abiotic and biotic interactions of the soil and have important ecological, as well as environmental, effects (Peñuelas et al., 2014). This part of the chapter focuses on the BVOCs emitted from plant litter, plant roots and rhizomes, and microorganisms and the role they play in mediating interactions in the soil.

BVOCs are universally known to mediate interactions between biotic environments – these can be plant-to-plant, plant-to-animal, plant-to-microbe or microbe-to-microbe interactions. They do so by acting as important info-chemicals for intra- and inter-organismic communication and have growth manipulating agents. Hung *et al.* (2013) used *Arabidopsis thaliana* as a model to test the effects of VOCs from the soil fungus *Trichoderma viride*. They found that plants grown in the presence of *T. viride* (absence of any physical contact) volatiles had increased total biomass and chlorophyll concentration, besides growing taller and bigger. Their analysis of *T. viride* revealed 51 compounds with isobutyl alcohol, isopentyl alcohol and 3-methylbutanal as the most abundant species. Some BVOCs act as plant growth promoters by suppressing fungal growth (Kai *et al.*, 2007).

2.3.1 Soil as a source of BVOCs

Multiple biotic and abiotic factors result in non-methane BVOC emission from soil. Among biotic processes, microbial decomposition of soil organic matter is an important contributor. Such BVOC emissions are likely to have strong influences on soil ecology and terrestrial biogeochemistry (Leff and Fierer, 2008). Microbial BVOCs are either released as end-products or intermediates in aerobic/anaerobic respiratory processes or other metabolic pathways like fermentation. Above- and belowground dead plant biomass and root exudates contribute highly to soil organic matter (Kögel-Knabner, 2002). Microbial degradation of these plant-derived substrates forms one of the most important sources of soil VOCs, such as acetone (bare soil without litter) and methanol (soil covered with leaf litter) (Schade and Goldstein, 2001). Physical processes like evaporation of VOCs from plant litter or Maillard-type reactions also contribute to bursts of VOCs from soil after a rain or dew episode (Greenberg *et al.*, 2012). Roots are a good source of soil VOCs but their contribution to overall soil VOCs has not been studied in detail (Peñuelas *et al.*, 2014). Exudates from roots and root litter increase the microbial activity in soil, which may in turn increase or decrease the root-rhizosphere fluxes of VOCs (Asensio *et al.*, 2007; Rinnan *et al.*, 2013). Both abiotic and biotic processes are responsible for soil VOC emissions. However, no study has directly measured the contribution of these processes (Peñuelas *et al.*, 2014). Studies have, nonetheless, shown that different VOCs can have preferential mechanisms of release; for example, methanol and

terpenes prefer abiotic processes, whereas acetone is associated with biotic processes (Warneke *et al.*, 1999; Schade and Custer, 2004; Gray *et al.*, 2010).

2.3.2 Soil as a sink of BVOCs

Soils can also act as VOC sinks. Many microbes consume VOCs as a source of carbon (Owen *et al.*, 2007; Ramirez *et al.*, 2010). A number of studies (Arsenio *et al.*, 2007; Greenberg *et al.*, 2012; Aaltonen *et al.*, 2013) have also reported deposition of atmospheric VOCs in soil. Additionally, humic substances and mineral particles present in the soil trap VOCs through adsorption (Diamadopoulos *et al.*, 1998; Ruiz *et al.*, 1998). Nitrate and hydroxyl radicals, ozone and hydrogen peroxides increase the sink potential of soils via physicochemical degradation of VOCs (Insam and Seewald, 2010). Although experiments show that soils can absorb 80% of the VOCs produced by litter (Ramirez *et al.*, 2010), little information is available on the strength of soil as a sink of VOCs and needs further investigation. Tables 2.1 and 2.2 show the types of BVOCs from different ecosystems with their emission rates; as can be seen, terpenes and oxygenated VOCs dominate emissions occurring from soil. A pattern of lower BVOC emissions from soil compared to aboveground vegetation can be observed generally; discrepancies over significance of soil BVOC contribution occur while accounting for overall ecosystem fluxes. Differences in BVOC emissions from soil or aboveground vegetation may be attributed to the choice of methodology, temperature/moisture conditions or seasonal variations, etc. (Hellén *et al.*, 2006; Aaltonen *et al.*, 2011; Oderbolz *et al.*, 2013).

2.3.3 VOCs from soil microorganisms

There is an entire kingdom of microbial organisms in the soil, which secrete diverse chemicals including VOCs. Detection of microbial VOCs (mVOCs) depends upon several factors, such as the presence of substrates, growth conditions, detection techniques used, etc. Therefore, at any given time, a measured VOC profile might not reflect the complete diversity or volatile emission

Table 2.1. BVOCs measured in field conditions.

Ecosystem	Season	Source/sink of VOCs	Type of VOC	Emission rate ($\mu g/m^2/hr$)	Method used	References
Ponderosa Pine (PP) Plantation	Summer and autumn	Post-shower	Acetone	806	Dynamic chamber/GC-FID	Schade and Goldstein, 2001
Mediterranean Shrubland	Winter–summer	Shrubland soil	Methanol	144	Dynamic chamber/PTS-MS	Asensio et al., 2008
Agriculture (cereal)	Summer	Bare soil	Methanol	533	EC/PTR-MS	Schade and Custer, 2004
Scots Pine Forest	Spring–autumn	Forest floor	Monoterpenes	373	Static chamber/GC-MS and FID	Hellén et al., 2006
Scots Pine Forest	Spring–autumn	Forest floor	Methanol	194	Dynamic chamber/PTR-MS	Aaltonen et al., 2013
Norway Spruce Stand	Summer–autumn	Forest soil	Monoterpenes	47	Variant of dynamic chamber/HSGC-FID	Smolander et al., 2006
Sitka Spruce Plantation	Summer	Forest soil	Monoterpenes (30% limonene, 20% α-pinene, 20% myrcene, 20% camphene)	38	Dynamic chamber/GC-FID	Hayward et al., 2001
PP Plantation	Summer	Forest soil	Acetaldehyde	1.7	PTR-MS	Greenberg et al., 2012
Silver Birch Stand	Summer–autumn	Forest floor	Total monoterpenes	1.6	Variant of dynamic chamber/HSGC-FID	Smolander et al., 2006

Notes: GC-FID: gas chromatography with flame ionization detector; GC-MS: gas chromatography with mass spectrometry; HSGC-FID: head space gas chromatography with flame ionization detector; PTR-MS: proton transfer reaction mass spectrometry.

Table 2.2. BVOCs measured in laboratory experiments.

BVOCs source/sink	Type of VOC	Emission rate	Units	Method	References
Fagus **species** litter	Acetone	0.6	µg/gm DW/hr	Dynamic glass cell PTR-MS	Warneke *et al.*, 1999
Fagus **species** litter	Acetaldehyde	0.4	µg/gm DW/hr	Dynamic glass cell PTR-MS	Warneke *et al.*, 1999
Fagus **species** litter	Methanol	0.35	µg/gm DW/hr	Dynamic glass cell PTR-MS	Warneke *et al.*, 1999
Populus deltoides litter	Methanol	7.0	µmol/gm DW/hr	Dynamic jar headspace PTR-MS	Gray *et al.*, 2010
Pinus ponderosa litter	Methanol	2.5	µmol/gm DW/hr	Dynamic jar headspace PTR-MS	Gray *et al.*, 2010
Centaurea maculosa litter	Methanol	0.6	µmol/gm DW/hr	Dynamic jar headspace PTR-MS	Gray *et al.*, 2010
Rhododendron maximus litter	Methanol	1.6	µmol/gm DW/hr	Dynamic jar headspace PTR-MS	Gray *et al.*, 2010
Eucalyptus **species** litter	Methanol	5.9	µmol/gm DW/hr	Dynamic jar headspace PTR-MS	Gray *et al.*, 2010
Eucalyptus **species** litter	Monoterpene	0.4	µmol/gm DW/hr	Dynamic jar headspace PTR-MS	Gray *et al.*, 2010
Eucalyptus **species** litter	Propanol/ acetone	0.3	µmol/gm DW/hr	Dynamic jar headspace PTR-MS	Gray *et al.*, 2010

potential of a microorganism, but rather present only a snapshot. Identification of all VOCs emitted by microorganisms is also limited because the available database (NIST, Wiley) primarily contains VOCs from plants and animals. In other words, it would not be wrong to assume that new structures may be discovered in the future (Peñuelas *et al.*, 2014). Existing literature suggests that bacterial VOC profiles consist more of alkenes, ketones, pyrazines and terpenes; those identified from fungal species contain benzenoids, aldehydes, arsenics, chlorides, nitriles, thiofurans and bromides. Differences occur not only at the level of genus, but also species and strains (Peñuelas *et al.*, 2014).

Fermentation, amino-acid degradation, terpenoid biosynthesis and aerobic carbon metabolism are some of the processes producing VOCs in microorganisms. Primary metabolism and energy generation pathways, such as the Embden–Meyerhof pathway, heterolactic/ homolactic pathway, etc., result in intermediaries like pyruvate, glyceraldehyde 3-phosphate, lactate and acetate that act as precursors for VOC biosynthesis (Peñuelas *et al.*, 2014). Commonly known examples are synthesis of ethanol from *Saccharomyces cerevisiae* (Sniegowski *et al.*, 2002), lactic acid, ethanol by *Lactococcus* (Klijn *et al.*, 1995), etc. C_6 to C_{16} hydrocarbon blends, consisting of alkenes, aliphatic alcohols and ketones, are important mVOCs – typical fatty acid metabolism products. Shikimate pathways or degradation of L-phenylalanine or L-tyrosine generate aromatic compounds in microbes. Lactic acid bacteria contribute to the formation of volatile sulfur compounds like H_2S, dimethyl sulfide (DMS), methanethiol, dimethyl trisulfide (DMTS) through degradation of methionine (Schulz and Dickschat, 2007). These sulfur compounds play an important role in the global biogeochemical cycle as well.

Various interactions are mediated by mVOCs. Bacterial VOCs are known for their inhibitory effects on fungal growth and spore germination. In fact, bacterial VOC profiles are determined by environmental conditions, such as the presence of neighbouring plants, other bacteria, fungi, etc. (Kai et al., 2009). Soil fungi stasis, inhibition of fungal growth, is a phenomenon largely mediated by VOCs. Many recent studies have a documented growth inhibition effect of bacterial VOCs on fungi (Zou et al., 2007; Kai et al., 2009). Only a few tested fungal species remain unaffected by bacterial VOCs (Kai et al., 2007). Rhizoctonia solani is an important phytopathogen which causes serious damage to agricultural systems and natural forests. Several antagonistic bacterial VOCs are effective growth inhibitors of R. solani; it would therefore be interesting to develop VOC-based biological methods of fungal control (Kai et al., 2007).

Fungal VOCs, likewise, are known to possess anti-fungal as well as anti-bacterial properties. Saprobiont Schizophyllum commune releases VOCs that inhibit growth of Botrytis cinerea and Mucor miehei – important food pathogens (Campos et al., 2010). In addition, fungus-to-fungus interactions are mediated through VOCs (Hynes et al., 2007). However, how fungal VOCs affect the ecology of soil systems remains largely unexplored.

mVOCs interact with plants and animals in several ways, and the effects can be growth-stimulating, inhibitory or general communication. Microbial VOCs can affect the development of various animals. VOCs of Bacillus subtilis, Pseudomonas fluorescens, etc. impair the growth of protozoans like Acanthamoeba castellanii and are even lethal to Paramecium caudatum (Kai et al., 2008). Bacterial VOCs are important in triggering many plant responses. A microarray study by Zhang et al. (2007), revealed that more than 600 genes change their gene expression on exposure to bacterial VOCs.

Similarly, fungal VOCs also affect plant and animal fitness – volatiles of Fusarium oxysporum are known to possess nematicidal effects (Campos et al., 2010). The mycorrhizal symbiotic relationship, well known for its growth-promoting effects, can also activate immune responses in plants via influences on plant VOCs (Jung et al., 2012). Minerdi et al. (2009), in an experiment to study the effect of fungal VOCs on plants, exposed lettuce plants to airborne F. oxysporum

MSA 35 VOC. This stimulated shoot growth and increased overall biomass and chlorophyll content.

2.3.4   VOCs from plant roots and rhizomes

Like aboveground tissues, underground plant parts such as roots and rhizomes produce diverse VOCs. Terpenes are the most common belowground VOC and a common component of many extracts and essential oils of aromatic plants. A variety of monoterpenes and sesquiterpenes can be obtained from vetiver grass, rhizome of ginger, turmeric, etc. (Champagnat et al., 2006; Koo and Gang, 2012). One of the smallest VOC emitted by both above- and belowground plant tissues is methanol. While in aboveground plant parts, methanol is associated with leaf expansion and cell elongation (Hüve et al., 2007); in belowground parts, it causes root elongation and separation of root border cell caps (Driouich et al., 2007). Methanol is also known for inducing adventitious root formation in Arabidiopsis (Guénin et al., 2011). Root-produced methanol is used by methylotropic symbionts as a carbon source, which induces nodule formation (Sy et al., 2005). The role of methanol emissions in belowground herbivory is not clearly understood; however, pest attack of roots induces VOC release. C_6 volatiles have been detected in volatile blends in response to phylloxera, root borer feeding in plant roots. These C_6 root volatiles are assumed to possess bactericidal and fungicidal properties as observed in aboveground tissues as well (Prost et al., 2005). Schaller and Stintzi (2009) hint at the capability of C_6 volatiles to serve as short-range attractive cues for herbivores, parasites or plant signals. Table 2.3 gives an account of the diversity of VOCs produced by roots and rhizomes.

2.4   Why Study BVOC Emissions?

It is widely known and accepted that biological processes of the terrestrial ecosystem affect the atmosphere and the climate of the Earth. Biological processes related to carbon assimilation, its release as CO_2 or sequestration in organic material has been of particular interest. Nonetheless, it is also

Table 2.3. BVOCs from roots and rhizomes of various plants.

Plant species	VOCs	Method	References
Arabidopsis thaliana	Ethanol, ethyl acetate, 1,8-cineole	SPME, PTR-MS, solvent extraction	Chen *et al.*, 2004
Brassica nigra	Glucosinolate breakdown products, sulfides	PTR-MS	Tytgat *et al.*, 2013
Circuma longa	Monoterpenoids, α-zingiberene and other sesquiterpenoids	Organic solvent extraction	Koo and Gang, 2012
Lycopersicon esculentum	β-phellandrene	SPME	van Schie *et al.*, 2007
Pinus pinea	Monoterpene, sesquiterpene hydrocarbons	Passive diffusion method	Lin *et al.*, 2007

important to note that gases other than CO_2 are exchanged between the biosphere and atmosphere. BVOCs comprises a large variety of these gases, which differ in size, metabolic origin and physicochemical properties, and are released in the atmosphere from the vegetation-covered land masses (Laothawornkitkul *et al.*, 2009). Synthesis of BVOC and their properties, such as solubility, volatility and diffusivity, determine their emission rates. In other words, several internal and external factors modulate BVOC emission (Peñuelas and Llusià, 2003; Niinemets *et al.*, 2004). Because many processes affect BVOC synthesis and release compared to carbon assimilation in plants, changes in community structure and function, and land use change can lead to significant changes in BVOC emissions, even if the standing biomass and net primary productivity of the ecosystem remain unchanged. On entering the atmosphere, BVOC molecules gradually split and convert to CO_2 or other intermediate products. Chemical degradation of these intermediate compounds is more relevant to air quality and climate. Ecologically, BVOC emission plays an important role in growth, reproduction, defence and communication in the living system, and alterations arising from global and climate changes will improve our understanding of BVOC-mediated positive–negative feedback loops in the biosphere–atmosphere–climate system (Peñuelas and Staudt, 2010).

2.4.1 BVOC emissions in a changing world

A global mean temperature increase of 0.76°C has been recorded for the 20th century and a further increase of 1.8–4°C is projected for the 21st century (IPCC, 2013). There can be direct or indirect effects on trends of BVOC emission. Global models used frequently for BVOC emission response to temperature estimate an increase in BVOC emission by 10% over the past 30 years. Models predict a 30–40% increase in emissions from biological sources, with an additional increase of 2–3°C temperature (Keenan *et al.*, 2009). A number of factors are responsible for this rise. An increase in temperature enhances the enzymatic activity of VOC synthesis, as well as raising the VOC vapour pressure and decreasing the resistance to diffusion. A warmer climate means warmer average winter temperatures, this implies longer plant activity, leading to an increase in emissions. Analysis of model-based studies point to the importance of incorporating key temperature-sensitive coefficients in order to avoid overestimation of regional emissions (Staudt *et al.*, 2003; Keenan *et al.*, 2009). Research focusing on the study of medium- to long-term response of BVOC emissions is warranted to validate the expected increase in emissions (Peñuelas and Staudt, 2010). Components of global change such as land use change, tropospheric ozone, elevated CO_2 concentrations, nutrient availability, increased incidence of UV radiations and temperature can limit or promote constitutive or induced BVOC emission.

2.4.1.1 Land use change

For several constitutive BVOCs, such as wound-induced green leaf VOCs, genotype is an important determinant of their emission. Induced BVOCs involved in plant-to-plant or plant-to-insect interactions, depend upon the genetic make-up of plant species and sometimes on the insect inducer as well (De Moraes *et al.*, 1998). Changes in land use/land cover dramatically affect BVOC emissions because of such species specificities (Niinemets and Peñuelas, 2008). Examples can be drawn from the under-examined Arctic (see Box 2.1), temperate and boreal ecosystems or regions in the tropics. Increasing areas of tropical rainforest are being replaced by plantations such as rubber in China and oil palm in Malaysia – both are strong emitters of isoprenoid. Some of their compounds respond strongly to warming as well (Wilkinson *et al.*, 2006). Evergreens such as oaks, eucalyptus and pine are frequently used in afforestation programmes in temperate areas. These tree species are known for being strong BVOC emitters through the year (Keenan *et al.*, 2009). Elevated global temperatures increase the number of species that can thrive in colder climates, leading to a shift in vegetation types. For example, warming is leading to a progressive shift from a cold temperate ecosystem to a Mediterranean ecosystem in Spain (Peñuelas and Boada, 2003). A 5°C rise in minimum winter temperatures in freezing winters is speculated to increase the number of species that can thrive in an area by 7–20% (Niinemets and Peñuelas, 2008). Another important indirect effect is more leaf litter, which brings extra nutrients to the soil, providing an additional source of BVOC (Box 2.1). A more rapid change in species distribution is occurring because of profit-driven globalized trade in exotic plants, agri-products, etc., which alleviates biological and ecological constraints of dispersal (Niinemets and Peñuelas, 2008).

2.4.1.2 Tropospheric ozone

Concentrations of ozone in the troposphere are likely to increase in the future (Heald *et al.*, 2009). A strong variability occurs in the emission of various VOC types, depending upon plant species and season. Stress conditions such as wounding, water, ozone, etc. result in altered VOC emissions that may increase up to several orders of magnitude compared to non-stressed plants (Loreto and Sharkey, 1993; Sharkey and Loreto, 1993; Loreto and Velikova, 2001). A study conducted by Llusià *et al.* (2002) on the effect of ozone stress on Mediterranean woody plants reveals the above effects. For illustration, under similar ozone exposures, *Olea europaea* emitted less α-pinene as compared to *Quercus ilex ilex*, which showed elevated emissions of α-pinene and limonene. On considering all the species under study, there was an increase in net photosynthetic activity, total VOCs (45%) and limonene (95%) under ozone stress averaged over all seasons. An important point to note is that BVOCs are important precursors of ozone (see section 2.2.1 above), and an increase in BVOCs may lead to positive feedback in ozone formation.

Box 2.1. Effect of climate change on BVOC emission – example from the sub-Arctic.

The area north of 60 degrees is commonly referred to as the Arctic, and contributes 1–2% of the total isoprene and monoterpene emissions annually. An IPCC (2013) report projects an increase in temperature by 4–5°C by the year 2100 in the area north of 60°N – twice as much as the global mean temperature increase. Peñuelas and Fillela (2001) suggest that warmer conditions with a prolonged growing season will significantly change sub-Arctic ecosystem dynamics. The ongoing warming has led to expansion of shrubs to higher altitudes (Michelsen *et al.*, 2012) and an increase in BVOC emissions via changing biomass or species composition (e.g. Kramshoj *et al.*, 2016). Increasing abundance of deciduous plant species increases the leaf litter, generating new feedback loops. Addition of leaf litter increases microbial activity and the amount of nutrients returning to the soil and made available to plants (Rinnan *et al.*, 2008). Warming supports deciduous and evergreen shrubs and disfavours bryophytes, changing the vegetation structure and hence community-level BVOC emission (Valolahti *et al.*, 2015). Studies conducted in Abisko, northern Sweden (385m amsl) suggest that climate change in the sub-Arctic gradually changes the vegetation structure and composition of volatiles. Warming increases terpenoid emission resulting from increased volatility of stored compounds as well as by triggering *de novo* synthesis (Loreto and Velikova, 2001). Soudzilovskaia *et al.* (2013) suggest that warming may also change the genetic traits of plants and thus affect BVOC composition. An increased contribution to BVOC release by the sub-Arctic in the future is estimated by Valolahti *et al.* (2015).

2.4.1.3 Carbon dioxide concentration

Rising atmospheric CO_2 has many important environmental manifestations. Elevated CO_2 concentrations could increase productivity and standing biomass of vegetation, at least in the short term (Körner, 2006). This could affect the formation and release of BVOCs; however, it is not entirely clear if an elevated CO_2 level increases the emission of VOCs from plants (Peñuelas and Llusià, 1997), though studies have reported a reduction in isoprene emission in response to an increase in atmospheric CO_2 (Monson et al., 2016). Results vary depending upon the VOC species and plant species considered and their environmental and physiological conditions (Heald et al., 2009).

Similarly, limited information is available on the effects of increasing availability of nutrients, such as nitrogen. Eutrophication is known for its effects on changing biodiversity composition and function; its role in BVOC emission, therefore, cannot be overlooked (Peñuelas and Staudt, 2010). Increased incidence of UV radiations might increase the BVOC emission from the Arctic substantially (Box 2.1). Herbivory, too, acts as a stressor and alters BVOC emissions in plants (Himanen et al., 2009).

In general, given the paucity of precise and complete data on the effects of the many global change components, it is difficult to ascertain general trends in BVOC emission. Given the variability and complexity in the interactions between such components – biotic and abiotic – a large variability is introduced to BVOC emissions. For example, considering the direct and indirect effects of global temperature increase, how BVOC scenarios will change with long-term warming is still open to debate (Peñuelas and Staudt, 2010). Emission of BVOC is an unavoidable result of their volatile nature and most of the BVOCs have developed important physiological and ecological functions (Himanen et al., 2009). If these emissions increase under the changing climate, whether we will have a world with BVOCs as dominant ecological and evolutionary factors is an important point to consider (Peñuelas et al., 2014).

2.5 Conclusion

Based on the current review, it can be said that BVOCs will have an important atmospheric implication given the changing scenarios of global change components. The overall impact of an increase in BVOC emissions will affect their ecological, physiological and environmental role. While immense progress has been made in understanding the controls and effects of BVOCs, our knowledge is still insufficient to draw reliable quantitative predictions given the complexity of BVOC interactions with the physical and biological world.

References

Aaltonen, H., Pumpanen, J., Pihlatie, M., Hakola, H., Hellén, H., Kulmala, L. et al. (2011) Boreal pine forest floor biogenic volatile organic compound emissions peak in early summer and autumn. Agricultural and Forest Meteorology 151(6), 682–691.

Aaltonen, H., Aalto, J., Kolari, P., Pihlatie, M., Pumpanen, J. et al. (2013) Continuous VOC flux measurements on boreal forest floor. Plant and Soil 369(1–2), 241–256.

Arneth, A., Harrison, S.P., Zaehle, S., Tsigaridis, K., Menon, S. et al. (2010) Terrestrial biogeochemical feedbacks in the climate system. Nature Geoscience 3, 525–532.

Asensio, D., Peñuelas, J., Filella, I. and Llusià, J. (2007) On-line screening of soil VOCs exchange responses to moisture, temperature and root presence. Plant and Soil 291(1–2), 249–261.

Asensio, D., Peñuelas, J., Prieto, P., Estiarte, M., Filella, I. and Llusià, J. (2008) Interannual and seasonal changes in the soil exchange rates of monoterpenes and other VOCs in a Mediterranean shrubland. European Journal of Soil Science 59(5), 878–891.

Blande, J.D., Tiiva, P., Oksanen, E. and Holopainen, J.K. (2007) Emission of herbivore-induced volatile terpenoids from two hybrid aspen (Populus tremula × tremuloides) clones under ambient and elevated ozone concentrations in the field. Global Change Biology 13(12), 2538–2550.

Campos, V.P., Silva, R., de Pinho, C. and Souza Freire, E. (2010) Volatiles produced by interacting microorganisms potentially useful for the control of plant pathogens. Ciência e Agrotecnologia 34(3), 525–535.

Champagnat, P., Figueredo, G., Chalchat, J.-C., Carnat, A.-P. and Bessiere, J.-M. (2006) A study on the composition of commercial *Vetiveria zizanioides* oils from different geographical origins. *Journal of Essential Oil Research* 18(4), 416–422.

Chen, F., Ro, D.-K., Petri, J., Gershenzon, J., Bohlmann, J. *et al.* (2004) Characterization of a root-specific Arabidopsis terpene synthase responsible for the formation of the volatile monoterpene 1,8-cineole. *Plant Physiology* 135(4), 1956–1966.

Daum, P.H., Kleinman, L., Imre, D.G., Nunnermacker, L.J., Lee, Y.N. *et al.* (2000) Analysis of the processing of Nashville urban emissions on July 3 and July 18, 1995. *Journal of Geophysical Research: Atmospheres* 105(D7), 9155–9164.

De Moraes, C.M., Lewis, W.J., Pare, P.W., Alborn, H.T. and Tumlinson, J.H. (1998) Herbivore-infested plants selectively attract parasitoids. *Nature* 393(6685), 570.

Diamadopoulos, E., Sakellariadis, D. and Koukouraki, E. (1998) The effect of humic acid on the transport of volatile chlorinated hydrocarbons in soil. *Water Research* 32(11), 3325–3330.

Driouich, A., Durand, C. and Vicré-Gibouin, M. (2007) Formation and separation of root border cells. *Trends in Plant Science* 12(1), 14–19.

Fehsenfeld, F., Calvert, J., Fall, R., Goldan, P., Guenther, A.B. *et al.* (1992) Emissions of volatile organic compounds from vegetation and the implications for atmospheric chemistry. *Global Biogeochemical Cycles* 6(4), 389–430.

Fowler, D., Amann, M., Anderson, F., Ashmore, M., Cox, P. *et al.* (2008) Ground-level ozone in the 21st century: future trends, impacts and policy implications. *Royal Society Science Policy Report* 15(08). DOI: 10.1021/es4054434.

Gray, C.M., Monson, R.K. and Fierer, N. (2010) Emissions of volatile organic compounds during the decomposition of plant litter. *Journal of Geophysical Research: Biogeosciences* 115(G3), 2156–2202.

Greenberg, J.P., Asensio, D., Turnipseed, A., Guenther, A.B., Karl, T. and Gochis, D. (2012) Contribution of leaf and needle litter to whole ecosystem BVOC fluxes. *Atmospheric Environment* 59, 302–311.

Grosjean, D. (1995) Atmospheric chemistry of biogenic hydrocarbons – relevance to the Amazon. *Quimica Nova* 18(2), 184–201.

Guénin, S., Mareck, A., Rayon, C., Lamour, R., Ndong, Y.A. *et al.* (2011) Identification of pectin methylesterase 3 as a basic pectin methylesterase isoform involved in adventitious rooting in *Arabidopsis thaliana*. *New Phytologist* 192(1), 114–126.

Guenther, A.B., Jiang, X., Heald, C.L., Sakulyanontvittaya, T., Duhl, T. *et al.* (2012) The model of emissions of gases and aerosols from nature version 2.1 (MEGAN2.1): an extended and updated framework for modeling biogenic emissions. *Geoscientific Model Development* 5, 1471–1492.

Hayward, S., Muncey, R.J., James, A.E., Halsall, C.J. and Hewitt, C.N. (2001) Monoterpene emissions from soil in a Sitka spruce forest. *Atmospheric Environment* 35(24), 4081–4087.

Heald, C.L., Wilkinson, M.J., Monson, R.K., Alo, C.A., Wang, G. and Guenther, A. (2009) Response of isoprene emission to ambient CO_2 changes and implications for global budgets. *Global Change Biology* 15(5), 1127–1140.

Hellén, H., Hakola, H., Pystynen, K.-H., Rinne, J. and Haapanala, S. (2006) C_2–C_{10} hydrocarbon emissions from a boreal wetland and forest floor. *Biogeosciences* 3(2), 167–174.

Himanen, S.J., Nerg, A.M., Nissinen, A., Pinto, D.M., Stewart, C.N. *et al.* (2009) Effects of elevated carbon dioxide and ozone on volatile terpenoid emissions and multitrophic communication of transgenic insecticidal oilseed rape (*Brassica napus*). *New Phytologist* 181(1), 174–186.

Hung, R., Lee, S. and Bennett, J.W. (2013) *Arabidopsis thaliana* as a model system for testing the effect of *Trichoderma* volatile organic compounds. *Fungal Ecology* 6(1), 19–26.

Hüve, K., Christ, M.M., Kleist, E., Uerlings, R., Niinemets, Ü. *et al.* (2007) Simultaneous growth and emission measurements demonstrate an interactive control of methanol release by leaf expansion and stomata. *Journal of Experimental Botany* 58(7), 1783–1793.

Hynes, J., Müller, C.T., Jones, T.H. and Boddy, L. (2007) Changes in volatile production during the course of fungal mycelial interactions between *Hypholoma fasciculare* and *Resinicium bicolor*. *Journal of Chemical Ecology* 33(1), 43–57.

Insam, H. and Seewald, M.S.A. (2010) Volatile organic compounds (VOCs) in soils. *Biology and Fertility of Soils* 46(3), 199–213.

IPCC (2013) *Climate Change 2013: The Physical Science Basis. Contribution of Working Group I to the Fifth Assessment Report of the Intergovernmental Panel on Climate Change*, ed. Stocker, T.F., Qin, D., Plattner, G.-K., Tignor, M., Allen, S.K. *et al.* Cambridge University Press, Cambridge and New York. DOI: 10.1017/CBO9781107415324.

Joutsensaari, J., Loivamäki, M., Vuorinen, T., Miettinen, P., Nerg, A.-M. *et al.* (2005) Nanoparticle formation by ozonolysis of inducible plant volatiles. *Atmospheric Chemistry and Physics* 5(6), 1489–1495.

Jung, S.C., Martinez-Medina, A., Lopez-Raez, J.A. and Pozo, M.J. (2012) Mycorrhiza-induced resistance and priming of plant defenses. *Journal of Chemical Ecology* 38(6), 651–664.

Kai, M., Effmert, U., Berg, G. and Piechulla, B. (2007) Volatiles of bacterial antagonists inhibit mycelial growth of the plant pathogen *Rhizoctonia solani*. *Archives of Microbiology* 187(5), 351–360.

Kai, M., Vespermann, A. and Piechulla, B. (2008) The growth of fungi and *Arabidopsis thaliana* is influenced by bacterial volatiles. *Plant Signaling & Behavior* 3(7), 482–484.

Kai, M., Haustein, M., Molina, F., Petri, A., Scholz, B. and Piechulla, B. (2009) Bacterial volatiles and their action potential. *Applied Microbiology and Biotechnology* 81(6), 1001–1012.

Keenan, T., Niinemets, Ü., Sabate, S., Gracia, C. and Peñuelas, J. (2009) Process based inventory of isoprenoid emissions from European forests: model comparisons, current knowledge and uncertainties. *Atmospheric Chemistry and Physics* 9, 4053–4076.

Kesselmeier, J. and Staudt, M. (1999) Biogenic volatile organic compounds (VOC): an overview on emission, physiology and ecology. *Journal of Atmospheric Chemistry* 33(1), 23–88.

Klijn, N., Weerkamp, A.H. and De Vos, W.M. (1995) Detection and characterization of lactose-utilizing *Lactococcus* spp. in natural ecosystems. *Applied and Environmental Microbiology* 61(2), 788–792.

Kögel-Knabner, I. (2002) The macromolecular organic composition of plant and microbial residues as inputs to soil organic matter. *Soil Biology and Biochemistry* 34(2), 139–162.

Koo, H.J. and Gang, D.R. (2012) Suites of terpene synthases explain differential terpenoid production in ginger and turmeric tissues. *PLoS One* 7(12), e51481.

Körner, C. (2006) Plant CO_2 responses: an issue of definition, time and resource supply. *New Phytologist* 172(3), 393–411.

Kramshoj, M., Vedel-Peterson, I., Schollert, M., Rinnan, A., Nymand, J. *et al.* (2016) Large increases in Arctic biogenic volatile emissions are a direct effect of warming. *Nature Geoscience* 9, 349–352.

Kroll, J.H., Ng, N.L., Murphy, S.M., Flagan, R.C. and Seinfeld, J.H. (2005) Secondary organic aerosol formation from isoprene photooxidation under high NO_x conditions. *Geophysical Research Letters* 32(18), 1869–1877.

Kulmala, M. (2003) How particles nucleate and grow. *Science* 302(5647), 1000–1001.

Laothawornkitkul, J. *et al.* (2009) Biogenic volatile organic compounds in the Earth system. *New Phytologist* 183, 27–51.

Leff, J.W. and Fierer, N. (2008) Volatile organic compound (VOC) emissions from soil and litter samples. *Soil Biology and Biochemistry* 40(7), 1629–1636.

Lin, C., Owen, S.M. and Peñuelas, J. (2007) Volatile organic compounds in the roots and rhizosphere of *Pinus* spp. *Soil Biology and Biochemistry* 39(4), 951–960.

Llusià, J., Peñuelas, J. and Gimeno, B.S. (2002) Seasonal and species-specific response of VOC emissions by Mediterranean woody plant to elevated ozone concentrations. *Atmospheric Environment* 36(24), 3931–3938.

Loreto, F. and Sharkey, T.D. (1993) Isoprene emission by plants is affected by transmissible wound signals. *Plant, Cell and Environment* 16, 563–570.

Loreto, F. and Velikova, V. (2001) Isoprene produced by leaves protects the photosynthetic apparatus against ozone damage, quenches ozone products and reduces lipid peroxidation of cellular membranes. *Plant Physiology* 127, 1781–1787.

Michelsen, A., Rinnan, R. and Jonasson, S. (2012) Two decades of experimental manipulations of heaths and forest understory in the subarctic. *Ambio* 41 Supplement 3, 218–230.

Minerdi, D., Bossi, S., Gullino, M.L. and Garibaldi, A. (2009) Volatile organic compounds: a potential direct long distance mechanism for antagonistic action of Fusarium oxysporum strain MSA 35. *Environmental Microbiology* 11(4), 844–854.

Monson, R.K. and Holland, E.A. (2001) Biospheric trace gas fluxes and their control over tropospheric chemistry. *Annual Review of Ecology and Systematics* 32(1), 547–576.

Monson, R.K., Trahan, N., Rosenstiel, T.N., Veres, P., Moore, D. *et al.* (2007) Isoprene emission from terrestrial ecosystems in response to global change: minding the gap between models and observations. *Philosophical Transactions of the Royal Society A* 365, 1677–1695.

Monson, R.K., Neice, A.A., Trahan, N.A., Shiach, I., McCorkel, J.T. and Moore, D.J.P. (2016) Interactions between temperature and intercellular CO_2 concentration in controlling leaf isoprene emission rates. *Plant, Cell and Environment* 39, 2404–2413.

Ng, N.L., Kroll, J.H., Keywood, M.D., Bahreini, R., Varutbangkul, V. *et al.* (2006) Contribution of first-versus second-generation products to secondary organic aerosols formed in the oxidation of biogenic hydrocarbons. *Environmental Science & Technology* 40(7), 2283–2297.

Niinemets, Ü. and Peñuelas, J. (2008) Gardening and urban landscaping: significant players in global change. *Trends in Plant Science* 13(2), 60–65.

Niinemets, Ü., Loreto, F. and Reichstein, M. (2004) Physiological and physicochemical controls on foliar volatile organic compound emissions. *Trends in Plant Science* 9(4), 180–186.

Oderbolz, D.C., Aksoyoglu, S., Keller, J., Barmpadimos, I., Steinbrecher, R. *et al.*. (2013) A comprehensive emission inventory of biogenic volatile organic compounds in Europe: improved seasonality and land-cover. *Atmospheric Chemistry and Physics* 13(4), 1689–1712.

Owen, S.M., Clark, S., Pompe, M. and Semple, K.T. (2007) Biogenic volatile organic compounds as potential carbon sources for microbial communities in soil from the rhizosphere of *Populus tremula*. *FEMS Microbiology Letters* 268(1), 34–39.

Peñuelas, J. and Boada, M. (2003) A global change-induced biome shift in the Montseny mountains (NE Spain). *Global Change Biology* 9(2) 131–140.

Peñuelas, J. and Filella, I. (2001) Responses to a warming world. *Science* 294(5543), 793–795.

Peñuelas, J. and Llusià, J. (1997) Effects of carbon dioxide, water supply, and seasonality on terpene content and emission by *Rosmarinus officinalis*. *Journal of Chemical Ecology* 23(4), 979–993.

Peñuelas, J. and Llusià, J. (2001) The complexity of factors driving volatile organic compound emissions by plants. *Biologia Plantarum* 44(4), 481–487.

Peñuelas, J. and Llusià, J. (2003) BVOCs: plant defense against climate warming? *Trends in Plant Science* 8(3), 105–109.

Peñuelas, J. and Staudt, M. (2010) BVOCs and global change. *Trends in Plant Science* 15(3), 133–144.

Peñuelas, J., Filella, I. and Gamon, J.A. (1995) Assessment of photosynthetic radiation use efficiency with spectral reflectance. *New Phytologist* 131(3), 291–296.

Peñuelas, J., Asensio, D., Tholl, D., Wenke, K., Rosenkranz, M. *et al.* (2014) Biogenic volatile emissions from the soil. *Plant, Cell & Environment* 37(8), 1866–1891.

Pichersky, E. and Gershenzon, J. (2002) The formation and function of plant volatiles: perfumes for pollinator attraction and defense. *Current Opinion in Plant Biology* 5(3), 237–243.

Pichersky, E., Noel, J.P. and Dudareva, N. (2006) Biosynthesis of plant volatiles: nature's diversity and ingenuity. *Science* 311(5762), 808–811.

Poisson, N., Kanakidou, M. and Crutzen, P.J. (2000) Impact of non-methane hydrocarbons on tropospheric chemistry and the oxidizing power of the global troposphere: 3-dimensional modelling results. *Journal of Atmospheric Chemistry* 36(2), 157–230.

Prost, I., Dhondt, S., Rothe, G., Vicente, J., Rodriguez, M.J. *et al.* (2005) Evaluation of the antimicrobial activities of plant oxylipins supports their involvement in defense against pathogens. *Plant Physiology* 139(4), 1902–1913.

Ramirez, K.S., Lauber, C.L. and Fierer, N. (2010) Microbial consumption and production of volatile organic compounds at the soil–litter interface. *Biogeochemistry* 99(1–3), 97–107.

Rinnan, R., Michelsen, A. and Jonasson, S. (2008) Effects of litter addition and warming on soil carbon, nutrient pools and microbial communities in a subarctic heath ecosystem. *Applied Soil Ecology* 39(3), 271–281.

Rinnan, R., Gierth, D., Bilde, M., Rosenørn, T. and Michelsen, A. (2013) Off-season biogenic volatile organic compound emissions from heath mesocosms: responses to vegetation cutting. *Frontiers in Microbiology* 4, 1–10.

Ruiz, J., Bilbao, R. and Murillo, M.B. (1998) Adsorption of different VOC onto soil minerals from gas phase: Influence of mineral, type of VOC, and air humidity. *Environmental Science & Technology* 32(8), 1079–1084.

Schade, G.W. and Custer, T.G. (2004) OVOC emissions from agricultural soil in northern Germany during the 2003 European heat wave. *Atmospheric Environment* 38(36), 6105–6114.

Schade, G.W. and Goldstein, A.H. (2001) Fluxes of oxygenated volatile organic compounds from a ponderosa pine plantation. *Journal of Geophysical Research: Atmospheres* 106(D3), 3111–3123.

Schaller, A. and Stintzi, A. (2009) Enzymes in jasmonate biosynthesis – structure, function, regulation. *Phytochemistry* 70(13), 1532–1538.

Schulz, S. and Dickschat, J.S. (2007) Bacterial volatiles: the smell of small organisms. *Natural Product Reports* 24(4), 814–842.

Sharkey, T.D. and Loreto, F. (1993) Water stress, temperature, and light effects on the capacity for iso-prene emission and photosynthesis of kudzu leaves. *Oecologia* 95, 328–333.

Smolander, A., Ketola, R.A., Kotiaho, T., Kanerva, S., Suominen, K. and Kitunen, V. (2006) Volatile mono-terpenes in soil atmosphere under birch and conifers: effects on soil N transformations. *Soil Biology and Biochemistry* 38(12), 3436–3442.

Sniegowski, P.D., Dombrowski, P.G. and Fingerman, E. (2002) *Saccharomyces cerevisiae* and *Saccharomyces paradoxus* coexist in a natural woodland site in North America and display different levels of reproductive isolation from European conspecifics. *FEMS Yeast Research* 1(4), 299–306.

Sonwani, S., Saxena, P. and Kulshrestha, U. (2016) Role of global warming and plant signaling in BVOC emissions. In: Kulshrestha, U. and Saxena, P. (eds) *Plant Responses to Air Pollution*. Springer, Singapore, pp. 45–57.

Soudzilovskaia, N.A., Elumeeva, T.G., Onipchenko, V.G., Shidakov, I.I., Salpagarova, F.S., Khubiev, A.B., Tekeev, D.K. and Cornelissen, J.H.C. (2013) Functional traits predict relationship between plant abun-dance dynamic and long-term climate warming. *Proceedings of the National Academy of Sciences USA* 110(45), 18180–18184.

Spielmann, F.M., Langebner, S., Ghirardo, A., Hansel, A., Schnitzler, J.-P. and Wohlfahrt, G. (2017) Iso-prene and α-pinene deposition to grassland mesocosms. *Plant and Soil* 410, 313–322.

Staudt, M., Joffre, R. and Rambal, S. (2003) How growth conditions affect the capacity of *Quercus ilex* leaves to emit monoterpenes. *New Phytologist* 158(1), 61–73.

Sy, A., Timmers, A.C.J., Knief, C. and Vorholt, J.A. (2005) Methylotrophic metabolism is advantageous for *Methylobacterium extorquens* during colonization of *Medicago truncatula* under competitive condi-tions. *Applied and Environmental Microbiology* 71(11), 7245–7252.

Trainer, M., Williams, E.J., Parrish, D.D., Buhr, M.P., Allwine, E.J. *et al.* (1987) Models and observations of the impact of natural hydrocarbons on rural ozone. *Nature* 329(6141), 705–707.

Tytgat, T.O.G., Verhoeven, K.J.F., Jansen, J.J., Raaijmakers, C.E., Bakx-Schotman, T. *et al.* (2013) Plants know where it hurts: root and shoot jasmonic acid induction elicit differential responses in *Brassica olracea*. *PLoS One* 8, e65502.

Valolahti, H., Kivimäenpää, M., Faubert, P., Michelsen, A. and Rinnan, R. (2015) Climate change induced vegetation change as a driver of increased subarctic biogenic volatile organic compound emissions. *Global Change Biology* 21(9), 3478–3488.

van Schie, C.C.N., Haring, M.A. and Schuurink, R.C. (2007) Tomato linalool synthase is induced in trichomes by jasmonic acid. *Plant Molecular Biology* 64(3), 251–263.

Warneke, C., Karl, T., Judmaier, H., Hansel, A., Jordan, A. *et al.* (1999) Acetone, methanol, and other par-tially oxidized volatile organic emissions from dead plant matter by abiological processes: significance for atmospheric HO_x chemistry. *Global Biogeochemical Cycles* 13(1), 9–17.

Wilkinson, M.J., Owen, S.M., Possell, M., Hartwell, J., Gould, P. *et al.* (2006) Circadian control of isoprene emissions from oil palm (*Elaeis guineensis*). *The Plant Journal* 47(6), 960–968.

Zhang, H., Kim, M.-S., Krishnamachari, V., Payton, P., Sun, Y. *et al.* (2007) Rhizobacterial volatile emissions regulate auxin homeostasis and cell expansion in *Arabidopsis*. *Planta* 226(4), 839.

Zou, C.-S., Mo, M.-H., Gu, Y.-Q., Zhou, J.-P. and Zhang, K.-Q. (2007) Possible contributions of volatile-producing bacteria to soil fungistasis. *Soil Biology and Biochemistry* 39(9), 2371–2379.

3 Transport of Air Pollutants

Naveen Chandra[1]* and Vineet Goswami[2]
[1]*Japan Agency for Marine-Earth Science and Technology, Yokohama, Japan;* [2]*Colorado State University, Fort Collins, USA*

Abstract

Rapid population and economic growth and continuing industrialization have led to a significant rise in the levels of air pollutants in the Earth's atmosphere. The emission of pollutants from localized sources can impact the regional and global air quality through both dynamical and chemical processes in the atmosphere. Here, we discuss some of the important dynamical processes in the atmosphere and their impacts on the burden of pollutants in the troposphere – a region of the atmosphere where the impacts are directly experienced by humans and ecosystems.

3.1 Introduction

A small (~0.07%) volume of the Earth's atmosphere constitutes a myriad of gases such as carbon dioxide (CO_2), carbon monoxide (CO), methane (CH_4), sulfur hexafluoride (SF_6), chlorofluorocarbons (CFCs), non-methane volatile organic compounds (NMVOCs), nitrogen oxides (NO_x), sulfur dioxide (SO_2) and ozone (O_3), commonly known as 'trace gases'. Though occurring in small proportions, they play a fundamental part in sustaining life on Earth. The natural greenhouse effect (process of trapping the Earth's surface-emitted long wave radiation by the greenhouse gases present in the atmosphere, which help to maintain the Earth's temperature) makes Earth a habitable place. Various atmospheric trace constituents such as CO_2, CH_4, O_3 and CFCs contribute to the Earth's greenhouse effect and are hence known as greenhouse gases (GHGs). Higher levels of surface O_3, CO, SO_2 etc., will significantly affect the future climate, human health and crop yields (Seinfeld and Pandis, 2006; Lal *et al.*, 2017). Hence, these atmospheric gases are treated as air pollutants. Over the past two centuries, the Earth's climate and air quality have significantly changed due to the large emissions of GHGs and air pollutants from various anthropogenic (human-induced) activities, particularly the use of fossil fuels for energy.

The Vostok ice core records show that the levels of GHGs, particularly CO_2 and CH_4, have fluctuated in the past, when there were no anthropogenic activities (Petit *et al.*, 1999). These fluctuations derived from natural factors. As a consequence of this, the Earth's climate system pulsed between glacial and interglacial states. The time frame of these changes was roughly 100,000 years. However, the Earth's system has dramatically left the bounds of natural variability over the last 250 years due to extensive

---

* Corresponding author: nav.phy09@gmail.com

© CAB International 2019. *Air Pollution: Sources, Impacts and Controls*
(eds P. Saxena and V. Naik)

anthropogenic activities (Jacob and Winner, 2009; Stocker *et al.*, 2013). Therefore, it is important to understand the processes that regulate the abundance and distribution of atmospheric trace constituents and pollutants.

The evolution of the abundance of trace gases at different longitude (x), latitude (y) and altitude (z) with time in the Earth's atmosphere depends on their production (emission from direct sources + chemical production), loss (chemical loss + deposition), and transport (convection + advection + diffusion), which can be represented by the following continuity equation.

$$\frac{\partial n}{\partial t} = P - L - \nabla.(nU) \qquad \text{(Eq. 3.1)}$$

where:

n = atmospheric number density of trace gases in three dimensional (x,y,z) frame of reference fixed to the Earth

P = sources inside the hypothetical atmospheric volume (dx, dy, dz)

L = sinks inside the volume from chemical loss and deposition

U = wind vector (having three components: u, v, w)

$\nabla \cdot (nU)$ = transport of atmospheric trace species

Emission sources of pollutants are either localized (e.g. industrial chimneys, traffic along highways) or confined to a small region (e.g. biomass burning related to deforestation in tropical areas; an industrialized urban environment). The dense urban regions are the major 'hotspots' of air pollutants. The interplay between emission and transport controls the levels of pollutants at local, regional and global scales. Through the transport of atmospheric pollutants, the human-induced activities have global consequences, which is one of the most important issues for global climate change and health studies. One of the most prominent pieces of evidence for the impact of atmospheric pollution emission and transport on the composition of the global atmosphere is the Antarctic's 'ozone hole': a large reduction in the levels of stratospheric ozone over the southern polar cap each spring due to the reaction with chlorofluorocarbons (CFCs), which are transported from other polluted regions (Molina and Rowland, 1974; Crutzen and Arnold, 1986; Solomon, 1990; Solomon *et al.*, 1994). The Arctic

haze (the phenomenon of a visible reddish-brown springtime haze in the atmosphere) caused by industrial pollution from Europe is also another example of the transport-driven impact of atmospheric pollutants on the remote regions (Barrie, 1986).

The distribution of trace constituents in the atmosphere is a result of emission from their sources, their chemical interactions, drawdown by various sinks (driven by various energetic (thermodynamic) processes with major influence of solar radiation (photochemistry)) and transport of air driven by atmospheric dynamics. A comprehensive understanding of these processes is very important in order to understand the fate of trace gases in the atmosphere. This chapter will discuss the role of transport in the distribution of air pollutants in the atmosphere.

3.2 Spatial and Temporal Scale of Transport

The time taken by an air parcel to cross the globe depends on the different zonal (latitudinal) and meridional (longitudinal) wind regimes. For example, under the zonal (west to east) wind regimes, the air parcel can take less than 8 days to over a month to circulate zonally under the influence of the subtropical winter jet (40 m s^{-1}) and trade winds (easterlies << 10 m s^{-1}), respectively. The transport is comparatively slow in the meridional (south to north) direction because of low wind speed. In general, the zonal, intra-hemispherical and inter-hemispherical mixing rates of air parcels are about 2 weeks, 1–2 months and ~1.5 yrs, respectively. Trace gases exhibit an enormous range of spatial and temporal variability depending on their atmospheric lifetimes or residence times (Fig. 3.1).

The short-lived trace gases can be transported only over a limited distance due to their relatively shorter lifetime, resulting in large spatial heterogeneity in their distributions, while the long-lived species can be transported over larger distances, resulting in a spatial uniformity of these species in the atmosphere. For example, with an average lifetime of less than 0.01 s, the OH radical has a spatial transport scale of only about 1 cm. In contrast, CFCs, with a lifetime of

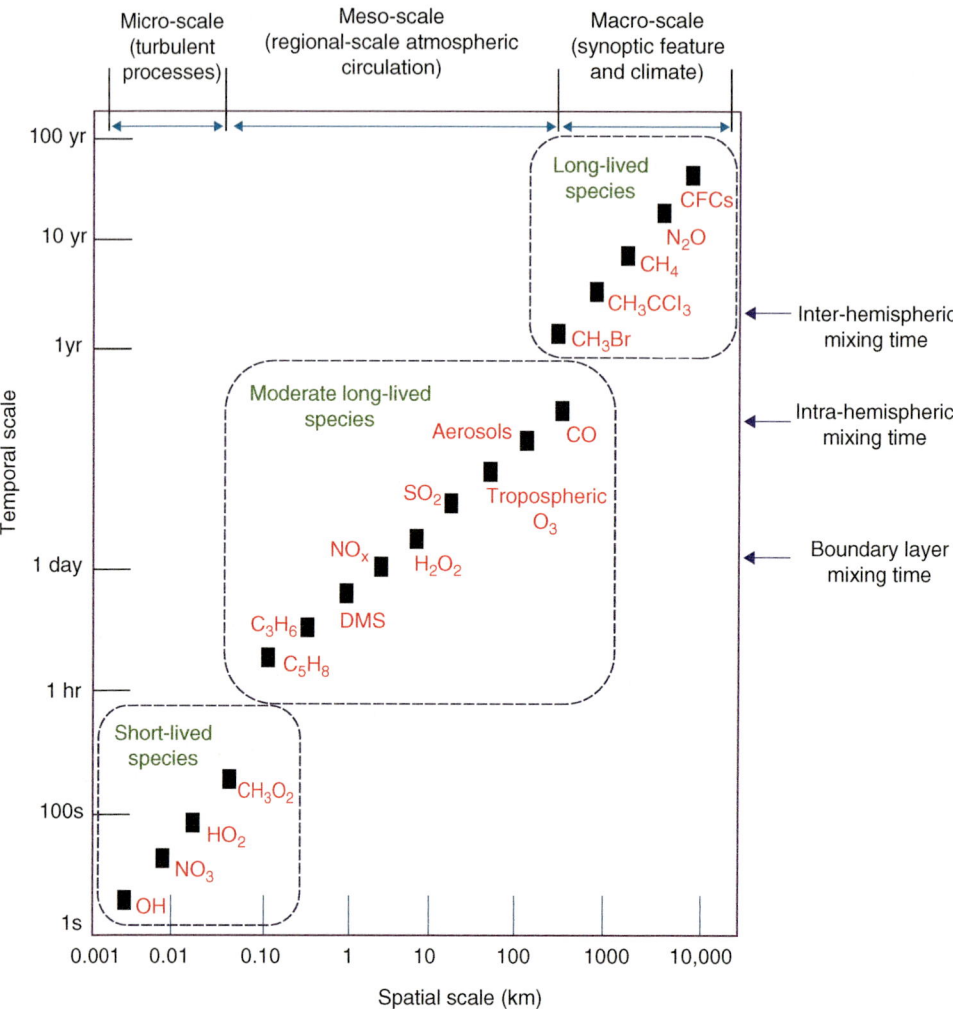

Fig. 3.1. Characteristic scales (spatial and temporal) for the variability of various trace constituents in the atmosphere. (Reproduced from Seinfeld and Pandis, 2006.)

more than 100 years, have a uniformly mixed (homogeneous) distribution throughout the Earth's atmosphere. The long average lifetime of CFCs even allows their mixing from the troposphere to the stratosphere.

The various processes acting on a wide range of scales have the potential to transport atmospheric pollutants over an extensive range of spatial scales (spanning eight orders of magnitude: Fig. 3.1). In particular, the spatial scale transport is broadly divided into three categories: (i) microscale; (ii) meso-scale; and (iii) macro-scale.

The micro-scale process, namely dispersion and dilution through atmospheric turbulence,

occurs in the boundary layer (~2–3 km above the Earth's surface) and plays a major role in the distributions of the pollutant in the boundary layer. This region of the boundary layer is directly influenced by biogenic and anthropogenic emissions. The day and night (diurnal) variation in temperature governs the dynamics of the atmospheric boundary layer. The formation of the convective boundary layer and inversion layer during daytime and nighttime, respectively, controls the dilution and accumulation of pollutants near the Earth's surface. Details about the formation of the boundary layer and its impact on the distribution of pollutants are given in section 3.3.

The meso-scale processes, occurring on scales of tens to hundreds of kilometres, include land–sea breeze circulation, mountain–valley winds and migratory high- and low-pressure fronts. Sea breeze circulation can transport the polluted continental air mass over the pristine marine environments. The synoptic scale motions, on a scale of hundreds to thousands of kilometres, dominate mainly in the free troposphere (above the boundary layer) and can transport trace constituents over long distances, even across continents.

The macro-scale processes, which operate on the largest spatial scale, result in the incorporation of air pollutants into the planetary-scale atmosphere through various mixing and transport mechanisms. The macro-scale transport processes are responsible for global impact/influence of atmospheric pollutants (trace gases) that are emitted by localized zones. The enhancement of GHG's forcing of climate and stratospheric depletion of ozone is one of the best examples of the consequences and impacts of the macro-scale transport.

3.3   Transport Processes in the Troposphere

The balance among gravitational, pressure-gradient and Coriolis forces mainly induces the transport of air masses in the atmosphere. The transport of air masses are governed by various dynamical processes such as advection, convection, etc. These processes are briefly summarized and their impacts on the distribution of air pollutants discussed in the following sections.

3.3.1   Advection: horizontal transport

The process of horizontal transport is generally referred to as advection. The horizontal pressure-gradient and Coriolis forces mainly control the movement of air mass in the horizontal direction, since gravity doesn't operate in this direction. These forces originate from the differential heating of the Earth's surface and rotation of the Earth, respectively. As a result of pressure gradient force, the air mass moves from high pressure to a low-pressure region. Further, the Coriolis force deflects the direction of air mass to the right in the northern hemisphere and left in the southern hemisphere (Jacob, 1999). Horizontal transport is fastest in the longitudinal direction and it takes only a few weeks for the circulation of the air parcel globally in a particular latitudinal band. Transport is slow in the latitudinal direction and the circulation of air masses takes 1–2 months from mid-latitude to polar or tropical regions (Jacob, 1999). The sluggishness of the inter-hemispheric transport results in around a year for exchange of air masses between southern and northern hemispheres. The chemical lifetime of air pollutants like CO ranges from a few weeks to a couple of months in the middle and upper troposphere, which is less than the inter-hemispheric mixing time (~1 yr); consequently, they are not well mixed in the troposphere. The global transport of CO from source regions makes it an excellent passive tracer to study the tropospheric transport and circulation of global and regional pollutants emitted from industrial activities and large-scale biomass burning (Lelieveld *et al.*, 2001; Lal *et al.*, 2007, 2013, 2014; Lawrence and Lelieveld, 2010; Mallik *et al.*, 2013a; Chandra *et al.*, 2016a; Bhardwaj *et al.*, 2017).

3.3.2   Convection: vertical transport

The upward or downward transport of air parcels is derived from buoyancy. Buoyancy is the net force exerted on the air parcel due to the difference between pressure-gradient force and gravity. If the air parcel is lighter than the surrounding environment, it will be uplifted by buoyancy, and if it is heavier than the surrounding environment then it will be accelerated back to its position by buoyancy. The cooling of the rising air by adiabatic expansion results in an increase in the relative humidity, resulting in the formation of clouds and drawdown of water. This phenomenon is generally associated with shady weather. In contrast, the sinking air gets heated up due to compression; resulting in lowering of the relative humidity, leading to fair weather (sunny and dry conditions). The wind speeds, associated with the vertical transport, are very low (in the range of 0.001–0.001 ms^{-1}) as compared to the horizontal wind speeds (in the range of $1-10 \, \mathrm{ms^{-1}}$) (Jacob, 1999). Therefore, the vertical transport of air parcels takes about 3 months to reach the tropopause (the uppermost part of the troposphere) from the surface.

Vertical transport could be amplified by local buoyancy; driven by localized meteorological parameters such as temperature, humidity, etc. The rapid buoyant motion is referred to as convection. Convection is an important force in the atmosphere that results in rapid vertical mixing of surface air with the upper tropospheric air in a very short time period (~hrs). The presence of moisture in the air parcel affects its stability conditions. As the air parcel ascends due to the convective force, it cools down and moisture condenses, releasing the latent heat in the atmosphere. The release of the latent heat raises the temperature of the air parcel, resulting in an unstable atmosphere, allowing the transport of air parcels up to large vertical distances (Brasseur *et al.*, 1999). This process is known as the moist convection or deep convection. Atmospheric convection is called *deep* when it extends from near the surface to above the 500 hPa level. The strength of the convective activity is determined by the available convective potential energy (CAPE, J kg⁻¹). Vertical mixing varies with season and latitude and is supposed to be fastest in the lower latitudes during periods of deep convective activity. Most atmospheric deep convection occurs in the tropics during the summer monsoon season (June to September) as the rising branch of the Hadley circulation, a strong local coupling between the surface and the upper troposphere. The changes in vertical distribution of temperature and moisture can impact the concentration of trace gases within the layer via dynamical processes (Brasseur *et al.*, 1999).

Equivalent potential temperature (EPT) can be used as a proxy to study the moist convective processes, as it is conserved in the condensation and evaporation of water (Emanuel, 1994). EPT is the temperature of an air parcel after condensing the moisture in the air parcel, releasing its latent heat, and bringing back adiabatically to the 1000 hPa level. EPT can be denoted mathematically as:

$$\mathrm{EPT} = \left[T + \frac{q \times L}{C_p} \right] \times \left(\frac{p_o}{p} \right)^{\frac{R}{C_p}} \qquad \text{(Eq. 3.2)}$$

where T is the temperature in Kelvin, q is specific humidity in g Kg⁻¹, p_o is pressure at the surface, p is altitude pressure, L is latent heat (2.54×10^6 J Kg⁻¹), R is universal gas constant (287 J K⁻¹) and C_p is specific heat capacity (1004 J K⁻¹).

The primary source of EPT in the atmosphere is the surface sensible and latent heat flux, driven by solar heating; and the primary sink is the radiative cooling of the troposphere (Betts, 1992). The increasing and decreasing value of EPT with height reflect the convective stable and unstable atmospheric system, respectively. The tropical atmosphere is characterized by a decrease in EPT with height (Sahu *et al.*, 2013), which shows vertical transports by deep convection are a prominent mechanism over this region. Many researchers have used EPT as a tracer to track the upward and downdraft of air (Betts, 1973, 1976; Sahu *et al.*, 2013, 2014).

3.3.3 Boundary layer dynamics

Figure 3.2(a) shows a schematic of the atmospheric boundary layer (ABL) change from day to night. The ABL is the lowest part of the atmosphere that is directly influenced by the presence of the Earth's surface and responds to surface forcing with a time scale of around an hour or less (Stull, 1988).

The top of ABL is expressed by the 'mixing depth' or 'ABL height' and can range from a few metres to about 3000 metres. The depth of the ABL depends on the stability of the air. If a hot air mass resides beneath a cold air mass, the air is unstable and it tends to overturn, resulting in an intense vertical mixing. This phenomenon takes place during the daytime when the radiative heating of the Earth's surface elevates the air temperature close to the surface. The daytime mixing layer is characterized by strong convection and is known as the convective boundary layer (CBL). The convection process results in a rapid vertical mixing of the heat, momentum and air pollutants, leading to their dilution over the source regions. On the other hand, a hotter air mass on top of a cooler air mass results in relatively stable conditions. The upward vertical mixing is hindered under these stable conditions. Such a condition occurs during the nighttime when radiative cooling results in the cooler air mass in the proximity

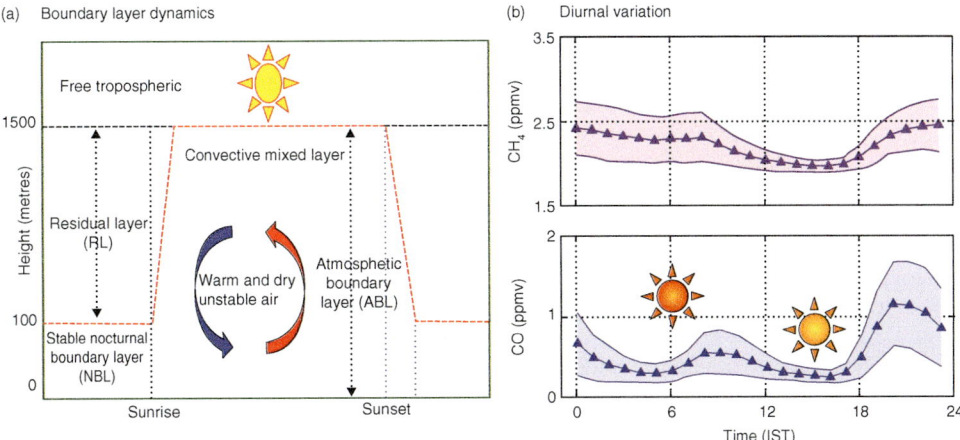

(a) Boundary layer dynamics

(b) Diurnal variation

Fig. 3.2. (a) Schematic of diurnal evolution of the atmospheric boundary layer (ABL) over land. (b) Diurnal variation of CH_4 and CO over Ahmedabad (23.1°N , 73.5°E, 50 m AMSL), average for the period of July 2014–June 2015. (From Chandra *et al.*, 2016b.)

of the Earth's surface forming a layer known as the nocturnal boundary layer (NBL). The boundary layer is shallow during this time and the mixing depth is small, often as low as 100 metres. The layer above the NBL is non-turbulent and known as the residual layer. The concentrations of pollutants in this layer are the same as the daytime mixed layer. The pollutants in the NBL remain trapped until sunrise. After sunrise, the stable conditions start to erode due to surface heating and the trapped polluted air mass in the NBL starts mixing with the residual layer air mass.

Figure 3.2(b) illustrates the effects of changes in ABL on the diurnal variation of pollutants along with their emission near the Earth's surface in an urban area. The figure shows the yearly averaged diurnal variation of CO and CH_4 at an urban region (Ahmedabad) (Lal *et al.*, 2015; Chandra *et al.*, 2016b). Low values during the daytime are a consequence of the dilution of both gases (CO and CH_4) in the high ABL height, while high values during nighttime are a result of the accumulation of pollutants in the shallow NBL. The peak during morning and evening hours appears due to the high vehicular emissions in this period. The difference in CO peak is due to the evolution of the boundary layer along with the emissions; high/low dilution of vehicular emissions during morning/evening rush hours is accentuated by higher/lower ABL height. The diurnal variation, discussed in Fig. 3.2(b), is only true for urban regions. In contrast to the urban

regions, over the high-altitude sites where the local emissions are negligible, the peak occurs during daytime, and relatively low concentration occurs during the rest of the day. The enhanced values during daytime are caused by upslope winds from the surrounding areas transporting pollution to the high-altitude regions through large-scale advection (Zellweger *et al.*, 2009; Zhang, 2010; Fang *et al.*, 2016).

The ABL height and wind speed collectively play a crucial role in governing the ambient levels of pollutants over the source region. This is expressed in terms of the ventilation coefficient (VC = ABL height × average wind speed near the ground (~10 m); Srivastava *et al.*, 2010; Lu *et al.*, 2012). The low and high VC indicates the accumulation and dispersion of local pollutants, respectively. The low/high VC determines the suppressed/facilitated ventilation of the city during the nighttime/daytime. In general, for surface sources, the more stable the air, the worse the dispersion conditions. The VC over the land surface has important implications for the concentration of urban air pollutants and their spatial distribution. Poor ventilation results in several episodic air pollution events.

3.3.4 Land–sea breeze dynamics

Land–sea breeze is the most studied meso-scale phenomenon. Its characteristics have been the subject of numerous studies, particularly focused on the impacts on air pollution levels. The land and sea breeze circulations are commonly encountered near shorelines, lakes or bays (Miller *et al.*, 2003).

The schematic diagram shown in Fig. 3.3 illustrates the air circulation pattern during the land–sea breeze. The differential heating over land and water, due to the different specific heat capacity (the amount of heat needed to raise the system's temperature by 1°C), creates air pressure differences, which is the main driving force behind the land–sea breeze circulation. The land surface heats up faster than the oceanic surface during daytime, as land has lower specific heat than water. As a result of this, surface air is lifted over the land relative to that over the sea; consequently a shallow low-pressure system is set up over the land. The thermal contrast creates a pressure-gradient force directed from the sea

(high-pressure area) to land (low-pressure area), and a shallow layer of marine air mass flows towards the land. This process is known as the sea breeze. Meanwhile, some of the diverging air lifted over land returns toward the sea. The surface air pressure increases over oceans due to the convergence of the air aloft over the ocean, which results in the more prominent flow of air from ocean to land. This completes the basic sea breeze circulation cell. A reverse circulation occurs at night because the ocean cools more slowly than the land, resulting in the process with contrasting air movement, known as the land breeze.

Pollutants can disperse in various directions via the sea breeze circulation (Hewson and Olsson, 1967; Melas *et al.*, 1995, 1998, 2000). Pollutants can be carried tens of kilometres offshore by the seaward return flow followed by a landward re-circulation, resulting

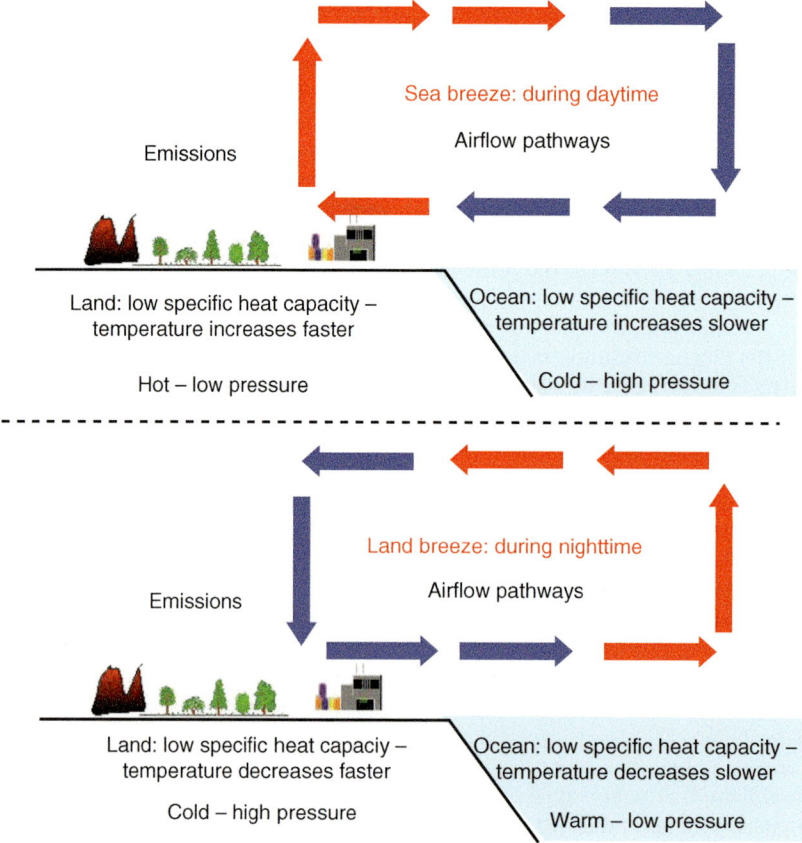

Fig. 3.3. Schematic of airflow pattern during the land–sea breeze circulation.

in an increase in the concentration of pollutants (Shair *et al.*, 1982; Abbs and Physick, 1992). Pollutants can also be transported many kilometres inland by the sea-breeze circulation (Abbs and Physick, 1992). It has been shown that the impact of the prevailing meteorological conditions and sea breeze on the pollution levels in a medium-sized coastal city is more important than the pollutants' transport from the neighbouring industrial areas (Papanastasiou and Melas, 2009).

3.4 Transport of Pollutants

In the above section, we discussed mainly the main transport processes (horizontal and vertical) in the troposphere. Here, we discuss their impacts on the level of air pollutants in the troposphere.

3.4.1 Impact of horizontal transport on the pollutants

Large-scale advection can transport the pollutants large distances in the atmosphere from their local source locations (Browell *et al.*, 1996; Blake *et al.*, 1999; Forster *et al.*, 2001; Zhang, 2010). As a result of this, the trend of emissions over the source region can influence the trend of pollutants over the receptor locations. For example, O_3 levels have increased by 0.6 ppb/yr during the years 1995–1998 over the mid-tropospheric (3.0–8.0 km) region of the United States during the spring season due to the transport of South and East Asia's anthropogenic emissions (Cooper *et al.*, 2010). There is also evidence, based on balloon-borne ozonesondes, that suggests the transport of pollutants from Europe and North America to East Asia (Naja and Akimoto, 2004), and in particular to the South Asian region (Srivastava *et al.*, 2012; Lal *et al.*, 2013). The transport of European emissions contributes to about 40% of the tropospheric O_3 over East Asia (Newell and Evans, 2000). Further, the model results suggest that the transport of O_3 emitted from the European region contributes to about 4 ppbv of O_3 over central Asia during the spring season (Wild *et al.*, 2004). A study based on the balloon-borne

ozonesondes of vertical distribution of O_3 over the city of Ahmedabad in western India revealed that the lower troposphere (below 3 km) is affected by the transport from Southern Europe and North Africa (Srivastava *et al.*, 2012). Further, a detailed study of atmospheric transport, using the FLEXPART retroplume technique, shows that the upper tropospheric air over Ahmedabad during the winter season is affected by transport from North Africa, the North Atlantic Ocean, the central US and the North Pacific Ocean (Lal *et al.*, 2014). The long-range transport of atmospheric SO_2 from Africa can affect its columnar values by a factor of 10–20 higher than the background levels in some places over northern India (Mallik *et al.*, 2013b). Further, it has also been shown that the advection of anthropogenic emissions of SO_2 from China can significantly affect the air quality over Japan (Igarashi *et al.*, 2004, 2006) and Europe (Fiedler *et al.*, 2009a,b).

The anthropogenic emissions of trace gases have been increasing sharply over the Asian region and are projected to increase further (Akimoto, 2003; Ohara *et al.*, 2007; Tanimoto *et al.*, 2009). Several campaign-based studies have been conducted with an aim to observe the influence of transport of pollutants from the Asian region (Naja, 1997; Chand *et al.*, 2003; Lal *et al.*, 2006; Sahu *et al.*, 2006; Lal *et al.*, 2007). The Indian Ocean Experiment (INDOEX) in 1999 concluded that the transport of polluted Asian continental air mass significantly influences the atmospheric composition of the northern Indian Ocean (Lelieveld *et al.*, 2001). Further, the surface studies of trace gases (O_3, CO, CH_4, non-methane hydrocarbons (NMHCs)) during the field campaigns over the marine region of the Bay of Bengal (BoB) – e.g. the Bay of Bengal Experiments (BOBEX) and Bay of Bengal Process Studies (BOBPS) (Sahu *et al.*, 2006; Lal *et al.*, 2007) – also reveal that the transport of major pollutants from the highly polluted Indo-Gangetic Plain (IGP) significantly influences their distribution over the northern part of the bay. The vertical studies of O_3 during the Integrated Campaign for Aerosols, Gases and Radiation Budget (ICARB) in 2006 showed that the outflow from the polluted regions of the atmosphere over northern India, particularly over the Indo-Gangetic Plain, resulted in an enhancement of about 13 ± 6 ppbv to the mixing ratio of ozone over the northern

BOB (Srivastava *et al.*, 2011). Thus, there is enough evidence to see that the intercontinental and hemispheric transport of air pollutants can affect the atmospheric burden of pollutants in a great manner over the clean remote region.

3.4.2 Asian summer monsoon and its impact on the transport of air pollutants

The Asian summer monsoon (ASM) circulation is one of the dominant climatological features of the global circulation during the Northern summer (June to September). The strong anticyclonic flow and divergence in the upper troposphere, along with the cyclonic flow and convergence in the lower troposphere, are the main characteristics of ASM circulation (Hoskins and Rodwell, 1995; Highwood and Hoskins, 1998; Scheeren *et al.*, 2003; Li *et al.*, 2005; Randel and Park, 2006; Park *et al.*, 2007). The anticyclone is centred over the Tibetan Plateau and the deep convection is dominant over the Bay of Bengal, northern India and the South China Sea (e.g. Tzella and Legras, 2011; Bergman *et al.*, 2013; Chandra *et al.*, 2016a, 2017). Along with this peculiar climatological feature, the Asian region is the hotspot of the emission of several anthropogenic pollutants (Akimoto, 2003; Ohara *et al.*, 2007; Kar *et al.*, 2010). Hence, the Asian region in the tropics has been an important concern for transport, dispersal and transformation of air pollutants due to the coexistence of deep convection and large emissions of pollutants from several anthropogenic sources.

The exchange of pollutants can take place between the boundary layer and upper troposphere by the deep convective process during the summer monsoon season. In this process, vertical transport of pollutants takes place very rapidly (within an hour), where the strong upper tropospheric anticyclone further traps them. These processes cause a significant enhancement in the abundance of several air pollutants such as CH_4, NO_x, CO, etc. at the upper tropospheric region (Park *et al.*, 2004; Li *et al.*, 2005; Randel and Park, 2006; Baker *et al.*, 2011, 2012; Patra *et al.*, 2011; Schuck *et al.*, 2012; Chandra *et al.*, 2016a, 2017). The updraft of pollutants also build up the Asian tropopause aerosol layer (ATAL; Vernier *et al.*, 2011), which causes a negative regional radiative forcing (-0.1 W m^{-2}: Vernier *et al.*, 2015), resulting in the cooling of the Earth's surface. Hence, transport in the Asian summer monsoon is plausibly an important factor for climate change. The upper tropospheric anticyclonic circulation is also coupled with the lower stratosphere and hence influences the composition of lower stratospheric air during the summer monsoon season (Dethof *et al.*, 1999). The modelling studies show that the lower-stratospheric water vapour may be strongly influenced by the upward extension of the monsoon circulation (Bannister *et al.*, 2004; Dessler and Sherwood, 2004; Gettelman *et al.*, 2004; Fu *et al.*, 2006). Hence the ASM monsoon provides a gateway for the entrance of the pollutants in the upper troposphere and lower stratosphere (Randel *et al.*, 2010), which may have harmful effects on tropospheric and stratospheric chemistry as well as the climate.

3.4.3 A case study for the CO transport (vertical and horizontal) in the troposphere

As discussed in the previous section, the dynamics over the Asian region play an important role in distributing the local pollutants throughout the troposphere. In this section, we attempt to understand the role of transport (vertical as well as horizontal) in the tropospheric distribution of CO over two selected regions in Asia: the northern Indian region (north of 25°N) and southern Indian region (south of 25°N). Figure 3.4 shows the monthly distribution of CO with height, averaged for the period of Jan 2001–Dec 2014 from Measurement of Pollution in the Troposphere (MOPITT) satellite, over slabs of southern (5–25°N, 70–85°E) and northern (26–35°N, 70–85°E) Indian regions (Fig, 3.4).

A strong enhancement in the CO mixing ratios has been observed in the upper tropospheric height (300–200 hPa) over both of the regions during the summer monsoon season (June–September). The elevated mixing ratios in the upper troposphere, as discussed in the former section, can be explained by transport (vertical and horizontal) of high CO emissions from the surface, as the vertical and horizontal transport time scales in the tropical region are much shorter

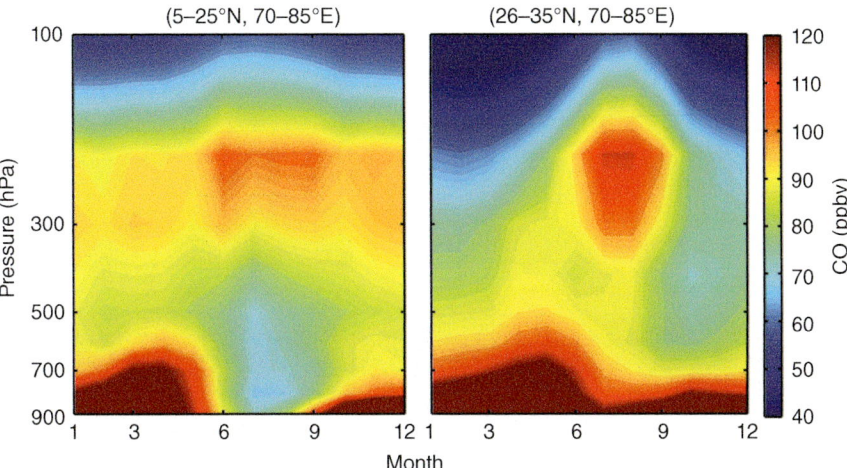

Fig. 3.4. Monthly CO distribution in the troposphere averaged for the period of 14 years (2001–2014) using MOPITT (v6) observations. The colours denote the range of CO mixing ratios in ppb.

than the chemical lifetime of CO (of the order of 1–2 months). The northern region shows comparatively higher CO levels than the southern region during the summer monsoon season. This is because: (i) surface CO emission sources are more prominent over the northern region than the southern region (Kar *et al.*, 2010), resulting in the uplifted air mass being relatively more enriched in CO over the northern region; and (ii) the transported air gets trapped within the strong anticyclonic circulation over the northern region, resulting in the prominent enhancement of CO (Chandra *et al.*, 2016a, 2017). The anticyclone does not cover the southern Indian region during the active phase of the southwest monsoon.

A relatively weaker CO plume at upper tropospheric height (300–200 hPa) has also been observed during the remaining three seasons over southern India, while northern India does not show this feature. Biomass burning and fire activities emit a huge amount of CO and further enter into the upper tropospheric region via the deep convective process (Barret *et al.*, 2008; Chandra *et al.*, 2016a). Once the polluted air mass reaches the upper troposphere, it is advected globally by strong westerly winds. (Fig. 3.5). The global distribution of CO at 300 hPa (Fig. 3.5) clearly shows that the southern part of India is mostly affected by the long-range transport of CO from the active fire regions (southern America, South Africa, North Africa)

during all the seasons, except the summer monsoon season. Hence, the dynamic processes in the atmosphere largely drive the distribution of CO in the troposphere (Barret *et al.*, 2008; Pommrich *et al.*, 2014; Chandra *et al.*, 2016a). This is a standard textbook example for the convective and long-range transport of CO in the troposphere. Similar mechanisms may also be applied for other atmospheric pollutants that have average tropospheric lifetimes comparable to or larger than the mixing time scale.

3.5. Summary

Air pollutants are continually being emitted into the atmosphere from various kinds of stationary sources (power plants, industrial facilities, etc.) and mobile sources (such as automobiles, aircraft, etc.). The emitted air pollutants travel hundreds and thousands of kilometres from their source regions, depending on their atmospheric lifetime, and affect human health and the atmospheric environment. Transport processes, such as convection and advection, mediate these localized pollutants in the atmosphere. This chapter briefly discusses these processes as well as summarizing their impact on the burden of pollutants in the atmosphere. Along with this we also report a case study of CO transport in the troposphere using the MOPITT satellite data.

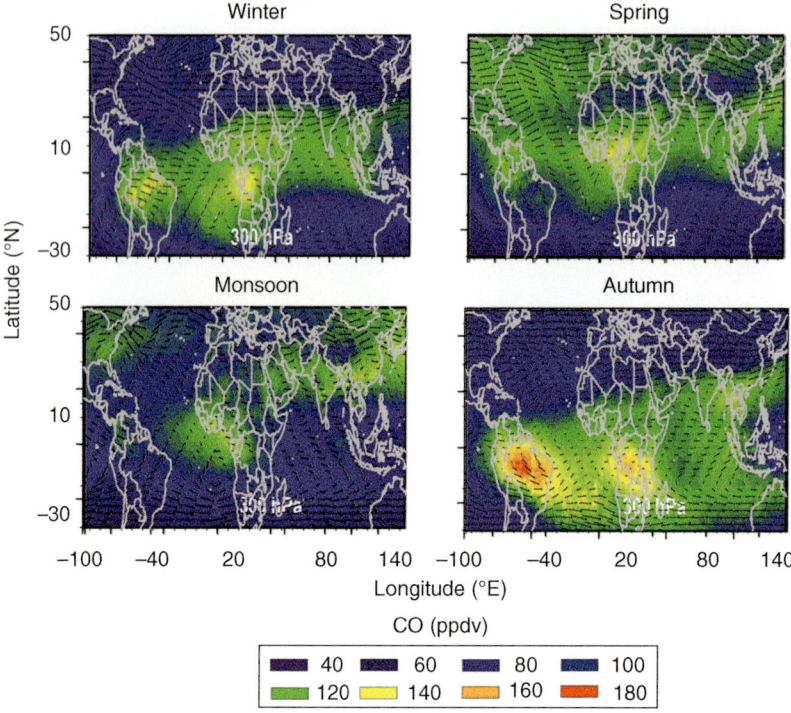

Fig. 3.5. Averaged (2001–2014) seasonal distributions of CO at the upper troposphere (300 hPa) observed by MOPITT satellite.

References

Abbs, D.J. and Physick W.L. (1992) Sea breeze observations and modeling: a review. *Australian Meteoro-logical Magazine* 41, 7–19.

Akimoto, H. (2003) Global air quality and pollution. *Science* 302, 1716–1719. DOI: 10.1126/science.1092666.

Baker, A.K., Schuck, T.J., Slemr, F., van Velthoven, P., Zahn, A. and Brenninkmeijer, C. (2011) Character-ization of non-methane hydrocarbons in Asian summer monsoon outflow observed by the CARIBIC aircraft. *Atmospheric Chemistry and Physics* 11(2), 503–518. DOI: 10.5194/acp-11-503-2011.

Baker, A.K., Schuck, T.J., Brenninkmeijer, C.A.M., Rauthe-Schöch, A., Slemr, F., van Velthoven, P.F.J. and Lelieveld, J. (2012) Estimating the contribution of monsoon-related biogenic production to methane emissions from South Asia using CARIBIC observations. *Geophysical Research Letters* 39, L10813. DOI: 10.1029/2012GL051756.

Bannister, R.N., O'Neill, A., Gregory, A.R. and Nissen, K.M. (2004) The role of the south-east Asian monsoon and other seasonal features in creating the 'tape-recorder' signal in the Unified Model. *Quarterly Journal of the Royal Meteorological Society* 130, 1531–1554.

Barret, B., Ricaud, P., Mari, C., Attié, J.-L., Bousserez, N. *et al.* (2008) Transport pathways of CO in the African upper troposphere during the monsoon season: a study based upon the assimilation of spaceborne observations. *Atmospheric Chemistry and Physics* 8, 3231–3246. DOI: 10.5194/acp-8-3231-2008.

Barrie, L. (1986) Arctic air pollution: an overview of current knowledge. *Atmospheric Environment* 20(4), 643-663.

Bergman, J.W., Fierli, F., Jensen, E.J., Honomichl, S. and Pan, L.L. (2013) Boundary layer sources for the Asian anticyclone: regional contributions to a vertical conduit. *Journal of Geophysical Research* 118, 2560–2575. DOI: 10.1002/jgrd.50142.

Betts, A.K. (1973) A composite mesoscale cumulonimbus budget. *Journal of the Atmospheric Sciences* 30, 597–610.

Betts, A.K. (1976) The thermodynamic transformation of the tropical subcloud layer by precipitation and downdrafts. *Journal of the Atmospheric Sciences* 33, 1008–1020.

Betts, A.K. (1992) FIFE atmospheric boundary layer budget methods. *Journal of Geophysical Research* 97(D17), 18523–18531. DOI: 10.1029/91JD03172.

Bhardwaj, P., Naja, M., Rupakheti, M., Panday, A.K., Kumar, R. *et al.* (2017) Variations in surface ozone and carbon monoxide in the Kathmandu Valley and surrounding broader regions during SusKat-ABC field campaign: Role of local and regional sources. *Atmospheric Chemistry and Physics Discussions.* DOI: 10.5194/acp-2017-306.

Blake, N.J., Blake, D.R., Wingenter, O.W., Sive, B.C., McKenzie, L.M. *et al.* (1999) Influence of southern hemispheric biomass burning on midtropospheric distributions of nonmethane hydrocarbons and selected halocarbons over the remote South Pacific. *Journal of Geophysical Research* 104, 16213–16232.

Brasseur, G.P., Orlando, J.J. and Tyndall, G.S. (1999) *Atmospheric Chemistry and Global Change.* Oxford University Press, New York.

Browell, E.V., Fenn, M.A., Butler, C.F., Grant, W.B., Clayton, M.B. *et al.* (1996) Ozone and aerosol distributions and air mass characteristics over the South Atlantic Basin during the burning season. *Journal of Geophysical Research* 101 (D19), 24043–24068.

Chand, D., Lal, S. and Naja, M. (2003) Variations of ozone in the marine boundary layer over the Arabian Sea and the Indian Ocean during the 1998 and 1999 INDOEX campaigns. *Journal of Geophysical Research* 108(D6), 4190. DOI: 10.1029/2001JD001589.

Chandra, N., Venkataramani, S., Lal, S., Sheel, V. and Pozzer, A. (2016a) Effects of convection and long-range transport on the distribution of carbon monoxide in the troposphere over India. *Atmospheric Pollution Research* 7, 775–785. DOI: 10.1016/j.apr.2016.03.005.

Chandra, N., Lal, S., Venkataramani, S., Patra, P.K. and Sheel, V. (2016b) Temporal variations of atmospheric CO_2 and CO at Ahmedabad in western India. *Atmospheric Chemistry and Physics* 16, 6153–6173. DOI: 10.5194/acp-16-6153-2016.

Chandra, N., Hayashida, S., Saeki, T. and Patra, P.K. (2017) What controls the seasonal cycle of columnar methane observed by GOSAT over different regions in India? *Atmospheric Chemistry and Physics* 17, 12633–12643. DOI: 10.5194/acp-17-12633-2017.

Cooper, O.R., Parrish, D.D., Stohl, A., Trainer, M., Nedelec, P., *et al.* (2010) Increasing springtime ozone mixing ratios in the free troposphere over western North America. *Nature* 463(7279), 344–48.

Crutzen, P.J. and Arnold, F. (1986) Nitric acid cloud formation in the cold Antarctic atmosphere: a major cause for springtime 'ozone hole'. *Nature* 324, 651–655.

Dessler, A.E. and Sherwood, S.C. (2004) Effect of convection on the summertime extratropical lower stratosphere. *Journal of Geophysical Research* 109, D23301. DOI: 10.1029/2004JD005209.

Dethof, A., O'Neill, A., Slingo, J.M. and Smit, H.G.J. (1999) A mechanism for moistening the lower stratosphere involving the Asian summer monsoon. *Quarterly Journal of the Royal Meteorological Society* 556, 1079–1106.

Emanuel, K.A. (1994) *Atmospheric Convection.* Oxford University Press, New York.

Fang, S., Tans, P.P., Steinbacher, M., Zhou, L., Luan, T. and Li, Z. (2016) Observation of atmospheric CO_2 and CO at Shangri-La station: results from the only regional station located at southwestern China. *Tellus B* [S.l.], v. 68.

Fiedler, V., Nau, R., Ludmann, S., Arnold, F., Schlager, H. and Stohl, A. (2009a) East Asian SO_2 pollution plume over Europe – Part 1: Airborne trace gas measurements and source identification by particle dispersion model simulations. *Atmospheric Chemistry and Physics* 9(14), 4717–4728.

Fiedler, V., Arnold, F., Schlager, H., Dornbrack, A., Pirjola, L. and Stohl, A. (2009b) East Asian SO_2 pollution plume over Europe – Part 2: Evolution and potential impact. *Atmospheric Chemistry and Physics* 9(14), 4729–4745.

Forster, C., Wandinger, U., Wotawa, G., James, P., Mattis, I. *et al.* (2001) Transport of boreal forest fire emissions from Canada to Europe. *Journal of Geophysical Research* 106(D19), 22887–22906.

Fu, R., Hu, Y., Wright, J.S., Jiang, J.H., Dickinson, R.E. *et al.* (2006) Short circuit of water vapor and polluted air to the global stratosphere by convective transport over the Tibetan Plateau. *Proceedings of the National Academy of Sciences USA* 103, 5664–5669.

Gettelman, A., Kinnison, D.E., Dunkerton, T.J. and Brasseur, G.P. (2004) Impact of monsoon circulations on the upper troposphere and lower stratosphere. *Journal of Geophysical Research* 109, D22101. DOI: 10.1029/2004JD004878.

Hewson, E.W. and Olsson, E.L. (1967) Lake effects on air pollution dispersion. *Journal of the Air Pollution Control Association* 17(11), 757–761. DOI: 10.1080/00022470.1967.10469069.

Highwood, E.J. and Hoskins, B.J. (1998) The tropical tropopause. *Quarterly Journal of the Royal Meteorological Society* 124, 1579–1604.

Hoskins, B.J. and Rodwell, M.J. (1995) A model of the Asian summer monsoon, I, The global scale. *Journal of the Atmospheric Sciences* 52, 1329–1340.

Igarashi, Y., Sawa, Y., Yoshioka, K., Matsueda, H., Fujii, K. and Dokiya, Y. (2004) Monitoring the SO_2 concentration at the summit of Mt. Fuji and a comparison with other trace gases during winter. *Journal of Geophysical Research* 109, D17304. DOI: 10.1029/2003JD004428.

Igarashi, Y., Sawa, Y., Yoshioka, K., Takahashi, H., Matsueda, H. and Dokiya, Y. (2006) Seasonal variations in SO_2 plume transport over Japan: observations at the summit of Mt. Fuji from winter to summer. *Atmospheric Environment* 40(36), 7018–7033.

Jacob, D.J. (1999) *Introduction of Atmospheric Chemistry.* Princeton University Press, Princeton, NJ.

Jacob, D.J. and Winner, D.A. (2009) Effect of climate change on air quality. *Atmospheric Environment* 43(1), 51–63. DOI: 10.1016/j.atmosenv.2008.09.051.

Kar, J., Deeter, M., Fishman, J., Liu, Z., Omar, A. *et al.* (2010) Wintertime pollution over the Eastern Indo-Gangetic Plains as observed from MOPITT, CALIPSO and tropospheric ozone residual data. *Atmospheric Chemistry and Physics* 10(24), 12273–12283.

Lal, S., Chand, D., Sahu, L.K., Venkataramani, S., Brasseur, G. and Schultz, M.G. (2006) High levels of ozone and related gases over the Bay of Bengal during winter and early spring of 2001. *Atmospheric Environment* 40, 1633–1644.

Lal, S., Sahu, L. and Venkataramani, S. (2007) Impact of transport from the surrounding continental regions on the distributions of ozone and related trace gases over the Bay of Bengal during February 2003. *Journal of Geophysical Research* 112(D14), 166, 187.

Lal, S., Venkataramani, S., Srivastava, S., Gupta, S., Mallik, C. *et al.* (2013) Transport effects on the vertical distribution of tropospheric ozone over the tropical marine regions surrounding India. *Journal of Geophysical Research: Atmospheres* 118, 1513–1524.

Lal, S., Venkataramani, S., Chandra, N., Cooper, O.R., Brioude, J. and Naja, M. (2014) Transport effects on the vertical distribution of tropospheric ozone over western India. *Journal of Geophysical Research: Atmospheres* 119, 10012–10026.

Lal, S., Chandra, N. and Venkataramani, S. (2015) A study of CO_2 and related trace gases using a laser based technique at an urban site in western India. *Current Science* 109, 2111–2116.

Lal, S., Venkataramani, S., Naja, M., Kuniyal, J.C., Mandal, T.K. *et al.* (2017) Loss of crop yields in India due to surface ozone: an estimation based on a network of observations. *Environmental Science and Pollution Research* 24, 20972–20981.

Lawrence, M. and Lelieveld, J. (2010) Atmospheric pollutant outflow from southern Asia: a review. *Atmospheric Chemistry and Physics* 10(22), 11017–11096.

Lelieveld, J., Crutzen, P.J., Ramanathan, V., Andreae, M.O., Brenninkmeijer, C.A.M. *et al.* (2001) The Indian Ocean experiment: widespread air pollution from South and Southeast Asia. *Science* 291, 1031–1036.

Li, Q., Jiang, J.H., Wu, D.L., Read, W.G., Livesey, N.J. *et al.* (2005) Convective outflow of South Asian pollution: a global CTM simulation compared with EOS MLS observations. *Geophysical Research Letters* 32, L14826. DOI: 10.1029/2005GL022762.

Lu, C., Deng, Q., Liu, W., Huang, B.-L. and Shi, L.-Z. (2012) *Journal of Central South University of Technology* 19, 615. DOI: 10.1007/s11771-012-1047-9.

Mallik, C., Lal, S., Venkataramani, S., Naja, M. and Ojha, N. (2013a) Variability in ozone and its precursors over the Bay of Bengal during post monsoon: transport and emission effects. *Journal of Geophysical Research* 118, 10190–10209.

Mallik, C., Lal, S., Naja, M., Chand, D., Venkataramani, S., Joshi, H. and Pant, P. (2013b) Enhanced SO_2 concentrations observed over northern India: role of long-range transport. *International Journal of Remote Sensing* 34(8), 2749–2762.

Melas, D., Ziomas, I.C. and Zerefos, C.H. (1995) Boundary layer dynamics in an urban coastal environment under sea breeze conditions. *Atmospheric Environment* 29, 3605–3617.

Melas, D., Ziomas, I.C., Klemm, O. and Zerefos, C.H. (1998) Anatomy of sea breeze circulation in Athens area under weak large scale ambient winds. *Atmospheric Environment* 32, 2223–2237.

Melas, D., Lavagnini, A. and Sempreviva, A. (2000) An investigation of the boundary layer dynamics of Sardinia Island under sea breeze conditions. *Journal of Applied Meteorology* 39, 516–524.

Miller, S.T.K., Keim, B.D., Talbot, R.W. and Mao, H. (2003) Sea breeze: structure, forecasting, and impacts. *Reviews of Geophysics* 41. DOI: 10.1029/2003RG000124.

Molina, J.S. and Rowland, F.S. (1974) Stratospheric sink for chlorofluoromethanes: chlorine atom-catalysed destruction of ozone. *Nature* 249, 810–812.

Naja, M. (1997) Tropospheric chemistry in the tropics. PhD thesis, Gujarat University, India.

Naja, M. and Akimoto, H. (2004) Contribution of regional pollution and long-range transport to the Asia-Pacific region: analysis of long-term ozonesonde data over Japan. *Journal of Geophysical Research* 109, 1306. DOI: 10.1029/2004JD004687.

Newell, R. and Evans, M. (2000) Seasonal changes in pollutant transport to the North Pacific: the relative importance of Asian and European sources. *Geophysical Research Letters* 27, 2509–2512. DOI: 10.1029/2000GL011501.

Ohara, T., Akimoto, H., Kurokawa, J., Horii, N., Yamaji, K. *et al.* (2007) An Asian emission inventory of anthropogenic emission sources for the period 1980–2020. *Atmospheric Chemistry and Physics* 7(16), 4419–4444.

Papanastasiou, D.K. and Melas, D. (2009) Climatology and impact on air quality of sea breeze in an urban coastal environment. *International Journal of Climatology* 29, 305–315. DOI: 10.1002/joc.1707.

Park, M., Randel, W.J., Kinnison, D.E., Garcia, R.R. and Choi, W. (2004) Seasonal variation of methane, water vapor, and nitrogen oxides near the tropopause: satellite observations and model simulations. *Journal of Geophysical Research* 109, D03302. DOI: 10.1029/2003JD003706.

Park, M., Randel, W.J., Gettelman, A., Massie, S.T. and Jiang, J.H. (2007) Transport above the Asian summer monsoon anticyclone inferred from Aura Microwave Limb Sounder tracers. *Journal of Geophysical Research* 112, D16309. DOI: 10.1029/2006JD008294.

Patra, P.K., Niwa, Y., Schuck, T.J., Brenninkmeijer, C.A.M., Machida, T. *et al.* (2011) Carbon balance of South Asia constrained by passenger aircraft CO_2 measurements. *Atmospheric Chemistry and Physics* 11, 4163–4175. DOI: 10.5194/acp-11-4163-2011.

Petit, J.R., Jouzel, J., Raynaud, D., Barkov, N.I., Barnola, J.M. *et al.* (1999) Climate and atmospheric history of the past 420,000 years from the Vostok ice core, Antarctica. *Nature* 399, 429–436.

Pommrich, R., Müller, R., Grooß, J.-U., Konopka, P., Ploeger, F. *et al.* (2014) Tropical troposphere to stratosphere transport of carbon monoxide and long-lived trace species in the Chemical Lagrangian Model of the Stratosphere (CLaMS). *Geoscientific Model Development* 7, 2895–2916. DOI: 10.5194/gmd-7-2895-2014.

Randel, W.J. and Park, M. (2006) Deep convective influence on the Asian summer monsoon anticyclone and associated tracer variability observed with Atmospheric Infrared Sounder (AIRS). *Journal of Geophysical Research* 111, D12314. DOI: 10.1029/2005JD006490.

Randel, W.J., Park, M., Emmons, L., Kinnison, D., Bernath, P. *et al.* (2010) Asian monsoon transport of pollution to the stratosphere. *Science* 328, 611–613. DOI: 10.1126/science.1182274.

Sahu, L., Lal, S. and Venkataramani, S. (2006) Distributions of O_3, CO and hydrocarbons over the Bay of Bengal: a study to assess the role of transport from southern India and marine regions during September–October 2002. *Atmospheric Environment* 40(24), 4633–4645.

Sahu, L.K., Sheel, V., Kajino, M. and Nedele, P. (2013) Variability in tropospheric carbon monoxide over an urban site in Southeast Asia. *Atmospheric Environment* 68, 243–255.

Sahu, L.K., Sheel, V., Kajino, M., Deushi, M., Gunthe, S.S. *et al.* (2014) Seasonal and interannual variability of tropospheric ozone over an urban site in India: a study based on MOZAIC and CCM vertical profiles over Hyderabad. *Journal of Geophysical Research: Atmospheres* 119, 3615–3641. DOI: 10.1002/2013JD021215.

Scheeren, H.A., Lelieveld, J., Roelofs, G.J., Williams, J., Fischer, H., *et al.* (2003) The impact of monsoon outflow from India and Southeast Asia in the upper troposphere over the eastern Mediterranean. *Atmospheric Chemistry and Physics Discussions* 3(3), 2285–2330.

Schuck, T.J., Ishijima, K., Patra, P.K., Baker, A.K., Machida, T. *et al.* (2012) Distribution of methane in the tropical upper troposphere measured by CARIBIC and CONTRAIL aircraft. *Journal of Geophysical Research* 117, D19304. DOI: 10.1029/2012JD018199.

Seinfeld, J.H. and Pandis, S.N. (2006) *Atmospheric Chemistry and Physics: From Air Pollution to Climate Change*, 2nd edn. Wiley, New York.

Shair, F.H., Sasaki, E.J., Carlan, D.E., Cass, G.R., Goodin, W.R. *et al.* (1982) Transport and dispersion of airborne pollutants associated with the land breeze–sea breeze system. *Atmospheric Environment* 16, 2043–2053.

Solomon, S. (1990) Progress towards a quantitative understanding of Antarctic ozone depletion. *Nature* 347, 347–354.

Solomon, S., Burkholder, J.B., Ravishankara, A.R. and Garcia, R.R. (1994) Ozone depletion and global warming potentials of CH_3I. *Journal of Geophysical Research* 99, 20929–20935.

Srivastava, S., Lal, S., Bala Subrahamanyam, D., Gupta, S., Venkataramani, S. *et al.* (2010) Seasonal variability in mixed layer height and its impact on trace gas distribution over a tropical urban site: Ahmedabad. *Atmospheric Research* 96, 79–87. DOI: 10.1016/j.atmosres.2009.11.015.

Srivastava, S., Lal, S., Venkataramani, S., Gupta, S. and Acharya, Y.B. (2011) Vertical distribution of ozone in the lower troposphere over the Bay of Bengal and the Arabian Sea during ICARB-2006: effects of continental outflow. *Journal of Geophysical Research* 116, D13301. DOI: 10.1029/ 2010JD015298.

Srivastava, S., Lal, S., Naja, M., Venkataramani, S. and Gupta, S. (2012) Influences of regional pollution and long range transport to western India: analysis of ozonesonde data. *Atmospheric Environment* 47, 174–182.

Stocker, T.F., Qin, D., Plattner, G.-K., Tignor, M., Allen, S.K. *et al.* (2013) Intergovernmental panel on climate change, 2013: summary for policymakers. In: *Climate Change 2013: The Physical Science Basis. Contribution of Working Group I to the Fifth Assessment Report of the IPCC.* 1, 4, 9.

Stull, R.B. (1988) *An Introduction to Boundary Layer Meteorology.* Kluwer Academic Press, Dordrecht, The Netherlands.

Tanimoto, H., Ohara, T. and Uno, I. (2009) Asian anthropogenic emissions and decadal trends in spring-time tropospheric ozone over Japan: 1998–2007. *Geophysical Research Letters* 36, L23802. DOI: 10.1029/2009GL041382.

Tzella, A. and Legras, B. (2011) A Lagrangian view of convective sources for transport of air across the Tropical Tropopause Layer: distribution, times and the radiative influence of clouds. *Atmospheric Chemistry and Physics* 11, 12517–12534. DOI: 10.5194/acp-11-12517-2011.

Vernier, J.P., Thomason, L.W. and Karl, J. (2011) CALIPSO detection of an Asian tropopause aerosol layer. *Geophysical Research Letters* 38, L07804. DOI: 10.1029/2010GL046614.

Vernier, J.P., Fairlie, T.D., Natarajan, M., Wienhold, F.G., Bian, J. *et al.* (2015) Increase in upper tropospheric and lower stratospheric aerosol levels and its potential connection with Asian pollution. *Journal of Geophysical Research* 120, 1608–1619. DOI: 10.1002/2014JD022372.

Wild, O., Prather, M.J., Hajime, A., Sundet, J.K., Isaksen, I.S.A. *et al.* (2004) CTM ozone simulations for spring 2001 over the western Pacific: regional ozone production and its global impacts. *Journal of Geophysical Research* 109, D15S02. DOI: 10.1029/2003JD004041.

Zellweger, C., Huglin, C., Klausen, J., Steinbacher, M., Vollmer, M. and Buchmann, B. (2009) Inter-comparison of four different carbon monoxide measurement techniques and evaluation of the long-term carbon monoxide time series of Jungfraujoch. *Atmospheric Chemistry and Physics* 9, 3491–3503. DOI: 10.5194/acp-9-3491-2009.

Zhang, L. (2010) Intercontinental transport of air pollution. *Frontiers of Environmental Science & Engineering in China* 4, 20–29. DOI: 10.1007/s11783-010-0014-7.

4 Methods for the Measurement of Air Pollutants

Shani Tiwari[1]* and Neha Mishra[2]

[1]*Nagoya University, Japan;* [2]*Indrapastha College, University of Delhi, India*

Abstract

Currently, air pollution is one of the most imperative problems and has a major role in global climate change and premature deaths throughout the world. Various types of air pollutants (e.g. particulate matter, NO_x, SO_x, CO and O_3) are present in the atmosphere and have significant effects on human health, agriculture, weather and climate. Accurate measurements of air pollutants on local, regional and global levels at high spatial and temporal resolution are needed to quantify their distribution and resulting impacts. Real-time ground-level measurements of air pollutants help to understand secondary aerosol formation through atmospheric chemical transformation. On the other hand, satellite measurements of air pollutants and trace gases provide a good platform to understand the current scenario of air quality and future climate change on a global scale. Measuring air pollutants with higher accuracy is challenging because of a variety of considerations, including their chemical composition, lifetime, emission sources and complex instrumentation. A continuous improvement in air pollutant measurement techniques and instrumentation has occurred during the last two decades; however, more efforts are needed to achieve more realistic data with higher accuracy. In this chapter, an overview of various measurement techniques of air pollutants (from ground-based and satellite-based measurements) are presented.

4.1 Introduction

Air pollution is one of the most important concerns of the 21st century because of its serious impacts on air quality, environment, human health and climate. During the last few decades, air pollution has continuously increased over South and East Asia (mainly India and China) due to a rapid increase in industrialization, population growth and related energy demands (Prasad *et al.*, 2006; Srivastava, 2017). Air pollution is broadly divided into two major types, ambient (outdoor) and indoor air pollution. The outdoor air pollution involves trace gases (e.g. ozone, SO_2, NO_x, CO and CH_4), all sizes of particulates, ranging from ultrafine to coarse, arising from various natural sources (dust storms, forest fires etc.), and anthropogenic emission sources (e.g. biomass burning, industrial emissions, vehicular traffic, gas-to-particle conversion). However, indoor environments predominantly gather mainly gaseous pollutants (e.g. CO, NO_2) and ultrafine and fine particles, mainly coming from combustion sources such as cooking and smoking. Air pollutants can be mainly classified as either primary or secondary. Primary air pollutants are directly emitted from a natural or anthropogenic process, such as ash from a volcanic eruption, carbon

---

* Corresponding author: pshanitiwari@gmail.com

monoxide (CO) gas from motor vehicle exhaust, or sulfur dioxide (SO_2) released from industries. Air pollutants formed in the air due to different chemical reactions are called secondary air pollutants (e.g. ozone, chlorofluorocarbons or CFCs, peroxyacetyl nitrate or PAN, NO_2, HNO_3). These are responsible for severe effects of air pollution such as smog, haze, eye irritation and damage to vegetation. Apart from the emission sources of air pollutants, regional meteorology and topography also play a crucial role in air pollution, which governs the pollutant transportation, dispersion and chemical transformation (Tiwari *et al.*, 2014).

According to the World Health Organization (WHO), air pollution is one of the world's biggest killers, causing around 3 million premature deaths worldwide every year (Lelieveld *et al.*, 2015). Among these, more than half of the deaths occur in developing countries (over half a million in India alone) (Chowdhury and Dey,

2016; Ghude *et al.*, 2016). However, some developed industrial nations (e.g. United States) also suffer with air pollution causing earlier deaths (Caiazzo *et al.*, 2013). Figure 4.1 shows the annual mean of fine particulate ($PM_{2.5}$ in $\mu g/m^3$) concentration in the major cities of South and East Asia in 2016. It clearly shows that most of the Asian cities have many times higher concentrations of fine particulate than the WHO target of $10 \mu g/m^3$. According to WHO, in 2016 Zabol (Iran) is the most polluted city of the major cities of Asia, followed by Agra, Allahabad (India) and Riyadh (Saudi Arabia).

Accurate measurement of air pollutants is necessary to understand the fundamental problems of air pollution, and provide solutions for policy makers at local, national and international levels (Bower and Mucke, 1999). The main objective of air pollution measurements is to provide quality-controlled scientific datasets for cost-effective control policies, and to understand the

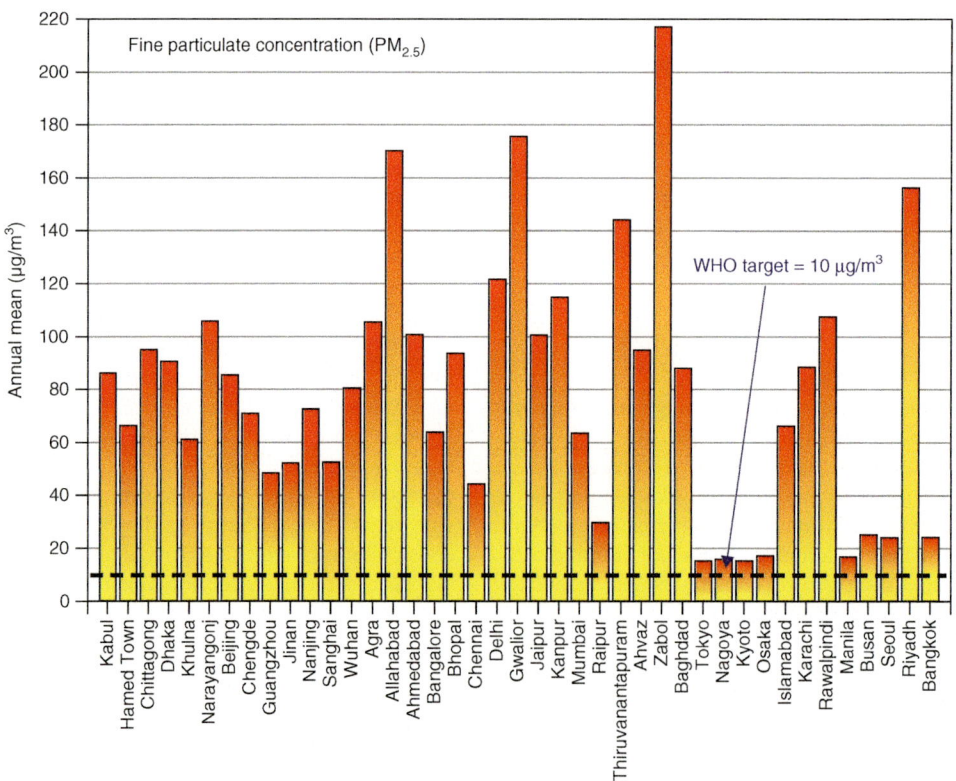

Fig. 4.1. Annual fine particulate concentration ($PM_{2.5}$) over South and East Asia. (Adapted from World Health Organization (2016).)

potential impacts of air pollution on human health, the ecosystem and climate change.

4.2 Types of Major Air Pollutants and Their Emission Sources

Below, we summarize key air pollutants relevant for human health, ecosystems and climate, and their emission sources.

Sulfur dioxides (SO_2)

Coal, petroleum and other fuels are often impure and contain sulfur as well as different organic compounds. The main anthropogenic source of sulfur dioxide emissions is the sulfur content of fossil fuels released by combustion. Relatively small amounts of sulfur are also emitted from forest fires, soils and vegetation (Simpson et al., 1999). Coal-fired power plants are the world's biggest source of sulfur dioxide, which contributes to smog, acid rain and health problems including lung and respiratory diseases (Greenberg et al., 2016; Saygın et al., 2017).

Carbon dioxides (CO_2)

In the atmosphere, CO_2 is the dominant carbon-bearing trace gas with a current concentration of approximately 405 ppm, which corresponds to a mass of 862 GtC (Quere et al., 2017). Atmospheric CO_2 presently contributes more than 80% of the gaseous radiative forcing responsible for anthropogenic climate change (IPCC, 2013). Anthropogenic sources (mainly from fossil-fuel combustion and industrial process) are the major emission sources of CO_2 (IPCC, 2013).

Carbon monoxides (CO)

Carbon monoxide (CO) is one of the most common and largely distributed air pollutants. CO is a highly dangerous gas and forms mainly when carbon fuel is not burned completely. Nearly two-thirds of CO emissions come from anthropogenic activities, such as crop residue burning, fossil fuel combustion, and methane oxidation by hydroxyl radical (OH) (Seinfeld and Pandis, 2010). Due to a relatively shorter life time (0.08–0.25 yr), CO is not well mixed in the troposphere. CO emission into the atmosphere can affect the life time of CH_4, photochemical production of CO_2 and tropospheric ozone, and hence may have a significant impact on global climate forcing.

Nitrogen oxides (NO_x)

Nitrogen dioxide (NO_2) and nitrogen oxide (NO), together termed NO_x, are pollutants produced when fuel is burned at high temperatures. Nitrogen oxide pollution comes from vehicle engines and power plants, and plays an important role in the formation of acid rain, ozone and smog. It also gets into the upper troposphere through lightning (Finney et al., 2016). Nitrogen oxides are also known as 'indirect greenhouse gases' since they play a significant role in global warming by producing ozone (Finney et al., 2016; Grewe et al., 2012) which has adverse effects on the respiratory systems of human beings and also causes damage to vegetation (Kim et al., 2006; Kampa and Castanas, 2008).

Ozone (O_3)

This is a type of oxygen gas whose molecules are made from three oxygen atoms joined together, instead of the two atoms in conventional oxygen (O_2). It is a reactive oxidant gas and very toxic for humans and vegetation at the surface level because it oxidizes biological tissues (Weschler, 2006; Kampa and Castanas, 2008; Liu and Peng, 2018). Nearly 90% of the Earth's atmospheric ozone is found in the stratosphere, which forms a layer (known as 'the ozone layer') around the Earth; however, the rest (10%) is found in the troposphere. This ozone layer absorbs the solar ultraviolet radiation, which is very harmful (mainly in the wavelength range from 290 to 320 nm) for the human skin, and can cause skin cancer (Eastham and Barrett, 2016). Stratospheric ozone is formed by the chemical reactions involving solar light (ultraviolet radiation) and oxygen molecules which separate the oxygen atoms. These oxygen atoms recombine with an oxygen molecule to produce ozone molecules. The ozone molecules can also be decomposed in the oxygen atom and an oxygen molecule by UV radiation. Thus, formation and loss of ozone is a cyclic process by UV radiation into the stratosphere. On the other hand, tropospheric ozone is mainly formed through chemical reactions, which involve hydrocarbon, nitrogen oxides and ozone itself.

Particulate matter (aerosol)

Particulate matter (PM) often also known as a part of aerosol particles, are tiny particles suspended in the atmosphere which range in size from a few hundred micrometres to nanometres. They are emitted into the atmosphere from both natural (e.g. dust storms) and anthropogenic sources (transportation and industry), through a gas-to-particles conversion. Particulate matter (PM or aerosols) is an important component of air pollution, having both long-term as well as short-term ill effects on human health, such as cardiovascular or lung and skin diseases, which sometimes lead to premature death (Krewski et al., 2000; Pope, 2000; HEI, 2004; Pope and Dockery, 2006), and on climate (Charlson et al., 1992; IPCC, 2013; Tiwari et al., 2016a) and vegetation (Mehta et al., 2017; Latha et al., 2018). Particulate matter assessment is of major concern around the world and many environmental protection agencies are working towards continuous monitoring and assessment of air pollution from surface-based stations. They have a significant potential to affect the Earth's climate through scattering and absorbing of solar radiation (Charlson et al., 1992; Ramanathan et al., 2001; Tiwari et al., 2015).

Methane (CH₄)

Methane is a main component of greenhouse gases and has an impact on a global scale. It has the potential to trap heat 34 times higher than carbon dioxide on a mass basis over a 100-year time scale (Myhre et al., 2013). Methane can come from many sources, both natural and human made. It is mainly emitted from the industrial emission of the oil and gas industries, crop residue burning (mainly paddy residue burning) and wetlands (Gedney et al., 2004; Jain et al., 2014). A recent study reported that up to about 8% of global CH$_4$ emissions occur in South Asia, covering less than 1% of global land (Chandra et al., 2017). The main sink of methane is through its reaction with ·OH radicals in the troposphere and bacterial uptake in soils (Fung et al., 1991; IPCC, 2013).

Volatile organic carbons (VOCs)

These carbon-based (organic) chemicals evaporate easily at ordinary temperatures and pressures, so they readily become gases. VOCs are emitted from a variety of both natural and anthropogenic sources, which show a significant spatial distribution from region to region (Goldstein and Galbally, 2007). The major anthropogenic sectors include emissions from chemical and petroleum industries, combustion of fossil fuels in automobile engines, and power plants (Sahu, 2012). They have chronic effects on people's health, such as inhalation risk, and eye, nose and throat irritation (Bolden et al., 2015; Dai et al., 2017; Guo et al., 2018; Soni et al., 2018). They also play a role in the formation of ozone and smog (Duan et al., 2008; Shao et al., 2009).

Chlorofluorocarbons (CFC)

Chlorofluorocarbons (CFCs) are completely anthropogenic, halogenated paraffin hydrocarbon compounds, and were widely used in refrigerators, aerosol cans, fire extinguishers and solvents (Metz et al., 2005), until it was discovered that they damage the Earth's ozone layer. Once CFCs are emitted into the atmosphere, they rise up to the stratosphere where they are broken down by UV radiation, releasing free chlorine atoms which cause ozone depletion.

4.3 Effects of Air Pollution on Health, Agriculture and Climate

Air pollution has significant effects on human health, agriculture and the environment, as well as climate change. London smog is a notorious example of the impact of air pollution on human health, and has killed more than 4000 people. Several studies have reported that most of the air pollutants (mainly particulate matter) directly affect the respiratory and cardiovascular system (Smith, 2000; Pope and Dockery, 2006; Pope *et al.*, 2009) and an increase in air pollutants is directly associated with mortality and morbidity (Vedal *et al.*, 1998; Pope and Dockery, 2006; Murray *et al.*, 2015; Chowdhury and Dey; 2016). $PM_{2.5}$ particles are more hazardous than PM_{10}, since they reach the human alveoli region. Several epidemiological studies reported that higher fine-particle exposure causes reduction in lung function and respiratory infections in children (Brauer *et al.*, 2007; Pope *et al.*, 2009; Balakrishnan *et al.*, 2011). In another study, Pope and Dockery (2006) estimated a mortality increase in the order of 4–6% with the increase in PM_{10}

and $PM_{2.5}$ concentrations by 20 and 10 µg/m³, respectively, in the ambient air of US cities.

Air pollution also has the potential to affect the growth of plants and agricultural production (Burney and Ramanathan, 2014). In heavy air pollution, air pollutants (mainly particulate matter) deposit on the leaves of plants (also known as aerosol shield) and reflect the solar radiation. Due to this shielding, the photosynthesis process of the plant is affected, which causes a reduction in crop production. On the other hand, air pollution has a significant impact on global climate change by modifying the Earth's radiation budget through the scattering and absorption of solar radiation (Wild, 2009; IPCC, 2013; Tiwari *et al.*, 2015). Figure 4.2 shows the net radiative

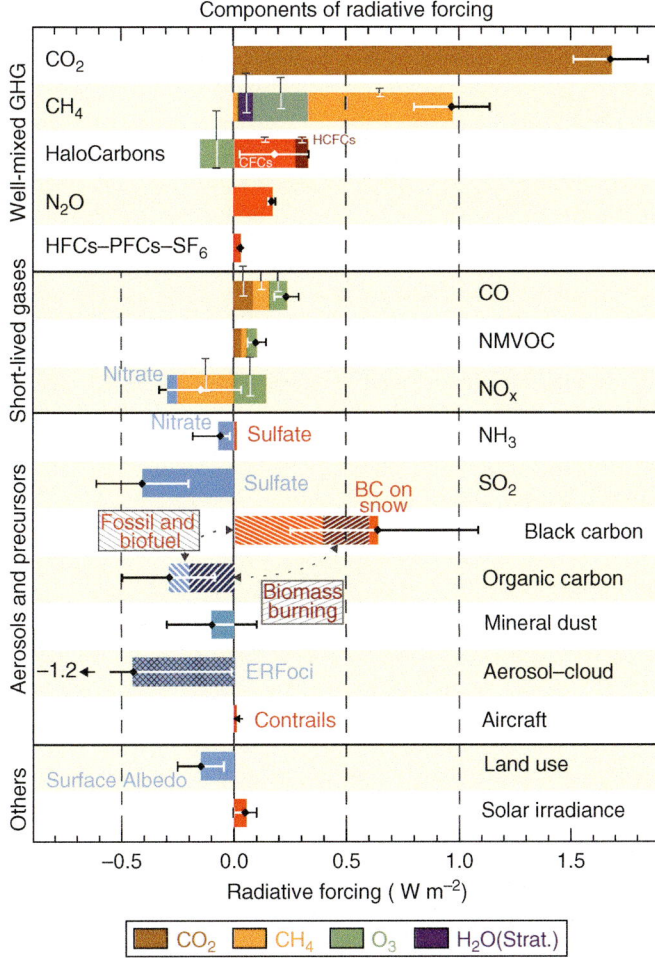

Fig. 4.2. Global radiative forcing of various air pollutants for the period 1750–2011. (Adapted from IPCC, 2013.)

forcing of various air pollutants reported by the Intergovernmental Panel on Climate Change (IPCC, 2013). Figure 4.2 reveals that there is net global warming (positive radiative forcing + 2.83 w/m^2) due to well-mixed greenhouse gases (WMGHG); among these, carbon dioxide has maximum forcing. Many studies reported that tropospheric ozone has positive radiative forcing, while stratospheric ozone has negative radiative forcing; however, aerosols have both positive and negative forcing, depending on aerosol types (IPCC, 2013; Tiwari *et al.*, 2015, 2016c).

4.4   Air Pollutant Measurement Techniques

4.4.1   Remote sensing techniques

The term 'remote sensing' refers to measurements made remotely without coming into physical contact with the subject of measurement. In this technique, information about the air pollutants is carried through electromagnetic radiation. In the last two or three decades, due to the rapid development in remote sensing methods, remote sensing measurements (mainly from satellites) have become one of the most effective tools to measure air pollutants. Remote sensing methods have the ability to provide information about the air pollutants with a high spatial–temporal resolution and also vertical profile measurements. Based on the emission sources and working principle, remote sensing is divided into two major categories as described below.

4.4.1.1   *Active remote sensing*

Active sensors provide their own energy source for illumination. The sensor emits radiation, which is directed toward the target to be investigated. The radiation reflected from that target is detected and measured by the sensor. In active remote sensing, the measuring device or sensor itself emanates radiation directed towards the object of interest, which is measured by the sensor after being reflected by the object. Examples of active remote sensing devices include Light Detection and Ranging (LIDAR) for the optical domain, and Radiowave Detection

and Ranging (RADAR) for the radio wave domain. Different types of LIDAR are widely used in the measurement of air pollutants throughout the world: elastic LIDAR is used for aerosol and dust measurement, Raman LIDAR is used for the measurement of ozone, and differential-absorption LIDAR (DIAL) is used for measurements of ammonia, methane and oxides of sulfur and nitrogen.

4.4.1.2   *Passive remote sensing*

In passive remote sensing, the device measures the energy emitted from the Earth's surface and atmosphere as a function of different physical characteristics related to atmosphere. This approach is widely used, particularly in satellite imagery. Examples of passive remote sensing devices are ground-based sun photometers (i.e. AERONET, MICROTOPS), and satellite observation (MODIS, OMI, CALIPSO, AIRS, etc.). A pictorial view of a satellite constellation observing the Earth's surface and atmosphere is shown in Fig. 4.3, in which the Aura satellite observes columnar aerosol optical depth, water vapour, greenhouse gases and other atmospheric pollutants. CALIPSO provides the vertical distribution of aerosol and their sub-types. However, Aura observed a high resolution vertical distribution of greenhouse gases and other atmospheric pollutants. A brief description about the various different remote sensing instruments is described in the following section.

AEROSOL ROBOTIC NETWORK (AERONET).   CIMEL sun/sky radiometer measurements are established globally under the Aerosol Robotic Network (AERONET) programme of NASA, USA (Fig. 4.4). It is a globally distributed network of automatic sun/sky scanning radiometers that measure solar radiation at several wavelengths, typically centred at 340, 380, 440, 500, 675, 870, 940 and 1020 nm, with triple observations per wavelength. Each band has a full width of approximately 10 nm at half maximum (FWHM). All of these spectral bands are utilized in direct sun measurements, while four of them (440, 675, 870 and 1020 nm) are used for sky radiance. Spectral aerosol optical depth (AOD) is obtained from direct sun measurements, while water vapour content in the atmosphere is retrieved from the direct measurements at 940 nm. However, inversion products of other aerosol optical properties, such

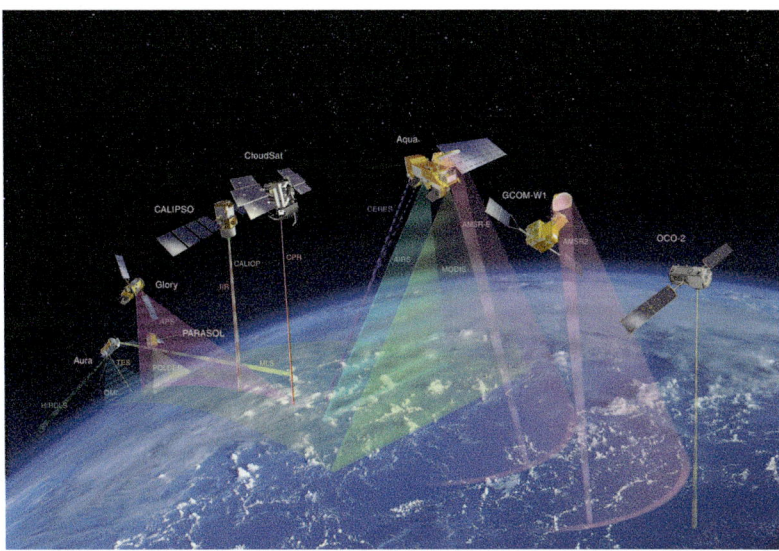

Fig. 4.3. A-train satellite constellation observing the Earth's atmosphere and surface at a broad swathe of wavelengths from space. (Adapted from National Aeronautics and Space Administration (n.d.))

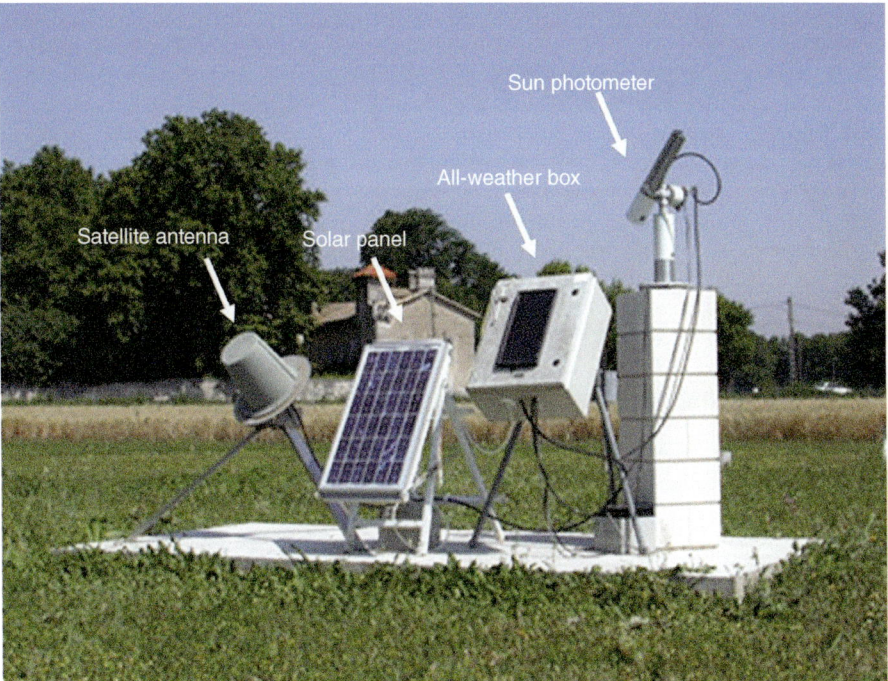

Fig. 4.4. The AERONET station with all its components.

as single scattering albedo (SSA), asymmetry parameter (AP), refractive indices and the column integrated aerosol size distributions are obtained from the sky radiance wavelengths (Holben *et al.*, 1998; Dubovik and King, 2000). The measurement protocols, calibration techniques and data processing have been described by Holben *et al.* (1998) and Eck *et al.*

(1999, 2005). The aerosols data are provided online in three different categories (Smirnov *et al.*, 2000), as a mixture of cloud contaminated (level 1.0), cloud screened (level 1.5) and quality assured (level 2.0) data.

The uncertainty in the calculation of AOD under cloud-free conditions is less than ± 0.01 for higher wavelengths (λ > 440 nm) and less than ± 0.02 for shorter wavelengths; however, it is less than ± 5% for the sky radiance measurements (Holben *et al.*, 1998; Eck *et al.*, 1999; Dubovik *et al.*, 2000). On the other hand, the estimated accuracy of AERONET for columnar water vapour (CWV) retrievals is within 10%. The errors in retrieved size distributions are not significant for the particles of radius (R) in the range 0.1–7 μm. The tendency of increasing errors in the retrieval of optical properties with the decrease in optical depth is higher in the case of refractive index and single scattering albedo than in the case of volume size distribution (Dubovik *et al.*, 2000). However, these errors do not significantly affect the important characteristic features of the size distribution (Dubovik *et al.*, 2000).

The long-term monthly mean spectral variability of aerosol optical depth (AOD) using a MICROTOPS sun photometer over Delhi is shown in Fig. 4.5 (Lodhi *et al.*, 2013). Figure 4.5 clearly reflects a high aerosol loading over Delhi during the last decade, which also shows a strong inter-/intra-seasonal as well as inter-annual variability. The monthly mean value of AOD

has a wide range from 0.25 to 1.75 for the different wavelengths. Figure 4.5 shows relatively steeper spectra of AOD during the winter/post-monsoon season, suggesting the dominance of fine mode aerosol particles which are mainly coming from the biomass/crop residue burning from northern India, mainly Haryana and Punjab (Kaskaoutis *et al.*, 2014; Tiwari *et al.*, 2016a). On the other hand, during the summer relatively less spectral dependence of AOD is observed than during the winter/post-monsoon season, indicating the dominance of coarse mode particles because of dust storms, which carry an abundance of dust particles from desert regions through long-range transport (Kaskaoutis *et al.*, 2012; Srivastava *et al.*, 2014; Singh *et al.*, 2016).

MODERATE RESOLUTION IMAGING SPECTRORADIOMETER (MODIS). The Moderate Resolution Imaging Spectroradiometer (MODIS) is a key instrument aboard the Terra and Aqua satellites for the measurement of aerosol and cloud properties, water vapour, etc. Terra was launched by NASA in December 1999 and has a southward orbit, while Aqua was launched in July 2002 and has an equator northward orbit (Levy *et al.*, 2007, 2010). MODIS has a swathe of 2330 km which makes it possible to observe global data in a single day and has an equilateral solar crossing time at ~10:30 am (Terra) and ~01:30 pm (Aqua). The MODIS sensor measures radiance at 36 spectral bands in the visible to thermal IR

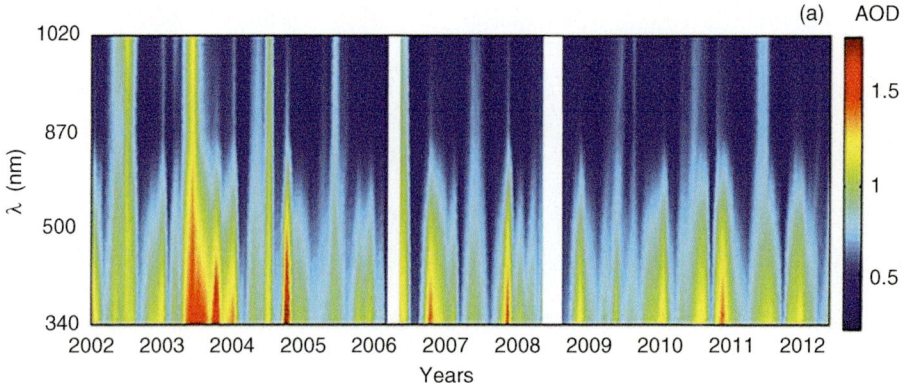

Fig. 4.5. Long-term monthly mean climatology (for a period of 11.5 years from December 2001 to May 2012) of spectral variation of aerosol optical depth (AOD) over Delhi. (Adapted from Lodhi *et al.*, 2013.)

spectral range of 0.41–14 µm, with 29 spectral bands at 1 km, five spectral bands at 500 m and two spectral bands at 250 m spatial resolution (Kaufman *et al.*, 1997). MODIS uses two different algorithms for land and ocean aerosol retrieval because of a large difference and heterogeneity in surface reflectance over land and ocean (Remer *et al.*, 2005; Levy *et al.*, 2010). It provides aerosol parameters at seven wavelength bands over oceanic regions and at three wavelength bands over land. Owing to the relatively stable and homogeneous reflectance of the ocean surface, the accuracy of AOD derived from the MODIS data over the ocean is better than that over land (Prijith *et al.*, 2013). The uncertainty associated in the MODIS retrieved AOD (τ) with AERONET is well documented as $\Delta\tau = \pm (0.05+ 0.15\tau_{MODIS})$ over land and $\Delta\tau = \pm (0.03+ 0.05\tau_{MODIS})$ over ocean (Remer *et al.*, 2008; Levy *et al.*, 2010).

Figure 4.6 shows 10 years (2005–2015) of monthly mean aerosol climatology over the Indian subcontinent, obtained from MODIS onboard the Terra satellite. The white patches in this figure show the absence of data. The figure reveals a large spatial variability in aerosol loading with a higher value of AOD over the entire Indo-Gangetic Basin (IGB) than the rest of India and its surroundings. Previous studies also reported higher aerosol concentration over the IGB, which also showed a negative gradient from western to eastern IGB (Lodhi *et al.*, 2013; Kaskaoutis *et al.*., 2013; Tiwari *et al.*, 2013, 2016a). Besides the IGB, coastal regions of the Indian subcontinent also have a relatively higher concentration of AOD than central and southern India, which may be due to dust outflow through the Arabian Sea and fossil fuel combustion from industries and ships (Satheesh *et al.*, 2006; Kedia and Ramachandran, 2008; Tiwari *et al.*, 2016b).

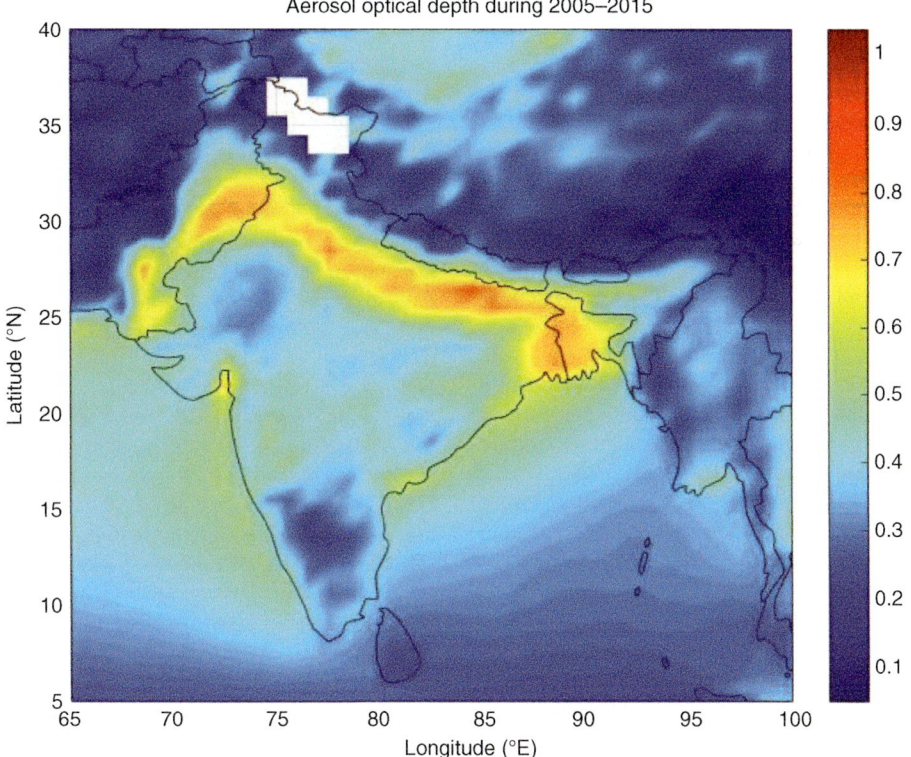

Fig. 4.6. Climatology of aerosol optical depth (AOD) at 550 nm over the Indian subcontinent for 2005–2015 obtained from MODIS on board the Terra satellite. The colour scale represents the aerosol optical depth (unitless).

CLOUD-AEROSOL LIDAR INFRARED PATHFINDER SATELLITE
OBSERVATION (CALIPSO).   The vertical distributions
of aerosols from passive remote sensing is a very
difficult problem, and crucial for the quantification of aerosol types and their radiative effects.
However, space-borne lidars can provide a global
view of the vertical structure of aerosol extinction from the Earth's surface through to the middle stratosphere. CALIPSO (Cloud-Aerosol Lidar
Infrared Pathfinder Satellite Observation), a part
of A-train sensor satellites (a series of afternoon
satellites) shown in Fig. 4.3, is widely used to obtain the vertical distribution of aerosol over
various regions (Mishra and Shibata, 2012a,b;
Prijith *et al.*, 2013). It is the first lidar satellite to
measure aerosol and cloud properties and was
launched on 28 April 2006 to study vertical
distributions and optical and physical properties.
Cloud-Aerosol Lidar with Orthogonal Polariza-
tion (CALIOP) is onboard the CALIPSO satellite,
which provides global vertically resolved measurements of aerosol and cloud distribution in
the atmosphere. CALIPSO's orbit has a 16-day
repeat cycle with equilateral crossing times at
1:30 and 13:30, which produce sub-satellite
tracks spaced longitudinally by 172 km at the
equator (Winker *et al.*, 2007). However, as
CALIPSO observations are limited to the sub-
satellite track, its spatial coverage is very poor
compared to MODIS. It measures the attenuated
backscatter intensity at two wavelengths, i.e. 532
and 1064 nm, with parallel and perpendicular
attenuated backscattered signals during the day
and night. The measured signals, processed
through a series of nested algorithms, can be
classified as clouds or aerosol and their subtypes
(clean continental, clean marine, dust, polluted
continental, polluted dust and smoke). Further
details about the CALIPSO algorithm and its description are provided by earlier studies (Labonne
et al., 2007; Winker *et al.*, 2007; Omar *et al.*,
2009; Tiwari *et al.*, 2016b). Aerosol extinction is
retrieved above clouds and below optically thin
clouds, as well as in the cloud-free columns
(Young and Vaughan, 2008). It also provides
information about the vertical distribution
of elevated aerosols and their transportation.
CALIPSO cannot provide information about the
aerosol and cloud at altitude regions that are
below optically thick clouds. Figure 4.7 shows the
monthly mean variation of aerosol backscatter
coefficient at 532 nm during 2006–2012 over:

(a) Bay of Bengal (BoB); and (b) Arabian Sea (AS)
obtained from CALIOP nighttime observations.
The aerosol backscatter is a measure of aerosol
loading. The maximum aerosol loading in the lower
part of the troposphere (boundary layer) is observed in May–August for both regions. A higher
aerosol loading is observed over AS in comparison to BoB. Apart from this, Fig. 4.7 clearly
reflects the presence of the elevated aerosol
layers (up to 2–4 km) during May–August over
both regions, which may be mainly due to the
long-range transportation of aerosol from the
desert regions at higher altitudes (2–3 km). During
October–November, another elevated aerosol
layer is found over both regions associated with
relatively higher altitude (> 4 km), which could
be because of long range transportation of biomass burning aerosol from the Indo-Gangetic
Basin. The elevated layers during May–August are
more pronounced over AS, while during October–
November over BoB.

OZONE MONITORING INSTRUMENT (OMI).   The Ozone
Monitoring Instrument (OMI) flies on NASA's
Aura satellite. Aura was launched on 15 July
2004, and OMI operations began on 1 October
2004. Aura has a sun-synchronous polar-orbit
with a local equator crossing time at the ascend-
ing overpass of about 13:38 hrs. OMI is a nadir
viewing instrument, which has 20 wide-field-
imaging spectrometers that collectively provide
daily global coverage with high spectral resolu-
tions and spatial resolution of 13 km × 24 km
(Levelt *et al.*, 2006). It measures the top of the
atmosphere upwelling radiances in the ultraviolet and visible (270–500 nm) regions of the
solar spectrum. Although the instrument was
designed primarily for retrieval of trace gases
like O_3, NO_2 and SO_2, it contains valuable information on aerosols. Further details of the OMI
sensors are available in Remer *et al.* (2005) and
Levelt *et al.* (2006), respectively. Due to higher
sensitivity, it can also provide the information
about the aerosol type, Extinction Aerosol Optical Depth (EAOD), Single Scattering Albedo
(SSA) and Absorbing Aerosol Optical Depth
(AAOD) using the OMAERUV algorithm (Torres
et al., 2007).

   In particular, the OMAERUV retrieval algorithm is highly sensitive to carbonaceous and
mineral aerosols and the aerosol layer height.
This algorithm assumes that the column aerosol

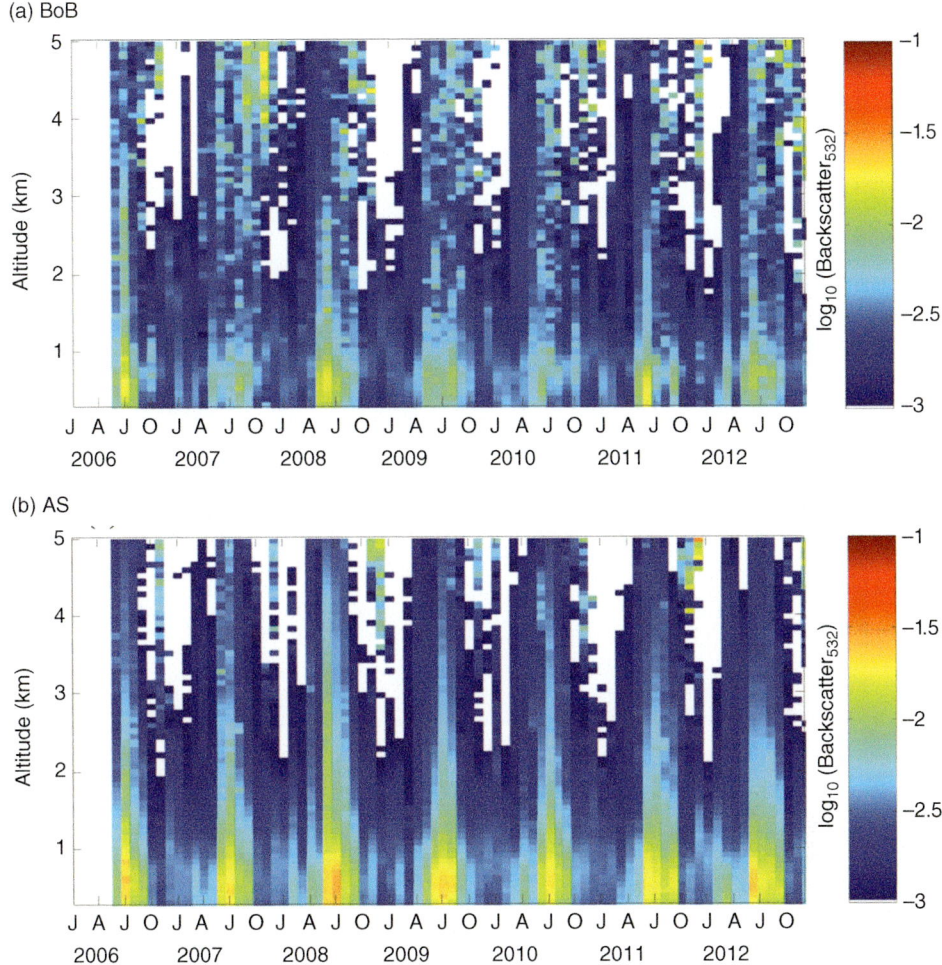

Fig. 4.7. Monthly mean variation of aerosol backscattering coefficient at 532 nm during 2006–2012 over (a) Bay of Bengal (BoB) and (b) Arabian Sea (AS) from CALIOP night-time observations. The aerosol backscatter coefficients are shown in logarithmic scale. (Adapted from Tiwari *et al.*, 2016b.)

loading can be represented by one of three types of aerosols and uses a set of aerosol models to account for the presence of these aerosols: carbonaceous aerosol from biomass burning, desert dust, and light-absorbing sulfate-based aerosols. Based on this, various aerosol types such as smoke, dust and sulfate can be easily distinguished. For carbonaceous and desert-dust particles, the aerosol load is assumed to be vertically distributed following a Gaussian function characterized by peak (aerosol layer height) and half-width (aerosol layer geometric thickness) values (Torres *et al.*, 2013). Apart from this, OMI also measures the cloud coverage which provides the data to derive the tropospheric ozone. Figure 4.8 shows the monthly mean climatology of total columnar ozone (in DU) over the Indian subcontinent for 10 years (2005–2015), indicating that the northern part experiences higher ozone than the southern part of the Indian subcontinent, which also shows a negative gradient from west to east over the IGB region. An earlier study also reported a large spatial variability in ozone concentration having higher value over northern India, particularly over the IGB (David and Nair, 2013).

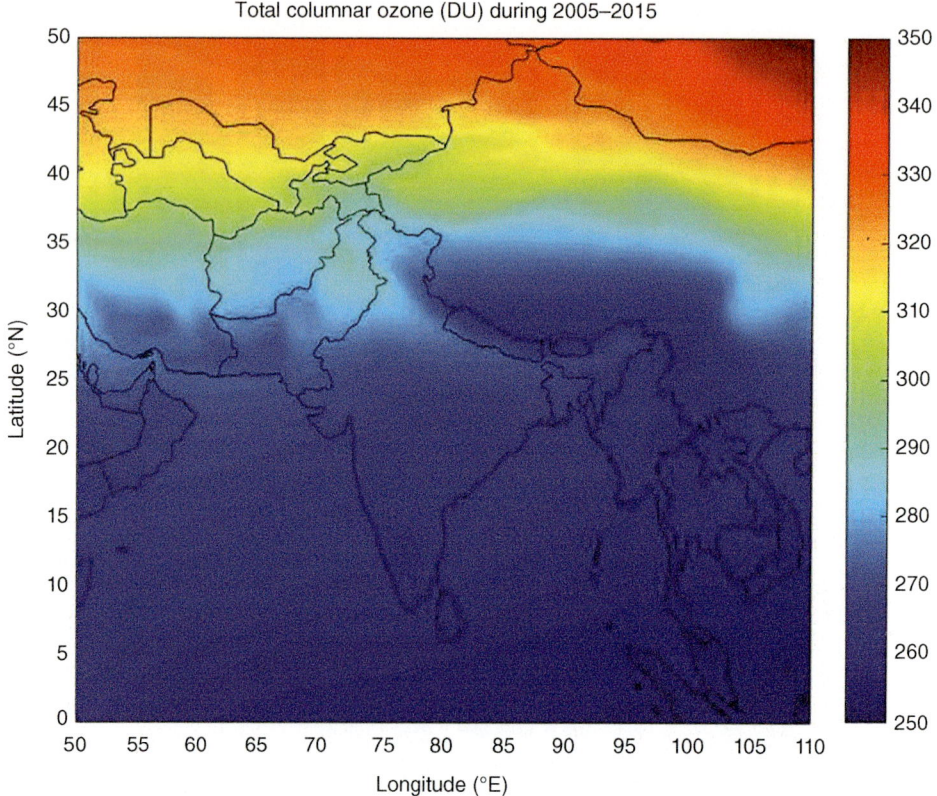

Fig. 4.8. Monthly mean climatology of total columnar ozone (in Dobson units) over the Indian subcontinent for 2005–2015, data obtained from OMI.

ATMOSPHERIC INFRARED SOUNDER (AIRS). Atmospheric Infrared Sounder (AIRS) is a high spectral resolution scanning spectrometer (http://disc.sci.gsfc.nasa.gov/AIRS/documentation/airs_instrument_guide.shtml) aboard the NASA's Aqua satellite, with 2378 bands in the thermal infrared (3.7–15.4 µm) and 4 bands in the visible (0.4–1.0 µm) (Aumann et al., 2003). Therefore, with such a large number of wavelength bands, AIRS provides us with a high resolution vertical structure, and therefore a sharp picture of the atmosphere. Radiances in the wavelength region 4.50–4.58 µm are utilized for AIRS carbon mono oxides retrieval (McMillan et al., 2005) following a radiative transfer model described in Susskind et al. (2003) and Strow et al. (2003). The instrument captures data with a spatial resolution of 13.5 km in the IR region and 2.3 km in the VIS/NIR region at nadir. The AIRS instrument measures air temperature and water vapour as a function of height, besides measuring abundances of

gaseous components such as ozone, carbon monoxide, carbon dioxide, methane and ammonia, etc. Besides these, AIRS can also track volcanic emissions and smoke from wildfires, and can indicate a region which may be heading for a drought. It is widely used in weather forecasting due to its hyperspectral observation. ECMWF studies have shown that in many circumstances, AIRS is responsible for reducing forecast errors by more than 10%, which is the largest forecast improvement of any single satellite instrument of the 2000s. Figure 4.9 shows the monthly mean climatology of daytime total columnar loading of carbon monoxide over the Indian subcontinent during 2005–2015. Figure 4.9 suggests a higher concentration of CO over the IGB and the western coast than the rest of the Indian subcontinent. Kar et al. (2008) also reported a higher concentration of CO over the IGB using MOPITT, while Girach and Nair (2014) reported a higher concentration of CO over the western coastal region of the

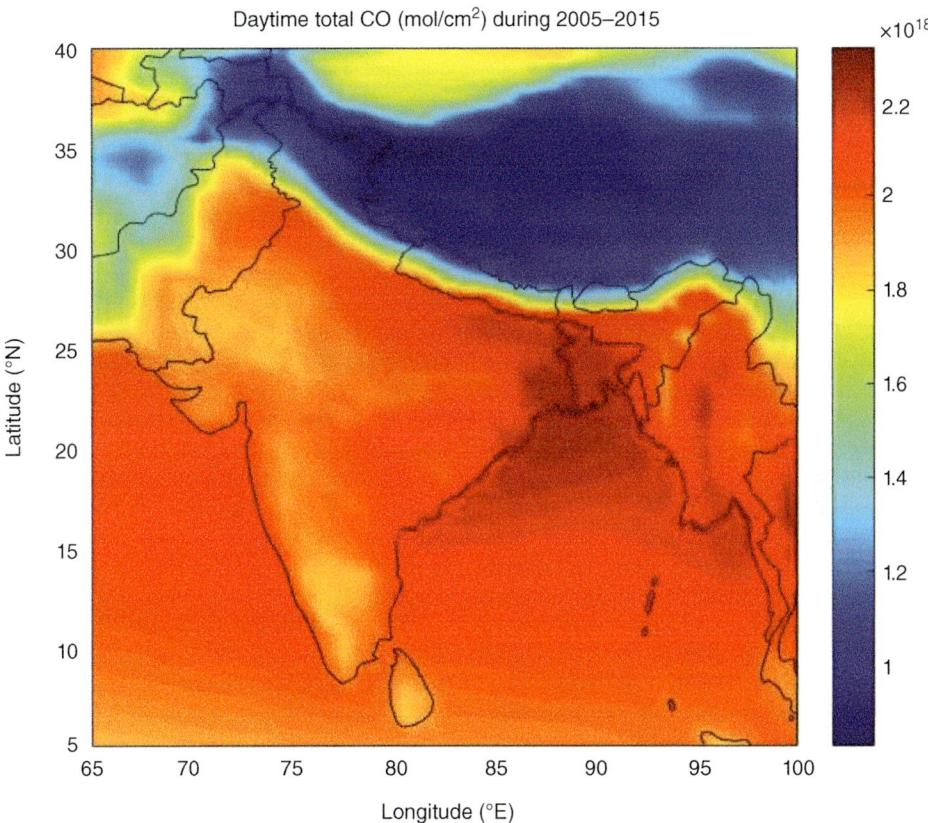

Daytime total CO (mol/cm^2) during 2005–2015

Fig. 4.9. Monthly mean climatology of daytime total column carbon monoxide (CO in mol/cm^2) over the Indian subcontinent for 2005–2015, data obtained from AIRS.

India subcontinent, which also shows a strong seasonal variability and is influenced by the synoptic meteorology.

4.5   Sampling techniques

Sampling techniques are one of the most fundamental techniques for the *in situ* measurements of air pollutants. There are two types of sampling method – active sampling and passive sampling. In active sampling, a pump is used which pulls the air through a collection device (like a filter or sensor) and is used mostly in concentration measurements. It is independent of wind speed and can collect a large sample volume; however, accurate and precise flow rate and sampling duration must be quantified (Watson *et al.*, 2013). On the other hand, passive sampling does not involve any pumping, and pollutants are collected through their own movement and diffusion. Passive sampling is a cost-effective tool but environmental factors can affect it (Zabiegała *et al.*, 2010). Once the samples are collected through different methods (volumetric, colorimetric, gravimetric, photometric), these samples are analysed by various established physical and chemical techniques such as the scanning electron microscope (SEM), or spectroscopic and chromatography techniques, which are discussed in upcoming sections. The sampling of air pollutants highly depends on the objective of the study, sampling location, local emission sources, frequency and duration of the sampling period. The sampling station should be selected to represent the neighbourhood (up to a few metres), to regional scales (up to a few hundred kilometres), and be at a height of about 3 to 10 m from ground level (Chow *et al.*, 2002). Nowadays, different instruments, based on sampling techniques (e.g. volume sampler for particulate matters, gas analyser, aethalometer, nephelometer) are used worldwide for the measurements of different air pollutant species which can be easily coupled with different analysis techniques. The day-to-day variability of particulate matter (PM) concentrations (PM_{10} and $PM_{2.5}$ in µg/m³) using a dust sampler (IPM-FDS, Instrumex) over Varanasi during January to March 2014 is shown in Fig. 4.10. The figure suggests a significant day-to-day variability of particulate matter over Varanasi during the study period, which continuously decreases from January to March because of calm meteorological conditions during March (transition month) (Kumar et al., 2015). A relatively higher concentration of particulate matter is observed over Varanasi than the national standards (PM_{10} = 100 µg m⁻³; $PM_{2.5}$ = 60 µg m⁻³; cpcb.nic.in) and United States Environmental Protection Agency standards (PM_{10} = 150 µg m⁻³; $PM_{2.5}$ = 35 µg m⁻³; USEPA, n.d.). The average mass concentrations of PM_{10} and $PM_{2.5}$ are recorded as 233 ± 58.37 and 138 ± 47.12 µg m⁻³ (mean ± sd), respectively. On the other hand, the monthly mean variation

of trace gases for 4 years over Kanpur using the gas analyser is shown in Fig. 4.11. The highest concentration of SO_2 is observed during winter (4.6 ± 4.1 ppb) while the minimum is in the monsoon season (0.5 ± 0.8 ppb) (Fig. 4.11a). In contrast to SO_2, O_3 is found to be at a maximum during summer (Fig. 4.11d), which may be because of the linear relationship with incoming solar radiation (Han et al., 2011; Gaur et al., 2014). Another peak in O_3 is also observed in winter, which may be because of long-range transportation and thermal decomposition of O_3 precursors (e.g. PAN) (Gaur et al., 2014). A similar pattern in monthly variation of NO_x and CO (Figs 4.11b and 4.11c, respectively) was also observed during the study period, which may be due to the combined effect of near-surface anthropogenic emissions and boundary layer processes.

4.6   Spectroscopic Techniques

The spectroscopic technique is highly sophisticated and widely used for air pollutant measurements.

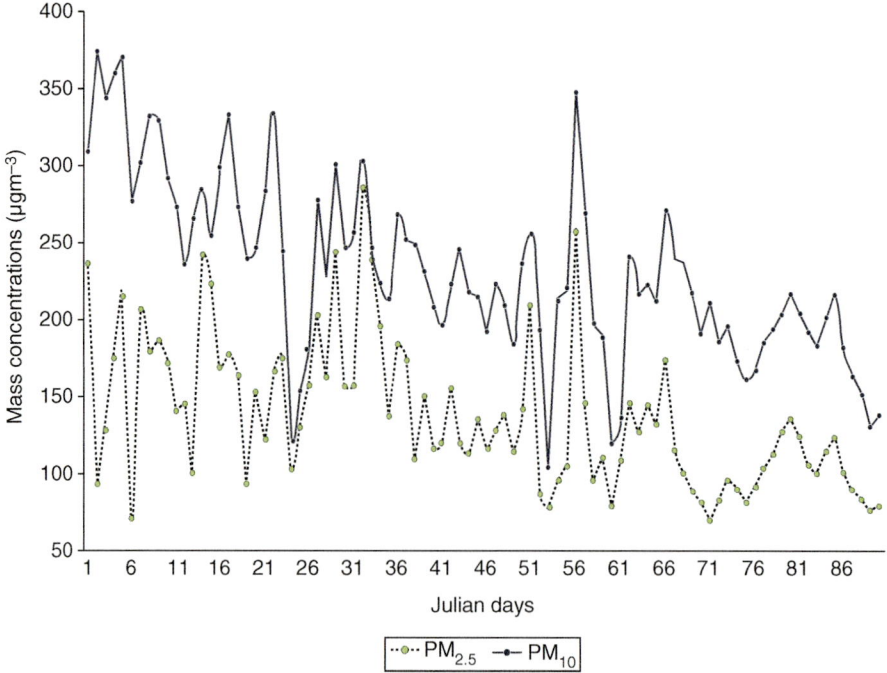

Fig. 4.10. Day-to-day variability of particulate matter concentrations (PM_{10} and $PM_{2.5}$ in µg/m⁻³) over Varanasi during January–March 2014. (Adapted from Kumar et al., 2015.)

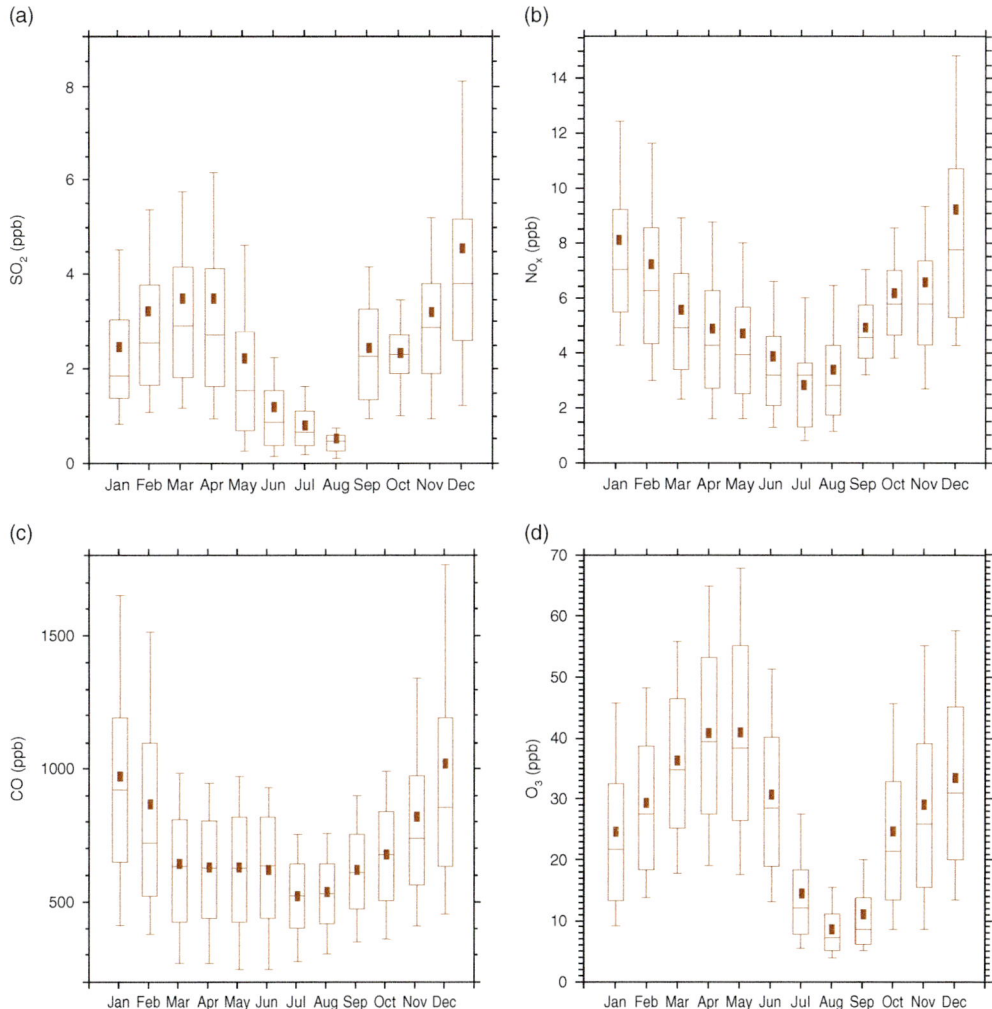

Fig. 4.11. Monthly mean variation of (a) SO_2, (b) NO_x, (c) CO, and (d) O_3 at Kanpur for 4 years (June 2009–May 2013). The horizontal solid line indicates the median, the filled square indicates the mean, the top and bottom of the box indicate the 75th and 25th percentile, respectively, the top and bottom whiskers indicate the 95th and 5th percentile, respectively. (Adapted from Gaur *et al.*, 2014.)

It is based on the principles of spectrometry, in which the interactions between light and matter use radiation of intensity and wavelengths. Spectrometers are the most important instruments of these techniques, which measure radiation types and wavelengths. These spectrometers are used in ground-based as well as satellite-based measurements. There are several spectroscopic techniques (e.g. absorption spectroscopy, mass spectroscopy, laser spectroscopy, Fourier Transfer Infrared spectroscopy (FTIR), ultraviolet-visible spectroscopy), which are used for the measurement of different atmospheric constituents. Mass spectroscopy sorts and measures the masses between samples within the sample through mass to charge ratio. Fourier Transform-Infrared Spectroscopy (FTIR) is mainly used to identify organic (and in some cases inorganic) materials and measures the absorption of infrared radiation by the sample material versus wavelengths. The spectroscopic techniques are easily coupled with other analytical methods (e.g. gas chromatography

with mass spectrometry (GC-MS), ion chromatography mass spectrometry (ICMS), time-of-
flight mass spectrometry (TOF-MS), proton
transfer reaction time-of-flight mass spectrometry (PTR-TOF-MS)) and provide accurate
information on the complex composition of air
pollutants.

The hourly and daily variations of VOCs over
Ahmedabad during 1 January to 1 February
2014 using PTR-TOF-MS are shown in
Fig. 4.12. The break in time series (7–9 and 24–
27 January) represents the absence of data because of the calibration process. The periodic
nature of VOC variation is observed on a diurnal

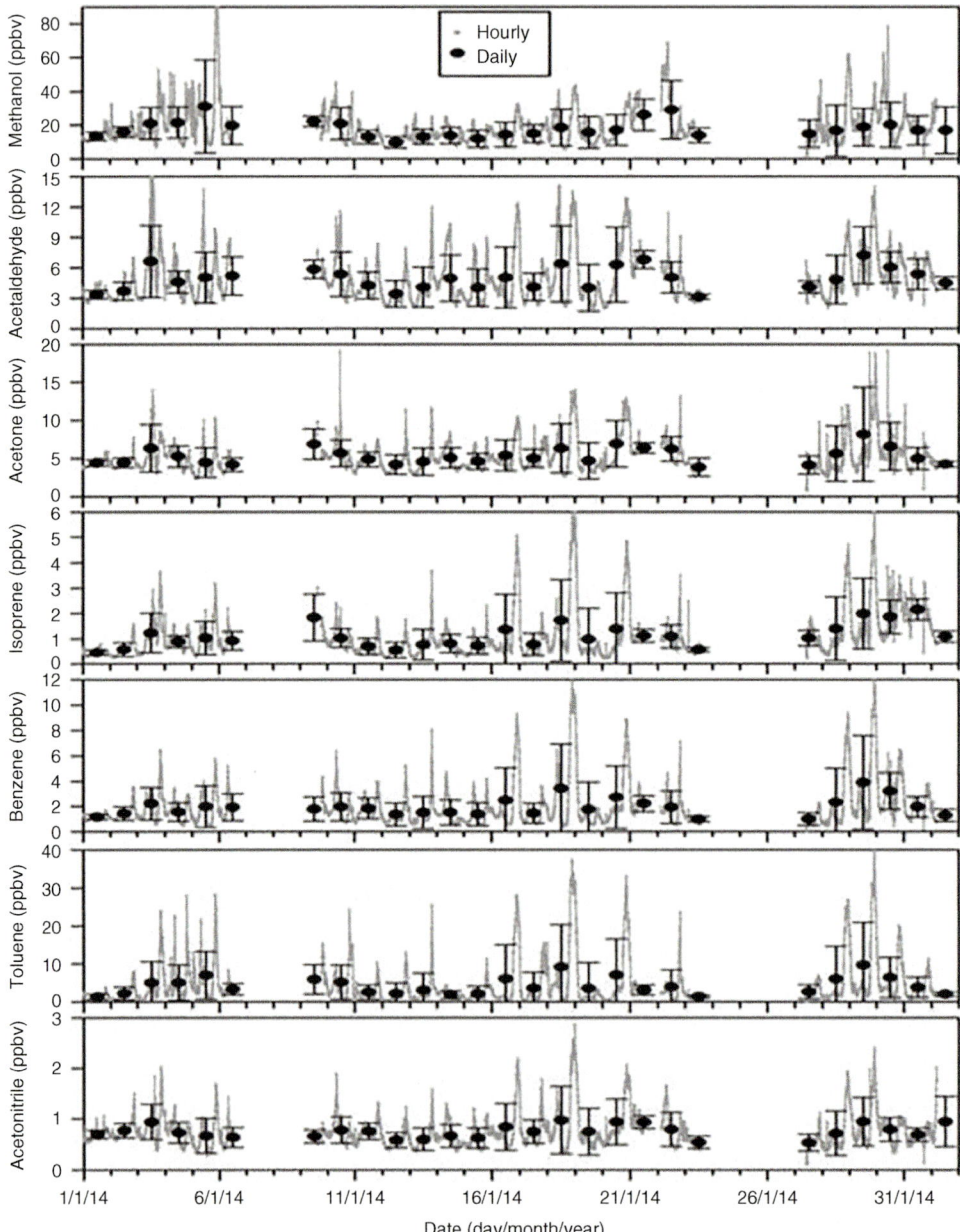

Fig. 4.12. The time series variations of VOCs using 10 min and daily means of VOCs during 1 January–1
February 2014. (Adapted from Sahu et al., 2016.)

basis; however, a significant day-to-day variability is also observed during the study period. Higher values of VOCs are observed during 29–31 January, which may due to the transport of pollutants from biomass burning over the IGB (Sahu et al., 2016). The average value of methanol, acetone and acetaldehyde were 17.85 ± 4.2 ppbv, 5.0 ± 1.1 ppbv and 5.4 ± 1.0 ppbv, respectively; however, the values of primary VOCs, such as benzene, toluene, acetonitrile and isoprene, were 1.98 ± 0.74 ppbv, 4.3 ± 2.3 ppbv, 0.76 ± 0.13 pbbv and 1.1 ± 0.47 ppbv, respectively.

4.7   Chromatography Techniques

Chromatography techniques are widely used in measurements of air pollutants. They work on the principle of the separation of the air pollutants' constituents, based on their speed (Schuck et al., 2009). In chromatography techniques, a dissolved air pollutant mixture in a mobile phase is moving in a fixed direction towards the stationary phase (which may be made from polar or nonpolar materials), where they interact and mixtures are separated. Different types of chromatography are used in air pollution measurement, which depends on the mobile phase; e.g. inert gases are widely used in gas chromatography (GC); buffered aqueous solution is used in ion chromatography (IC); liquid solvent is used in high pressure liquid chromatography (HPLC). GC is preferred for organic compound analysis which can be easily vaporized without decomposition. GC-based measurements are highly sensitive and provide an excellent chemical speciation at pptv level (Haskin, 2013). It can be easily coupled with mass-spectrometric and chemoluminescence detectors, which can easily identify a small fraction of the constituents in the mixture. GC measurements require several minutes sampling and a relatively longer analytical time, which causes the possibilities of error due to the rapid chemical reaction into the atmosphere. The diurnal variation of carbon monoxide (CO), methane (CH_4) and non-methane hydrocarbons (NMHCs) over Kolkata on a seasonal basis for 1 year (March 2012–February 2013) using gas chromatography, is shown in Fig. 4.13. This figure suggests that the values of CO and other hydrocarbons are relatively higher during the winter season than others, particularly in the morning (7.30 am) and evening (7.30 pm),

which may be mainly due to the atmospheric stability and mixing layer height during the winter season (Mallik et al., 2014). A relatively higher value of i-C_4H_{10} values is observed during peak traffic hours in the morning, suggesting gasoline evaporation.

On the other hand, ion chromatography (IC) is used for the separation of ionic constituents (mainly inorganic anions and cations) or polar molecules of the air pollutants mixture, based on their affinity to the ion exchanger. The IC analysis of 22 PM_{10} samples collected at Varanasi during April to July 2011 is shown as cumulative mass concentration of anions (Cl^-, SO_4^{2-}, PO_4^{2-} and NO_3^-) and Na^+, K^+, Mg^{2+}, NH^{4+} and Ca^{2+} in Fig. 4.14(a), and their percentage contribution of each ionic species in water soluble (WS) mass in Fig. 4.14(b). Figure 4.14 suggests that, among all chemical species, SO_4^{2-} have the highest concentration, followed by NO_3^-, Ca^{2+}, Na^+, Cl^-, K^+, $PO4^{2-}$ NH_4^+ and Mg^{2+}, respectively. The higher concentration of SO_4^{2-} is mainly because of SO_2 emission from industrial sectors, which suggests the formation of secondary particles due to gas-to-particle conversion (Behera and Sharma 2010a; Squizzato et al., 2013). Similar patterns of mean concentrations of water-soluble ions were also reported by Behera and Sharma (2010a) over those with relatively higher concentrations, which may be due to larger traffic density and industrial sectors, and coal-based power plants (Behera and Sharma, 2010b; Tiwari et al., 2016c).

Conclusions

Air pollutants have a significantly detrimental effect on the environment, climate and human beings, and are causes of several serious health problems and loss of crop production. The long-term effects are uncertain, due to their various physicochemical properties, regional meteorology and instrument limitation, though numerous studies have been carried out worldwide to understand the effects of these pollutants on human health, agriculture and climate. The main aim of this chapter is to provide updated information about the various measurement techniques of air pollutants used worldwide; various types of air pollutants and their possible emission sources are also discussed. It also provides information

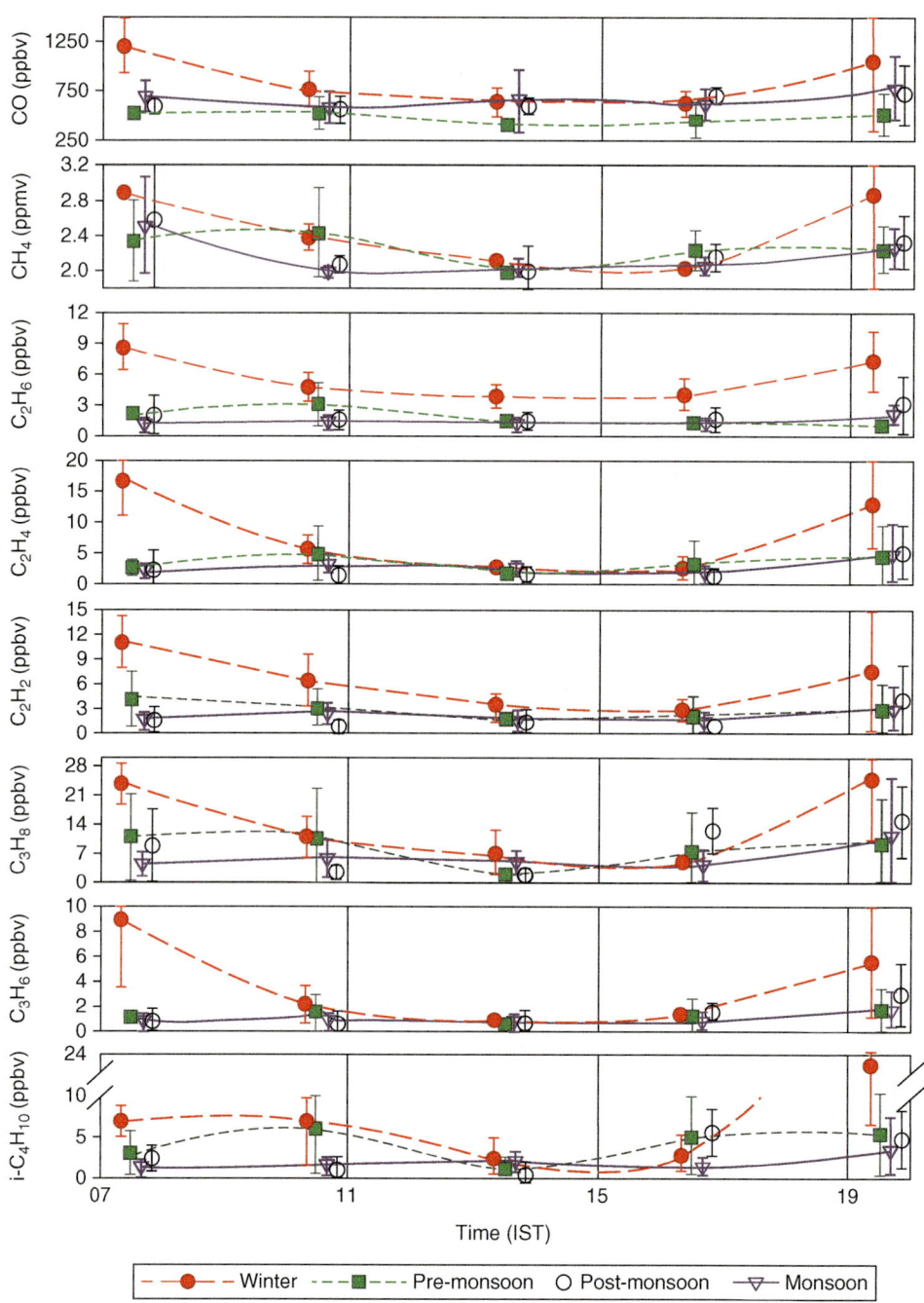

Fig. 4.13. Daytime variations of CO, CH_4 and non-methane hydrocarbons (NMHCs) over Kolkata during March 2012–February 2013. (Adapted from Mallik *et al.*, 2014.)

about air pollutant concentrations and their spatial distribution over the Indian subcontinent during the last decade. This chapter reveals that the air pollutants (e.g. aerosol, ozone, CO) over the Indo-Gangetic Basin are relatively high compared to the rest of the Indian subcontinent because of favourable meteorological conditions, various emission sources, higher pollution

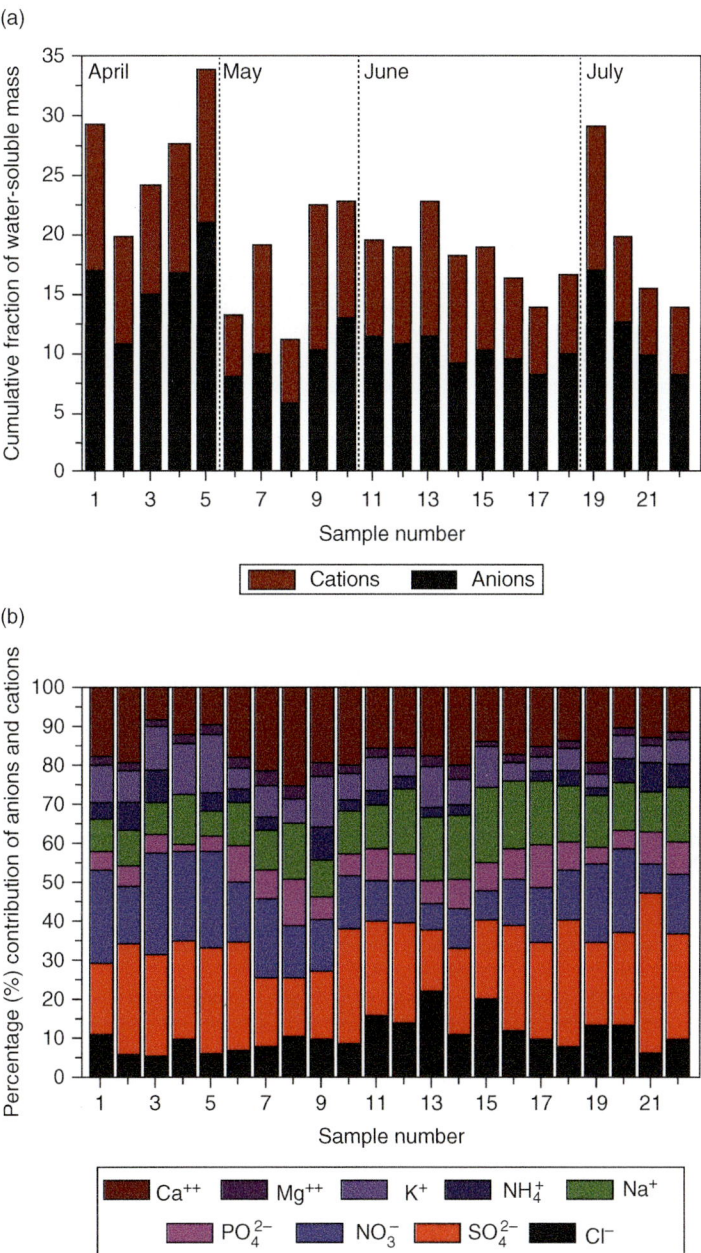

Fig. 4.14. (a) Cumulative fraction of anions and cations in the water soluble (WS) aerosol mass; and (b) percentage (%) contribution of various ionic species on WS mass in Varanasi during April–July 2011. (Adapted from Tiwari *et al.*, 2016c.)

density and related energy demands. Rapid urbanization and industrialization are the major causes of poor air quality and air pollution. To measure the pollutant concentration, several techniques (spectroscopic, gas chromatography, ion chromatography, remote sensing) are widely used. Nowadays, remote sensing tools have become the most effective and popular tools to measure the air pollutants and their transportation because of their ability to provide data with high spatial-temporal resolution, along with vertical distribution. Although the technology for air pollutant measurements has improved during the last two decades, it is still not at a satisfactory level. Therefore, cost-effective and quality-assured measurement techniques of air pollutants (particularly mean concentration of certain species) is greatly needed to help policy makers. Based on quality-assured data, policy makers could modify the current emission control scenario.

References

Aumann, H.H., Chahine, M.T., Gautier, C., Goldberg, M.D., Kalnay, E. *et al.* (2003) AIRS/AMSU/ HSB on the Aqua mission: design, science objectives, data products, and processing systems. *IEEE Transactions on Geoscience and Remote Sensing* 41, 253–264.

Balakrishnan, K., Ganguli, B., Ghosh, S., Sankar, S., Thanasekaran, V. *et al.* (2011) Short-term effects of air pollution on mortality: results from a time-series analysis in Chennai, India. Research report (Health Effects Institute), No. 157, pp. 7–44.

Behera, S.N. and Sharma, M. (2010a) Reconstructing primary and secondary components of $PM_{2.5}$ composition for an urban atmosphere. *Aerosol Science and Technology* 44, 983–992.

Behera, S.N. and Sharma, M. (2010b) Investigating the potential role of ammonia in ion chemistry of fine particulate matter formation for an urban environment. *Science of the Total Environment* 408, 3569–3575.

Bolden, A.L., Kwiatkowski, C.F. and Colborn, T. (2015) New look at BTEX: are ambient levels a problem? *Environmental Science and Technology* 49, 5261–5276.

Bower, J.S. and Mucke, H.-G. (1999) Design, operation and quality assurance and control in a monitoring system. *Monitoring Ambient Air Quality for Health Impact Assessment*. WHO Regional Publications, European Series, No. 85. World Health Organization, Copenhagen.

Brauer, M., Hoek, G., Smit, H.A., de Jongste, J.C., Gerritsen, J. *et al.* (2007) Air pollution and development of asthma, allergy and infections in a birth cohort. *European Respiratory Journal* 29, 879–888.

Burney, J. and Ramanathan, V. (2014) Recent climate and air pollution impacts on Indian agriculture. *Proceedings of the National Academy of Sciences USA* 111(46), 16319–16324.

Caiazzo, F., Ashok, A., Waitz, I.A., Yim, S.H.L. and Barrett, R.H. (2013) Air pollution and early deaths in the United States. Part I: quantifying the impact of major sectors in 2005. *Atmospheric Environment* 79, 198–208.

Chandra, N., Hayashida, S., Saeki, T. and Patra, P.K. (2017) What controls the seasonal cycle of columnar methane observed by GOSAT over different regions in India? *Atmospheric Chemistry and Physics* 17, 12633–12643.

Charlson, R.J., Schwartz, S.E., Hales, J.M., Cess, R.D., Coakley, J.A. *et al.* (1992) Climate forcing by anthropogenic aerosols. *Science* 255, 423–430.

Chow, J.C., Engelbrecht, J.P., Watson, J.G., Wilson, W.E., Frank, N.H. and Zhu, T. (2002) Designing monitoring networks to represent outdoor human exposure. *Chemosphere* 49(9), 961–978.

Chowdhury, S. and Dey, S. (2016) Cause-specific premature death from ambient $PM_{2.5}$ exposure in India: estimate adjusted for baseline mortality. *Environment International* 91, 283–290.

Dai, H., Jing, S., Wang, H., Ma, Y., Li, L. *et al.* (2017) VOC characteristics and inhalation health risks in newly renovated residences in Shanghai, China. *Science of the Total Environment* 577, 73–83.

David, L.M. and Nair, P.R. (2013) Tropospheric column O_3 and NO_2 over the Indian region observed by Ozone Monitoring Instrument (OMI): seasonal changes and long-term trends. *Atmospheric Environment* 65, 25–39.

Duan, J., Tan, J., Yang, L., Wu, S. and Hao, J. (2008) Concentration, sources and ozone formation potential of Volatile Organic Compounds (VOCs) during ozone episode in Beijing. *Atmospheric Research* 88, 25–35.

Dubovik, O. and King, M.D. (2000) A flexible inversion algorithm for retrieval of aerosol optical properties from sun and sky radiance measurements. *Journal of Geophysical Research* 105, 20673–20696.

Dubovik, O., Smirnov, A., Holben, B.N., King, M.D., Kaufman, Y.J. *et al.* (2000) Accuracy assessments of aerosol optical properties retrieved from AERONET sun and sky-radiance measurements. *Journal of Geophysical Research* 105, 9791–9806.

Eastham, S.D. and Barrett, S.R.H. (2016) Aviation-attributable ozone as a driver for changes in mortality related to air quality and skin cancer. *Atmospheric Environment* 144, 17–23.

Eck, T.F., Holben, B.N., Reid, J.S., Dubovik, O., Smirnov, A. *et al.* (1999) Wavelength dependence of the optical depth of biomass burning, urban, and desert dust aerosols. *Journal of Geophysical Research* 104, 31333–31349.

Eck, T.F., Holben, B.N., Dubovic, O., Smirnov, A., Goloub, P. *et al.* (2005) Columnar aerosol optical properties at AERONET sites in central eastern Asia and aerosol transport to the tropical mid-Pacific. *Journal of Geophysical Research* 110, D06202. DOI: 10.1029/2004JD005274.

Finney, D.L., Doherty, R.M., Wild, O., Young, P.J. and Butler, A. (2016) Response of lightning NO_x emissions and ozone production to climate change: insights from the atmospheric chemistry and climate model intercomparison project. *Geophysical Research Letters* 43, 5492–5500.

Fung, I., John, J., Lerner, J., Matthews, E., Prather, M. *et al.* (1991) 3-Dimensional model synthesis of the global methane cycle. *Journal of Geophysical Research – Atmospheres* 96, 13033–13065.

Gaur, A., Tripathi, S.N., Kanawade, V.P., Tare, V. and Shukla, S.P. (2014) Four-year measurements of trace gases (SO_2, NO_x, CO, and O_3) at an urban location, Kanpur, in Northern India. *Journal of Atmospheric Chemistry* 71, 283–301.

Gedney, N., Cox, P.M. and Huntingford, C. (2004) Climate feedback from wetland methane emissions. *Geophysical Research Letters* 31, L20503. DOI: 10.1029/2004GL020919.

Ghude, S.D., Chate, D.M., Jena, C., Beig, G., Kumar, R. *et al.* (2016) Premature mortality in India due to $PM_{2.5}$ and ozone exposure. *Geophysical Research Letters* 43, 4650–4658.

Girach, I.A. and Nair, P.R. (2014) Carbon monoxide over Indian region as observed by MOPITT. *Atmospheric Environment* 99, 599–609.

Goldstein, A.H. and Galbally, I.E. (2007) Known and unexplored organic constituents in the Earth's atmosphere. *Environmental Science and Technology* 41, 1514–1521.

Greenberg, N., Carel, R.S., Derazne, E., Bibi, H., Shpriz, M. *et al.* (2016) Different effects of long-term exposures to SO_2 and NO_2 air pollutants on asthma severity in young adults. *Journal of Toxicology and Environmental Health, Part A* 79, 342–351. DOI: 10.1080/15287394.2016.1153548.

Grewe, V., Dahlmann, K., Matthes, S. and Steinbrecht, W. (2012) Attributing ozone to NO_x emissions: implications for climate mitigation measures. *Atmospheric Environment* 59, 102–107.

Guo, T., Li, X., Li, J., Peng, Z., Xu, L. *et al.* (2018) On-line quantification and human health risk assessment of organic by-products from the removal of toluene in air using non-thermal plasma. *Chemosphere* 194, 139–146.

Han, S., Bian, H., Feng, Y., Liu, A., Li, X. *et al.* (2011) Analysis of the relationship between O_3, NO and NO_2 in Tianjin, China. *Aerosol and Air Quality Research* 11, 128–139.

Haskin, C. (2013) Determination of the concentration of atmospheric gases by gas chromatography. *McNair Scholars Research Journal* 6(1), Article 6.

HEI (2004) *Health Effects of Outdoor Air Pollution in Developing Countries of Asia: A Literature Review*, ed. Health Effects Institute, International Oversight Committee, Special Report No. 15. Health Effects Institute, Boston, MA.

Holben, B.N., Eck, T., Slutser, I., Tanre, D., Bais, J.P. *et al.* (1998) AERONET federated instrument network and data archive for aerosol characterization. *Remote Sensing of Environment* 66(1), 1–16.

IPCC (2013) *Fifth Assessment Report: Climate Change 2013*. Cambridge University Press, New York.

Jain, N., Bhatia, A. and Pathak, H. (2014) Emission of air pollutants from crop residue burning in India. *Aerosol and Air Quality Research* 14, 422–430.

Kampa, M. and Castanas, E. (2008) Human health effects of air pollution. *Environmental Pollution* 51, 362–367.

Kar, J., Jones, D.B.A., Drummond, J.R., Attié, J.L., Liu, J., Zou, J., Nichitiu, F., Seymour, M.D., Edwards, D.P. Deeter, M.N., Gille, J.C. and Richter, A. (2008) Measurement of low-altitude CO over the Indian subcontinent by MOPITT. *Journal of Geophysical Research* 113, D16307. DOI: 10.1029/2007JD009362.

Kaskaoutis, D.G., Gautam, R., Singh, R.P., Houssos, E.E., Goto, D. *et al.* (2012) Influence of anomalous dry conditions on aerosols over India: transport, distribution and properties. *Journal of Geophysical Research* 117, D09106. DOI: 10.1029/2011JD017314.

Kaskaoutis, D.G., Sinha, P.R., Vinoj, V., Kosmopoulos, P.G., Tripathi, S.N. *et al.* (2013) Aerosol properties and radiative forcing over Kanpur during severe aerosol loading conditions. *Atmospheric Environment* 79, 7–19.

Kaskaoutis, D.G., Kumar, S., Sharma, D., Singh, R.P., Kharol, S.K. *et al.* (2014) Effects of crop residue burning on aerosol properties, plume characteristics, and long-range transport over Northern India. *Journal of Geophysical Research – Atmospheres* 119, 5424–5444.

Kaufman, Y.J., Tanré, D., Remer, L.A., Vermote, E.F., Chu, A. and Holben, B.N. (1997) Operational remote sensing of tropospheric aerosol over land from EOS moderate resolution imaging spectroradiometer. *Journal of Geophysical Research* 102, 17051–17068.

Kedia, S. and Ramachandran, S. (2008) Features of aerosol optical depths over the Bay of Bengal and the Arabian Sea during pre-monsoon season: variabilities and anthropogenic influence. *Journal of Geophysical Research* 113, D11201. DOI: 10.1029/2007JD009070.

Kim, S.-W., Heckel, A., McKeen, S.A., Frost, G.J., Hsie, E.-Y. *et al.* (2006) Satellite-observed U.S. power plant NO$_x$ emission reductions and their impact on air quality. *Geophysical Research Letters* 33, L22812. DOI: 10.1029/2006GL027749.

Krewski, D., Burnett, R.T., Goldberg, M.S., Hoover, K., Siemiatycki, J. *et al.* (2000) *Reanalysis of the Harvard Six Cities Study and the American Cancer Society Study of Particulate Air Pollution and Mortality, A Special Report of the Institute's Particle Epidemiology Reanalysis Project.* Health Effects Institute, Cambridge, MA.

Kumar, M., Tiwari, S., Murari, V., Singh, A.K. and Banerjee, T. (2015) Wintertime characteristics of aerosols at middle Indo-Gangetic Plain: impacts of regional meteorology and long range transport. *Atmospheric Environment* 104, 162–175.

Labonne, M., Bréon, F.M. and Chevallier, F. (2007) Injection height of biomass burning aerosols as seen from a spaceborne lidar. *Geophysical Research Letters* 34, L11806. DOI: 10.1029/2007GL029311.

Latha, R., Vinayak, B. and Murthy, B.S. (2018) Response of heterogeneous vegetation to aerosol radiative forcing over a Northeast Indian station. *Journal of Environmental Management* 206, 1224–1232.

Lelieveld, J., Evans, J.S., Fnais, M., Giannadaki, D. and Pozzer, A. (2015) The contribution of outdoor air pollution sources to premature mortality on a global scale. *Nature* 525, 367–371.

Levelt, P.F., van den Oord, G.H.J., Dobber, M.R., Malkki, A., Huib, V. *et al.* (2006) The ozone monitoring instrument. *IEEE Transactions on Geoscience and Remote Sensing* 44, 1093–1101.

Levy, R.C., Remer, L.A., Mattoo, S., Vermote, E.F. and Kaufman, Y.J. (2007) Second-generation operational algorithm: retrieval of aerosol properties over land from inversion of Moderate Resolution Imaging Spectroradiometer spectral reflectance. *Journal of Geophysical Research* 112, D13211. DOI: 10.1029/2006JD007811.

Levy, R.C., Remer, L.A., Kleidman, R.G., Mattoo, S., Ichoku, C. *et al.* (2010) Global evaluation of the collection 5 MODIS dark-target aerosol products over land. *Atmospheric Chemistry and Physics* 10, 10,399–10,420.

Liu, J.C. and Peng, R.D. (2018) Health effect of mixtures of ozone, nitrogen dioxide, and fine particulates in 85 US counties. *Air Quality, Atmosphere & Health.* DOI: 10.1007/s11869-017-0544-2.

Lodhi, N.K., Beegum, S.N., Singh, S. and Kumar, K. (2013) Aerosol climatology at Delhi in the western Indo-Gangetic Plain: microphysics, long-term trends, and source strengths. *Journal of Geophysical Research* 118. DOI: 10.1002/jgrd.50165.

Mallik, C., Ghosh, D., Ghosh, D., Sarkar, U., Lal, S. and Venkataramani, S. (2014) Variability of SO$_2$, CO, and light hydrocarbons over a megacity in Eastern India: effects of emissions and transport. *Environmental Science and Pollution Research* 21, 8692–8706.

McMillan, W.W., Barnet, C., Strow, L., Chahine, M.T., McCourt, M.L. *et al.* (2005) Daily global maps of carbon monoxide from NASA's atmospheric infrared sounder. *Geophysical Research Letters* 32, L11801. DOI: 10.1029/2004GL021821.

Mehta, M., Dubey, S. and Prabhishini, V. (2017) Associative study of aerosol pollution, precipitation and vegetation in Indian region (2000–2013). In: Singh, V.P., Yadav, S. and Yadava, R.N. (eds) *Environmental Pollution: Select Proceedings of ICWEES-2016.* Springer, Singapore, pp. 147–152.

Metz, B., Kuijpers, L., Solomon, S., Anderson, S.O., Davidson, O. *et al.* (2005) *Safeguarding the Ozone Layer and the Global Climate System: Issues Related to Hydrofluorocarbons and Perfluorocarbons.* Cambridge University Press, Cambridge, UK.

Mishra, A.K. and Shibata, T. (2012a) Climatological aspects of seasonal variation of aerosol vertical distribution over central Indo-Gangetic Belt (IGB) inferred by the spaceborne lidar CALIOP. *Atmospheric Environment* 46, 365–375.

Mishra, A.K. and Shibata, T. (2012b) Synergistic analyses of optical and microphysical properties of agricultural crop residue burning aerosols over the Indo-Gangetic Basin (IGB). *Atmospheric Environment* 57, 205–218.

Murray, C., Forouzanfar, M.H., Alexander, L., Anderson, H.R., Bachman, V.F. *et al.* (2015) Global, regional and national comparative risk assessment of 76 behavioural, environmental, occupational and metabolic risks or clusters of risks in 188 countries 1990–2013: a systematic analysis for the GBD background. *Lancet* 6736, 1–27. DOI: 10.1016/S0140-6736(15)00128-2.

Myhre, G., Shindell, D., Bréon, F.-M., Collins, W., Fuglestvedt, J. *et al.* (2013) Anthropogenic and natural radiative forcing. In: Stocker, T.F., Qin, D., Plattner, G.-K., Tignor, M., Allen, S.K. *et al.* (eds) *Climate Change (2013). The Physical Science Basis. Contribution of Working Group I to the Fifth Assessment Report of the Intergovernmental Panel on Climate Change.* Cambridge University Press, Cambridge and New York, pp. 659–740. DOI: 10.1017/CBO9781107415324.018.

Omar, A.H., Winker, D.M., Vaughan, M.A., Hu, Y., Trepte, C.R. *et al.* (2009) The CALIPSO automated aerosol classification and lidar ratio selection algorithm. *Journal of Atmospheric and Oceanic Technology* 26(10), 1994–2014. DOI: 10.1175/2009JTECHA1231.1.

Pope, C.A. (2000) Epidemiology of fine particulate air pollution and human health: biologic mechanisms and who's at risk? *Environmental Health Perspectives* 104, 713–723.

Pope, C.A. and Dockery, D.W. (2006) Health effects of fine particulate air pollution: lines that connect. *Journal of the Air & Waste Management Association* 56, 709–742.

Pope, C.A., Ezzati, M. and Dockery, D.W. (2009) Fine-particulate air pollution and life expectancy in the United States. *The New England Journal of Medicine* 360, 376–386.

Prasad, A.K., Singh, R.P. and Kafatos, M. (2006) Influence of coal based thermal power plants on aerosol optical properties in the Indo-Gangetic Basin. *Geophysical Research Letters* 33, L05805. DOI: 10.1029/2005GL023801.

Prijith, S.S., Rajeev, K., Thampi, B.V., Nair, S.K. and Mohan, M. (2013) Multi-year observations of the spatial and vertical distribution of aerosols and the genesis of abnormal variations in aerosol loading over the Arabian Sea during Asian summer monsoon season. *Journal of Atmospheric and Solar-Terrestrial Physics* 105–106, 142–151.

Quere, C.L., Andrew, R.M., Friedlingstein, P., Sitch, S., Pongratz, J. *et al.* (2017) Global carbon budget 2017. *Earth System Science Data Discussions*. DOI: 10.5194/essd-2017-123.

Ramanathan, V., Crutzen, P.J., Kiehl, J.T. and Rosenfeld, D. (2001) Aerosols, climate, and the hydrological cycle. *Science* 5549, 2119–2124.

Remer, L.A., Kaufman, Y.J., Tanré, D., Mattoo, S., Chu, D.A. *et al.* (2005) The MODIS aerosol algorithm, products and validation. *Journal of the Atmospheric Sciences* 62, 947–973.

Remer, L.A., Kleidman, R.G., Levy, R.C., Kaufman, Y.J., Tanre, D. *et al.* (2008) Global aerosol climatology from the MODIS satellite sensors. *Journal of Geophysical Research* 113, D14S07. DOI: 10.1029/2007JD009661.

Sahu, L.K. (2012) Volatile organic compounds and their measurements in the troposphere. *Current Science* 102(10), 1645–1649.

Sahu, L.K., Yadav, R. and Pal, D. (2016) Source identification of VOCs at an urban site of Western India: effect of marathon events and anthropogenic emissions. *Journal of Geophysical Research – Atmospheres* 121, 2416–2433.

Satheesh, S.K., Moorthy, K.K., Kaufman, Y.J. and Takemura, T. (2006) Aerosol optical depth, physical properties and radiative forcing over the Arabian Sea. *Meteorology and Atmospheric Physics* 91, 45–62.

Saygın, M., Gonca, T., Öztürk, Ö., Has, M., Çalışkan, S. *et al.* (2017) To investigate the effects of air pollution (PM_{10} and SO_2) on the respiratory diseases asthma and chronic obstructive pulmonary disease. *Turk Toraks Dergisi* 18(2), 33–39.

Schuck, T.J., Brenninkmeijer, C.A.M., Slemr, F., Xueref-Remy, I. and Zahn, A. (2009) Greenhouse gas analysis of air samples collected onboard the CARIBIC passenger aircraft. *Atmospheric Measurement Techniques* 2, 449–464.

Seinfeld J.H. and Pandis S.N. (2010) *Atmospheric Chemistry and Physics: From Air Pollution to Climate Change*, 2nd edition. John Wiley, New York.

Shao, M., Zhang, Y., Zeng, L., Tang, X., Zhang, J. *et al.* (2009) Ground-level ozone in the Pearl River Delta and the roles of VOC and NO_x in its production. *Journal of Environmental Management* 90, 512–518.

Simpson, D., Winiwarter, W., Borjesson, G., Cinderby, S., Ferreiro, A. *et al.* (1999) Inventorying emissions from nature in Europe. *Journal of Geophysical Research* 104, 8113–8152.

Singh, A., Tiwari, S., Sharma, D., Singh, D., Tiwari, S. *et al.* (2016) Characterization and radiative impact of dust aerosols over Northwestern part of India: a case study during a severe dust storm. *Meteorology and Atmospheric Physics* 128(6), 779–792.

Smirnov, A., Holben, B.N., Eck, T.F., Dubovik, O. and Slutsker, I. (2000) Cloud-screening and quality control algorithms for the AERONET database. *Remote Sensing of Environment* 73(3), 337–349.

Smith, K.R. (2000) National burden of disease in India from indoor air pollution. *Proceedings of the National Academy of Sciences USA* 97(24), 13286–13293.

Soni, V., Singh, P., Shree, V. and Goel, V. (2018) Effects of VOCs on human health. In: Sharma, N., Agarwal, A., Eastwood, P., Gupta, T. and Singh, A. (eds) *Air Pollution and Control. Energy, Environment, and Sustainability*. Springer, Singapore, pp 119 – 142. DOI: 10.1007/978-981-10-7185-0_8

Squizzato, S., Masiol, M., Brunelli, A., Pistollato, S., Tarabotti E. *et al*. (2013) Factors determining the formation of secondary inorganic aerosol: a case study in the Po Valley (Italy). *Atmospheric Chemistry and Physics* 13, 1927–1939.

Srivastava, A.K., Yadav, V., Pathak, V., Singh, S., Tiwari, S., Bisht, D.S. and Goloub, P. (2014) Variability in radiative properties of major aerosol types: a year-long study over Delhi —an urban station in Indo-Gangetic basin. *Science of The Total Environment* 473–474, 659–666.

Srivastava, R. (2017) Trends in aerosol optical properties over South Asia. *International Journal of Climatology* 37, 371–380. DOI: 10.1002/joc.4710.

Strow, L., Hannon, S., De Souza-Machado, S. and Motteler, H. (2003) An overview of the AIRS radiative transfer model. *IEEE Transactions on Geoscience and Remote Sensing* 41(2), 303–313.

Susskind, J., Barnet, C. and Blaisdell, J. (2003) Retrieval of atmospheric and surface parameters from AIRS/AMSU/HSB data in the presence of clouds. *IEEE Transactions on Geoscience and Remote Sensing* 41, 390–409.

Tiwari, S., Srivastava A.K. and Singh, A.K. (2013) Heterogeneity in pre-monsoon aerosol characteristics over the Indo-Gangetic Basin. *Atmospheric Environment* 77, 738–747.

Tiwari, S., Bisht, D.S., Srivastava, A.K., Pipal, A.S., Taneja, A. *et al*. (2014) Variability in atmospheric particulates and meteorological effects on its mass concentrations over Delhi, India. *Atmospheric Research* 145–146, 45–56.

Tiwari, S., Srivastava A.K., Singh, A.K. and Singh, S. (2015) Identification of aerosol types over Indo-Gangetic Basin: implications to optical properties and associated radiative forcing. *Environmental Science and Pollution Research*. DOI: 10.1007/s11356-015-4495-6.

Tiwari, S., Tiwari, S., Hopke, P.K., Attri, S.D., Soni, V.K. and Singh, A.K. (2016a) Variability in optical properties of atmospheric aerosols and their frequency distribution over a mega city 'New Delhi', India. *Environmental Science and Pollution Research* 23, 8781–8793.

Tiwari, S., Mishra, A.K. and Singh, A.K. (2016b) Aerosol climatology over the Bay of Bengal and Arabian Sea inferred from space-borne radiometers and lidar observations. *Aerosol and Air Quality Research*. DOI: 10.4209/aaqr.2015.06.0406.

Tiwari, S., Dumka, U.C., Kaskaoutis, D.G., Ram, K., Panicker, A.S. *et al*. (2016c) Aerosol chemical characterization and role of carbonaceous aerosol on radiative effect over Varanasi in central Indo-Gangetic Plain. *Atmospheric Environment* 125, 437–449.

Torres, O., Tanskanen, A., Veihelmann, B., Ahn, C., Braak, R. *et al*. (2007) Aerosols and surface UV products from Ozone Monitoring Instrument observations: an overview. *Journal of Geophysical Research* 112, D24S47. DOI: 10.1029/2007JD008809.

Torres, O., Ahn, C. and Chen, Z. (2013) Improvements to the OMI near-UV aerosol algorithm using A-train CALIOP and AIRS observations. *Atmospheric Measurement Techniques* 6, 3257–270.

USEPA (n.d.) National Ambient Air Quality Standards Table. US Environmental Protection Agency. Available at: https://www.epa.gov/criteria-air-pollutants/naaqs-table (accessed 7 August 2018).

Vedal, S., Petkau, J., White, R. and Blair, J. (1998) Acute effects of ambient inhalable particulates in asthmatic and non-asthmatic children. *American Journal of Respiratory and Critical Care Medicine* 157, 1034–1043.

Watson, J.G., Chow, J.C., Tropp, R.J., Wang, X., Kohl, S.D. and Chen, L.W.A. (2013) Standards and traceability for air quality measurements: flow rates and gaseous pollutants. *MAPAN – Journal of Metrology Society of India* 28(3), 167–79.

Weschler, C.J. (2006) Ozone's impact on public health: contributions from indoor exposures to ozone and products of ozone-initiated chemistry. *Environmental Health Perspectives* 114, 1489–1496.

Wild, M. (2009) Global dimming and brightening: a review. *Journal of Geophysical Research* 114, D00D16. DOI: 10.1029/2008JD011470.

Winker, D.M., Hunt, W.H. and McGill, M.J. (2007) Initial performance assessment of CALIOP. *Geophysical Research Letters* 34, L19803. DOI: 10.1029/2007GL030135.

World Health Organization (WHO) (2016) Ambient Air Pollution Database, May 2016. Available at: http://www.who.int/airpollution/en/ (accessed 12 September 2018).

Young, S.A. and Vaughan, M.A. (2008) The retrieval of profiles of particulate extinction from Cloud Aerosol Lidar Infrared Pathfinder Satellite Observations (CALIPSO) data: algorithm description. *Journal of Atmospheric and Oceanic Technology* 26, 1105–1119.

Zabiegała, B., Kot-Wasik, A., Urbanowicz, M. and Namieśnik, J. (2010) Passive sampling as a tool for obtaining reliable analytical information in environmental quality monitoring. *Analytical and Bioanalytical Chemistry* 396(1), 273–296.

5 Air-Pollution Modelling Aspects: An Overview

Monojit Chakraborty,[1]* Sangeeta Bansal,[2] Renu Masiwal[3] and Amit Awasthi[4]
[1]ETH Zurich, Switzerland; [2]Guru Gobind Singh Indraprastha University, Delhi, India; [3]CSIR-National Physical Laboratory, New Delhi, India; [4]University of Petroleum and Energy Studies, Dehradun, India

---

Abstract

This chapter presents a brief review on air-pollution modelling and its techniques, i.e. methods in which a complex phenomenon in air-quality processes is transformed into a simple one, via computational simulation. It includes the description of models, both non-reactive (e.g. plume models) and reactive (e.g. photochemical models). Also discussed is the role of meteorological phenomena in the dispersion of pollutants in the atmosphere. The chapter is introduced through a philosophical and practical discussion on the implications of mathematical air-quality modelling. Pertinent mathematical modelling techniques, namely Box, Gaussian, Eulerian, Lagrangian and particle modelling approaches, are evaluated, and the mathematical air-pollution modelling approach is discussed.

5.1 Introduction

The atmosphere is composed of numerous gases in a reactive and complex system, in which physicochemical processes simultaneously occur. Monitoring of ambient quality provides only a glimpse of spatial and temporal conditions of the atmosphere. Such monitoring and measurements are often difficult to understand without a substantial concept of modelling for describing the atmospheric processes. Moreover, measurements alone could not describe atmospheric phenomena for the policy makers to execute an effective approach to minimize air pollution issues. In understanding completely the gaseous interactions within a system, a cumulative investigation of each atmospheric process (i.e. physical, chemical, transportation, scavenging, etc.) needs to be undertaken. The mathematical models provide such understanding in an integrated manner. The monitoring and measurement used in mathematical models is the best approach towards the complete understanding of inherent atmospheric processes.

5.1.1 Definition of air-pollution modelling

Models reflect a mathematical description of hypothesis, which has the ability to solve complex sets of equations derived through the behaviour of some physical or other processes.

5.1.2 Model formulation

There are several steps involved in the model design; its applications; performance testing, etc.

---

* Corresponding author: monojit.chakraborty@gmail.com

1. Defining and considering the problem of interest.
2. Assessing the dimension of the problem of interest.
3. Exercise on the dimension of the model.
4. Selecting the chemical, physical and dynamical processes for the simulation.
5. Selecting the variables.
6. Transforming into the computational domain.
7. Codifying and implementing algorithms.
8. Optimizing of model parameters.
9. Setting up initial conditions.
10. Setting up boundary layer conditions.
11. Procurement of input data.
12. Procurement of ambient monitoring data for evaluation.
13. Model predictions by interpolated input data.
14. Use of statistics and graphical techniques incorporation.
15. Comparing model results with monitored data.
16. Model sensitivity tests.
17. Improvement of algorithms (Jacobson, 2005).

5.1.3 Processes of chemical transformations

The mixing of ratios of chemical species in the troposphere is regulated by four types of processes: (i) *emissions of pollutants*: where pollutants are emitted into the atmosphere through numerous sources, for example, fossil fuel combustion, biogenic emissions, etc.; (ii) *chemical reactions*: photochemical reactions, hydrochemical reactions, etc., lead to the formation and removal of species in the atmosphere; (iii) *transport*: transportation by winds drive away those pollutants or reacted species from their source of origin; (iv) *deposition*: all those species floating in the troposphere either are finally deposited back to the Earth's surface or escape from the atmosphere to outer space (insignificant amount). Deposition of floating materials in the troposphere takes in two forms: 'dry deposition', occurring due to gravitation pull upon floating particulates; and 'wet deposition', where scavenging through precipitation is involved.

5.2 Air Pollution and Meteorology

During the release of pollutants into the ambient atmosphere, they are transported by air movements and subsequently the level of concentration of air pollutants in the atmosphere is diluted over time. Meteorological parameters include ambient temperature, humidity, wind speed, wind direction, mixing height, atmospheric stability, turbulence, temperature and inversion; these propagate the dilution process and dispersion of the pollutants and can be described as follows.

5.2.1 Wind vector

Wind is considered as a vector parameter, having both speed and direction. The horizontal movements of this vector are usually considered where the vertical movement is relatively low. Through the movement of wind, continuous pollutant releases are diluted at the point of emission, according to the direction of wind flow. Therefore, the concentrations in the plume and wind speed are inversely proportional to each other. It is a known fact that wind speed is a function of vertical height, and proportional to the power law vertical height, i.e. a relationship between the wind speeds at the different heights. The exponent (α) depends primarily on atmospheric stability, ranging from 0.07 (unstable) to 0.55 (stable) conditions (Turner, 1994).

When wind flows on the surface along a particular path, it creates friction while travelling, called roughness length (RL; z_0), which is dependent on height, space and wind speed gradient. The RL is an important factor for the modelling which varies between 0.001 and 0.3 m at the ground level, ~0.5 to 1 m for suburban areas and 1 and 3 m for urban areas (Grimmond *et al.*, 1998; Haq and Schwela, 2008).

5.2.2 Turbulence

This is also associated with wind oscillations in a circular motion (eddies) oriented generally to vertical or horizontal or in between all pathways. This is also a causative factor in which a polluted air parcel is diluted in the atmosphere. There are two kinds of turbulence: mechanical and buoyant. When wind speed increases with a faster air stream next to it, is called mechanical turbulence (Haq and Schwela, 2008). Buoyant turbulence is the result of heating or cooling of air near to the Earth's surface. The warm air goes upwards, creating a low-pressure zone, called positive buoyant turbulence, contrary to during the night temperature inversion, which results in the atmosphere becoming stable, inhibiting vertical motion, and is called negative buoyant turbulence (Keck *et al.*, 2013).

5.2.3 Mixing height

This is the maximum height for dispersion of tropospheric pollutants. Under unstable conditions, strong vertical mixing from the ground to approximately 1 km can be seen. This mixing height is much lower in stable atmospheric conditions. On a sunny day, a rising hot air stream is a general phenomenon which causes instability in the atmosphere. A series of upward and compensating downward motions will result in dispersion of the pollutants to the vertical and all other possible directions. In a clear night sky with light wind, buoyant turbulence shows a minimum value. In a neutral condition, the net heat flux on the ground is found to be around zero (Brankovic and Molteni, 1997; Seinfeld and Pandis, 2006).

5.2.4 Lapse rate

This is the rate of change in the ambient temperature while moving upward through the troposphere. The lapse rate (LR) is positive when the temperature drops with increasing height, zero when no change in temperature is observed, and negative when the temperature is proportional to the altitude (i.e. temperature inversion). The normal, or environmental LR rate, is referred to when an air parcel is stuck at a certain height. Further, the lapse rate is highly variable due to the influences of atmospheric phenomena like radiation, convection and condensation. It has been estimated that the average LR is about 6.5°C per km in the troposphere, while the adiabatic lapse rate involved in the change of ambient temperature is with respect to altitude, which affects whether an air parcel rises or sinks in the troposphere. Adiabatic lapse rates are usually distinguished by dryness and moistness.

5.3 Air Pollutant Emissions Sources

A number of diverse categories of pollution sources are responsible for the urban air-quality deterioration. For convenience, pollutant sources can be clustered into stationary, mobile or transportation sources.

- Point source – single, identifiable emission source, like a chimney through which gas is emitted from a combustible furnace.
- Line source – the emissions from vehicular transport on a roadway.
- Area source – the emissions from a larger area with multiple emission sources.

5.4 Plume Types and Stability

In neutral conditions, when the vertical air temperature gradient has been between dry adiabatic and isothermal, the turbulent currents disperse in horizontal and vertical directions resulting in a plume that looks like a cone, which is named 'coning'. This situation is most likely to occur during cloudy or windy periods (Fig. 5.1).

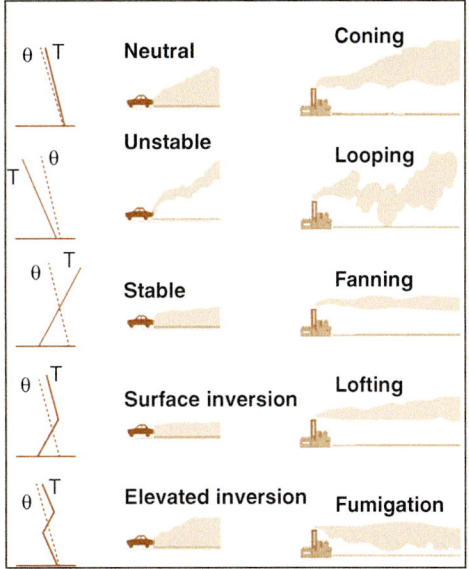

Fig. 5.1. Vertical dispersion under various conditions for low and high elevations of the source (T: Temperature, θ: Adiabatic lapse). (Adapted from Liptak, 1974.)

When the atmosphere is unstable, a large loop is formed by the continuously emitted plume, which is named 'looping' (Fig. 5.1). Looping occurs when the super-adiabatic lapse rate and solar heating generate unstable air, which brings the plume to ground level and again to the plume height, like a whip going up and down as the atmosphere mixes around.

During the temperature inversion, the plume with a lower mixing height spreads out horizontally but does not mix vertically; the plume is said to be 'fanning' (Fig. 5.1).

Under the atmospheric conditions where above the plume height is unstable and below is stable, a 'lofting' situation is created. The pollutants go up into the environment, therefore it has less impact on the ground level (Fig. 5.1).

When an air parcel is trapped in a temperature inversion, a super-adiabatic lapse rate causes the plume to reach the ground level along the length of the plume, this turbulence is defined as 'fumigation'. Solar heating causes the super-adiabatic lapse rate at the ground level. This condition has been favoured by clear skies and light winds (Fig. 5.1) (Baynton *et al.*, 1965).

5.5 Introduction of Atmospheric Modelling

Air-pollution measurements give important, quantitative information about ambient gases' mixing ratios, scavenging time, atmospheric deposition, etc., which means they can only describe air quality at specific locations and times, without giving clear guidance on the identification of the causes of the air-quality problem. Whereas, modelling on air pollution is a numerical tool used to describe the underlying relationship among the pollutants in the atmosphere. Thus, the results of air-pollution models of past and future scenarios will help in determining the effectiveness of abatement strategies against air pollution.

Atmospheric models can broadly be divided into two types: physical and mathematical. Physical models are used occasionally to simulate processes that are occurring in the atmosphere to form a prototype demonstration of the actual system, by making physical approximations and using equations that describe the phenomena of interest more directly, excluding or filtering out the phenomena that are less interesting. Mathematical models can be used immediately the decision over a suitable set of physical equations has been decided, though the mathematical methods could introduce errors while explaining the atmospheric processes. Mathematical models on atmospheric processes can broadly be categorized into two types: (i) models established on the fundamental description of tropospheric physico-chemical processes; and (ii) models established on statistical data analysis. These established models can explain atmospheric dispersion and chemical transport of atmospheric pollutants. The pollutant dispersion from the sources could be described by the Gaussian Plume Model, which commonly refers to the behaviour of gases and particles in turbulent flow as turbulent 'diffusion'. Whereas, a chemical transport model is an understanding of individual atmospheric processes, for example, chemistry, transport, removal, which could be described as a fixed coordinates system, i.e. the Eulerian approach, and where coordinates of a parcel of air pollutants can't be changed with the time. Another approach is the Lagrangian, in which concentration changes are described relative to the moving fluid or parcel of air pollutants. Both diffusion and chemical transport models are described below.

5.5.1 Point source modelling (Gaussian Plume Model)

A simple model, named the Gaussian Plume Model, was developed to understand the dispersion behaviour of plumes from industrial stacks. The application of this model is to calculate the maximum impact of plumes at the ground level and the maximum distance of impact away from the source. Equation 5.1 describes a mixing process from the results of a Gaussian concentration distribution both in crosswind and in vertical directions, while centring the downwind line from the origin (Fig. 5.2).

$$c(x, y, z) = \frac{Q}{2\pi\sigma_y\sigma_z u} \exp\left(\frac{-y^2}{2\sigma_y^2}\right)$$
$$\left(\exp\left(\frac{-(z-h)^2}{2\sigma_z^2}\right) + \exp\left(\frac{-(z+h)^2}{2\sigma_z^2}\right)\right)$$

(Eq. 5.1)

Where, c = ambient concentration at a given point.

Q = the emission rate.

y, x, z = are the crosswind, downwind and vertical wind direction, respectively.

u = wind speed at the 'h' height of the release.

σ_y, σ_z = deviations describing the crosswind and vertical mixing of the pollutant, and these are formulated on the horizontal and the vertical eddy diffusivity.

In Fig. 5.2, the H_s denoted here is the actual and H_e the effective stack height. The maximum stretch of crosswind and vertical deviation of the concentration profile are the key parameters in this model. To get the actual geometrical dimension of the plume by plotting the standard deviation of its concentration distribution, multiple experiments are required. Both the vertical and horizontal directions are required, in a function of the atmospheric stability and downwind distance from the source. The plotting is presented in Fig. 5.2.

5.5.2 Chemical transport model (modelling at urban and larger scales)

The classic Los Angeles-type smog is an example of a photochemical smog that drew attention to the severity of the air pollution over Europe (Fig. 5.3). This incident was similar to many episodes witnessed in large urban regions throughout the world in the early 1970s. Scientists began to realize that air pollution was not only a local phenomenon, but that long-range transport of pollutants from distant sources was also responsible. It became clear that the SO_2 and NO_x emissions from industrial tall chimneys could lead to acid formation or acidification in the atmosphere at long distances from its origin. It was also realized that ozone (O_3) was a problem in urbanized and industrialized areas. Therefore, scientists realized that those situations could not be tackled by simple Gaussian-plume-type modelling until some modifications were done.

Two different modelling approaches were followed: the Eulerian and Lagrangian models. In Eulerian modelling, while considering the ambient pollution concentration, the area is divided into fixed vertical and horizontal grid cells. In a Lagrangian modelling framework, an air parcel or 'puff' is tracked along a local wind trajectory by keeping its identity during its transport in certain directions. It is assumed that no mass exchange occurred between the air parcel and its surroundings, considering only that the surface emissions from the vehicles enter through the bottom side of the box. Thus the air parcel moves continuously, and it is therefore required that the model can simulate spatial variations in pollution concentrations at different

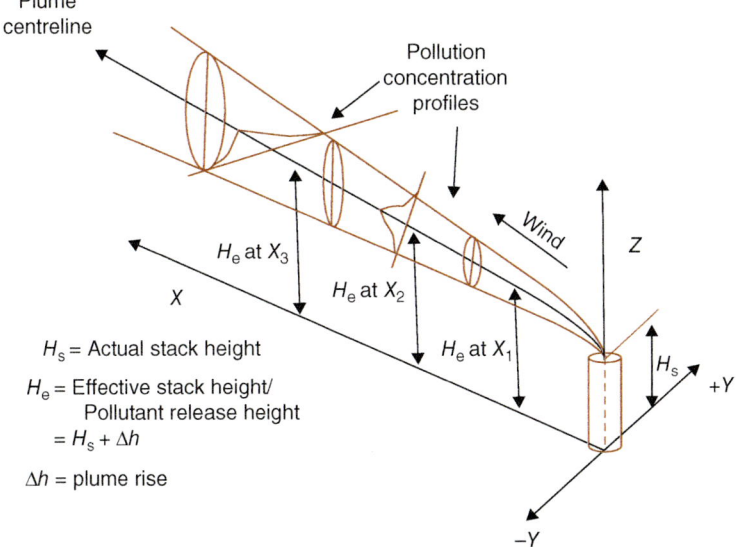

H_s = Actual stack height

H_e = Effective stack height/
    Pollutant release height
    = $H_s + \Delta h$

Δh = plume rise

Fig. 5.2. Gaussian plume dispersion – a schematic diagram. (From Boubel *et al.*, 1994.)

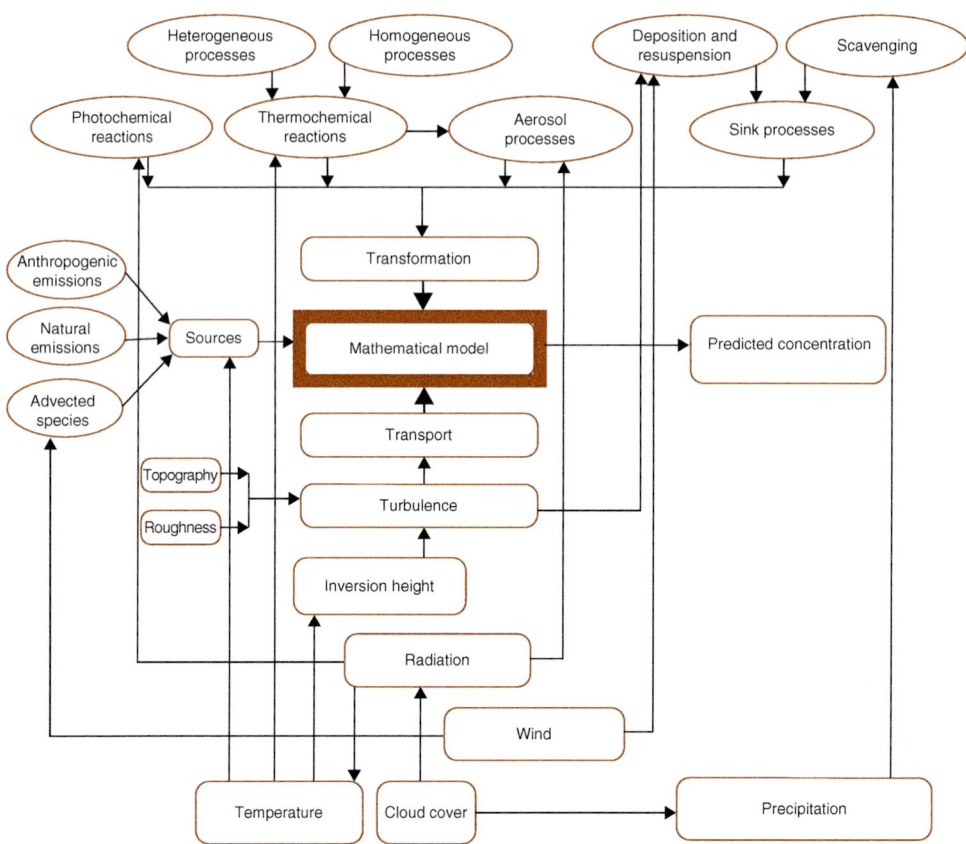

Fig. 5.3. Chemical transport model framework. (From Seinfeld and Pandis, 2006.)

times. In contrast, an Eulerian modelling approach relies on a fixed space or block with a particular dimension. It is believed the pollutants enter and leave each cell through its walls, and the model simulates the pollutants' concentrations at all locations as a function of time. Eulerian models are characterized in several dimensions. The simplest is the zero-dimensional box model, where the fixed space or block is represented by only one box (Fig. 5.4), wherein the model assumes that the pollution concentrations are the same everywhere, therefore concentration is only the functions of time; c_i (t). The next is a one-dimensional (1D) column model, which assumes that pollution concentrations are the functions of two vectors: height and time: c_i (z, t). Then the 2D model consists of two horizontally homogeneous layers, assuming that c_i are unchanged within the box in a dimension, depending

on the other two, height and time: c_i (x, z, t). The global atmospheric chemistry can be often described by the 2D model, where it can assume that pollution levels are functions only of latitude and altitude, but independent of longitude. Finally, the 3D model can simulate the cumulative concentration in a given area: c_i (x_i, y_i, z_i, t). Obviously, it is understandable that the model hierarchy gets more complex and accurate with the higher dimensions. The leading equations of these models will be given later.

5.5.2.1   Zero-dimensional box model

Gases and particles may enter or leave the box from any side, by assuming that all pollutants will be mixed well instantaneously and placed in a uniform manner in the box. The air parcel moves and adds up emissions in the box from different locations and times along with the direction of the wind. The box is orientated in such a way that the directional component of the wind velocity is both perpendicular and incident to one face of the box. The city is considered as a rectangle (box) where complete atmospheric turbulence is produced and total mixing of pollutants up to the ceiling height of the rectangle (Fig. 5.4). In the steady-state condition, it also considered that in the upwind direction, the velocity of air is strong enough so that the mixing ratio of the pollutants is uniformly distributed over the whole city, which is assumed as a box, and also needs to ensure that the concentration is not higher at the downwind side than the upwind side. With all these assumptions, the mass balance equation is simplified as below.

A crude but simple zero-dimension model for air pollution in a city is as follows (see Fig. 5.5). Suppose an urban area is surrounded by two parallel sides, L_1 and L_2, and mixing height is h. Also supposing that a pollutant is emitted at a constant rate Q (kg/s) with initial concentration C at time (t) (kg/m^3). A horizontal, uniform and constant wind with velocity v enters perpendicularly to the L_1 side, and glides alongside to L_2.

$$C = b + (Q * l)/(v * h) \qquad \text{(Eq. 5.2)}$$

$$dCV/dt = QA + uC_{in}Wh - uC_{out}Wh \quad \text{(Eq. 5.3)}$$

Where:

C = concentration (μg/m^3)
b = background pollution level (μg/m^3)
Q = emission rate per unit area (g/s.m^2)
v = wind velocity (m/s)
l = length (m)
h = mixing depth (m)
A = is the horizontal area of the box ($L \times W$), W is the width
V = is the volume described by the box
C_{in} = species concentration entering the box
C_{out} = species exiting the box

5.5.2.2   One-dimensional box model

The one-dimensional (1D) model, is a union of two box models which are typically piled one on top of the other vertically. The advantages of a vertical 1D model over a pure box model are that advection diffusion and vertical transport can be explained in a 1D model, which attenuates solar

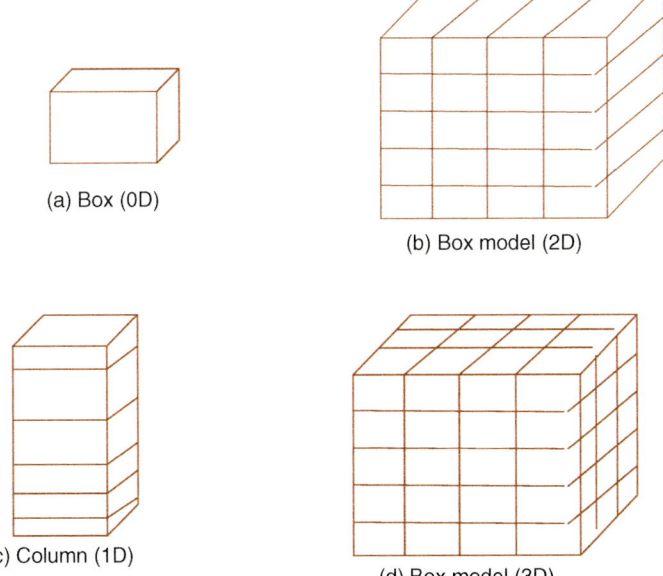

(a) Box (0D)

(b) Box model (2D)

(c) Column (1D)

(d) Box model (3D)

Fig. 5.4. Dimension-wise box model.

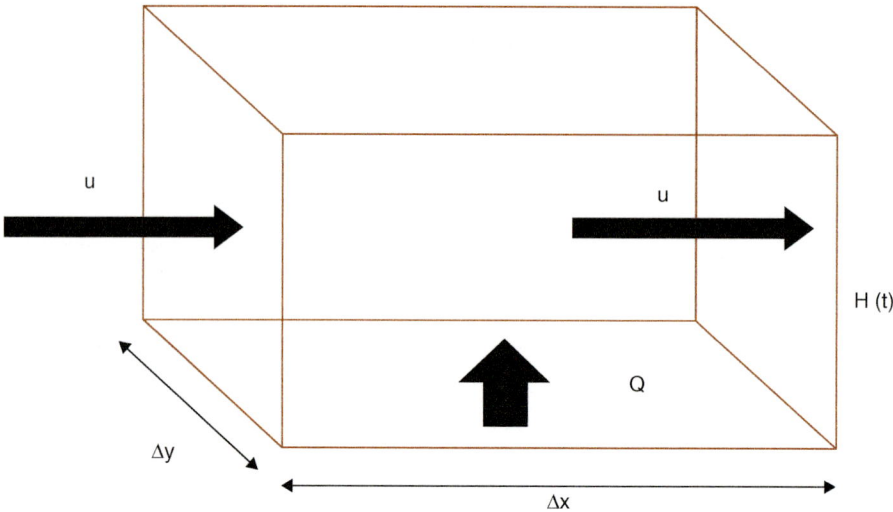

Fig. 5.5. Pollution generated in a simple box model.

radiation in each layer by the gases and particles when it travels through the column. The atmospheric transport equation (Eq. 5.4) is ideally suited for predicting the concentration distribution over extended areas; in many situations the atmosphere needs to be simulated only at a particular location. On the contrary, this model has a disadvantage compared to higher-D models, where vertical rise of wind is crudely estimated, while there is no scope to endorse the horizontal wind currents. Moreover, this model overlooks gases, particles, and potential temperature fluxes over the horizontal boundaries.

$$S(t) = S_0 - \int_t^{t_0} u(\tau)\, d\tau \qquad \text{(Eq. 5.4)}$$

where we have assumed that at time t_0 the trajectory ends at the desired location S_0. The location of the air parcel $S(t)$ for a given moment t, and τ is the residence time of the parcel on this backward trajectory can be calculated by a straightforward integration according to Eq. 5.5. This integration is carried backwards a couple of days until the air parcel is in a relatively clean area or in one with a well-characterized atmospheric composition.

$$\frac{dc_i}{dt} = \frac{Q_i}{H} + R_i - \frac{v_{d,i}}{H}c_i + \frac{u}{\Delta x}(c_i^o - c_i) \quad \text{(Eq. 5.5)}$$

Where Q_i is the pollution emission rate of i (kg h^{-1}), S_i the removal rate of i (kg h^{-1}) (the removal rate also can be described using dry deposition velocity of the species $v_{d,i}$, therefore $S_i = v_{d,i} c_i$ which replaces S_i), R_i, its chemical production rate (kg m^{-3} h^{-1}), c_i^o its background concentration, v is the atmospheric deposition velocity and u the wind speed with the wind assumed to have a specific direction.

5.5.2.3   Two-dimensional box model

The two-dimensional (2D) model is a combination of 1D models which are connected to each other by the sides. The model is laid in the three planes: x–y, x–z or y–z axis. In the 2D model, chemical transport, chemistry and dynamics can be simulated more accurately, which corresponds to a larger region up to a global scale (Garcia *et al.*, 1992). The 2D model assumes that during the daytime the pollutants are emitted at a constant rate from an area source, and mixing of gases takes place adjacent to the surface, which is spread over the city with a fixed downwind distance and large cross-wind dimensions. It also assumes that turbulent flux at the ground level is near to zero, and that the distance between the origin to the downwind x direction is l ($0 \le x \le$ l) and the emission source free region is not beyond the range, i.e. l$< x \le X_0$. where the crosswind concentration and direction remains unchanged. The basic governing equation (Eq. 5.6) of the primary pollutant (Pasquill and Smith, 1983) can be written as:

$$\frac{dc_p}{dt} + u(x,z)\frac{dc_p}{dx} + w(z)\frac{dc_p}{dz}$$
$$= \frac{d}{dz}\left(K_z(z)\frac{dc_p}{dz}\right) - (k + k_{wp})\,C_p \quad \text{(Eq. 5.6)}$$

Where:

$C_p = C_p\,(x,\,z,\,t)$ is the ambient mean concentration

u = mean wind speed in x-direction

w = mean wind speed in z-direction, K_z is the turbulent eddy diffusivity in z-direction

k = first order chemical reaction rate coefficient

k_{wp} = the first order rainout/washout coefficient

5.5.2.4   Three-dimensional chemical transport models

The 3D model is two parallel 2D models stacked on top of one another. The 3D model over a 2D model can be treated more realistically on air dynamics and transport of urban pollutants. This transport models is based on the mass balance approach to primary with the secondary pollutants. The leading equation is directed as the atmospheric diffusion equation (Eq. 5.7), which splits the gaseous transport period into an advection and turbulent transport contribution.

$$\frac{\partial c_i}{\partial t} + u_x\frac{\partial c_i}{\partial x} + u_y\frac{\partial c_i}{\partial y} + u_z\frac{\partial c_i}{\partial z}$$
$$= \frac{\partial}{\partial x}\left(K_{xx}\frac{\partial c_i}{\partial x}\right) + \frac{\partial}{\partial y}\left(K_{yy}\frac{\partial c_i}{\partial y}\right)$$
$$+ \frac{\partial}{\partial z}\left(K_{zz}\frac{\partial c_i}{\partial z}\right) + R_i(c_1, c_2, \ldots\ldots\ldots, c_n)$$
$$+ E_i(x,y,z,t) - S_i(x,y,z,t)$$
$$\text{(Eq. 5.7)}$$

$$u'c' = -K \cdot \nabla c \quad \text{(Eq. 5.8)}$$

Where:

$c_i(x, t)$ = concentration of i as a function of location x and time t

$u(x, t)$ = velocity vector

R_i = chemical production term for i species

$E_i(x, t)$ = emission fluxes

$S_i(x, t)$ = pollutants removal fluxes

u_x, u_y, u_z = x, y, and z components of the wind velocity

K = turbulent diffusivity

K_{xx}, K_{yy}, K_{zz} = corresponding eddy diffusivities

u' and c_i' = wind velocity turbulence fluctuations and concentration

The gradient transport theory, i.e. K theory (Eq. 5.8), is the easiest way out of the closure problem and is currently used in the majority of chemical transport models. It has been introduced and used to show its promising competitiveness in higher dimension closure approximation (Pai and Tsang, 1993). It is an important inference that the eddy diffusion coefficients to be used in a model are the link of the degree of performance when used as a meteorological model to calculate the inputs used in the transport model. The drawback of this model requires additional computational time and memory space than required in a lower-D model.

5.5.2.5   Multi-box models

A step forward from one- and two-box models is to a multi-box model, which can describe the behaviour of tropospheric pollutants more elaborately by summing up N number of boxes. In the multi-box model, the mass balance approach has also been applied, where each equation for each of the boxes is the same as in the single dimension model formulation, thus the equations combined through the pollutants in (W_{in}) and out (W_{out}) are considered. Let's assume that the generation and loss rates is 'X' with the emissions 'E', total chemical production 'P', cumulative loss 'L', and deposition 'D' within the box. The terms 'W_{in}', 'E' and 'P' are sources of 'X'; the terms 'W_{out}', 'L', and 'D' are sinks of 'X'. The mass of 'X' is time and again termed as an 'inventory' and the box itself is termed as a 'reservoir'. It is subsequently assumed that the box is well mixed in order to ease computation of the sources and all sinks termed as E_i, P_i, L_i, D_i, respectively. Where, a_i (kg) is symbolized as mass of 'X' in reservoir with species 'i'; and W_{ij} (kg s^{-1}) the transfer rate of 'X' from one reservoir to other. The mass balance equation as below (Eqs 5.9 and 5.10); if W_{ij} is first-order, so that $W_{ij} = k_{ij}a_i$ where k_{ij} is a transfer rate constant.

$$\frac{da_1}{dt} = E_1 + P_1 - L_1 - D_1 - F_1 + F_2 \quad \text{(Eq. 5.9)}$$

$$\frac{da_1}{dt} = E_1 + P_1 - L_1 - D_1 - k_1a_1 + k_2a_2 \quad \text{(Eq. 5.10)}$$

where (x) can be written and a similar equation can be written for a_2. We thus have two coupled first-order differential equations from which to calculate a_1 and a_2. The system at steady state $(da_1/dt = da_2/dt = 0)$ is described by two algebraic equations (Eqs 5.11 and 5.12). Consider the general 3-puff model for species X in Fig. 5.6.

The residence time of species X in box 1 is determined by summing the sinks from loss within the box and transfer to boxes 2 and 3:

$$\tau_1 = \frac{a_1}{L_1 + D_1 + W_1 + W_5} \quad \text{(Eq. 5.11)}$$

with similar expressions for t_2 and t_3. Often we are interested in determining the residence time within an ensemble of boxes. For example, we may be interested in knowing how long X remains in boxes 1 and 2 before it is transferred

to box 3. The total inventory in 'box' 1 and 2 is $a_1 + a_2$, and the transfer rates from these boxes to box 3 are W_3 and W_5, so that the residence time t_{1+2} of X in the ensemble of boxes 1 and 2 is given in the equation (Eq. 5.12). The terms W_1 and W_2 do not appear in the expression for t_{1+2} because they merely cycle X between boxes 1 and 2.

$$t_{1+2} = \frac{a_1 + a_2}{L_1 + L_2 + D_1 + D_2 + W_3 + W_5} \quad \text{(Eq. 5.12)}$$

5.5.2.6 Puff models

It has already been understood that box models are about the composition of the atmosphere within fixed domains or boxes through which the pollutants are being flown. Whereas, a puff model, on the contrary, describes the characteristics of one or more fluid elements like 'puffs' moving in a trajectory pathway (Fig. 5.7). A puff is a volume or parcel of air with dimensions ensuring that all points within it are transported by air blown in the same velocity, but a puff with a larger volume collective of molecules helps in understanding statistics because of its collective number of molecules (e.g. 1 cm^3 of air contents almost 10^{19} number of molecules).

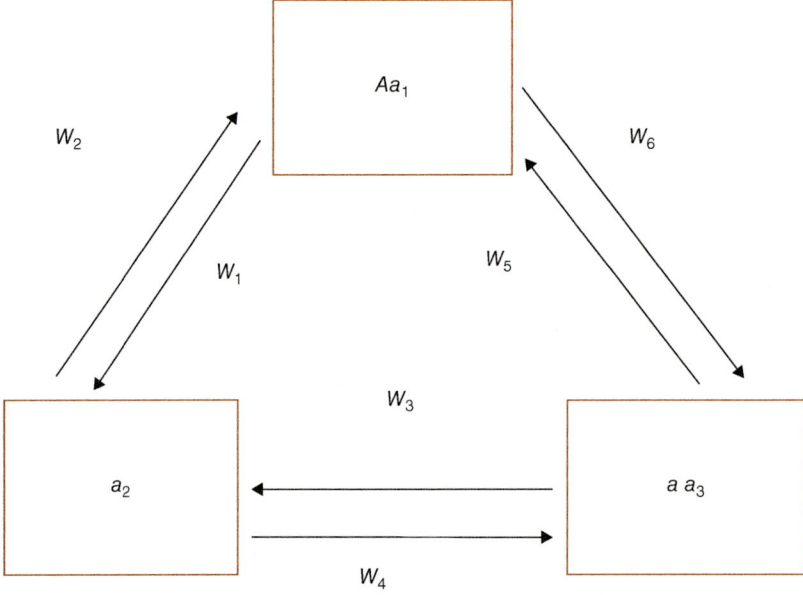

Fig. 5.6. Three-box model.

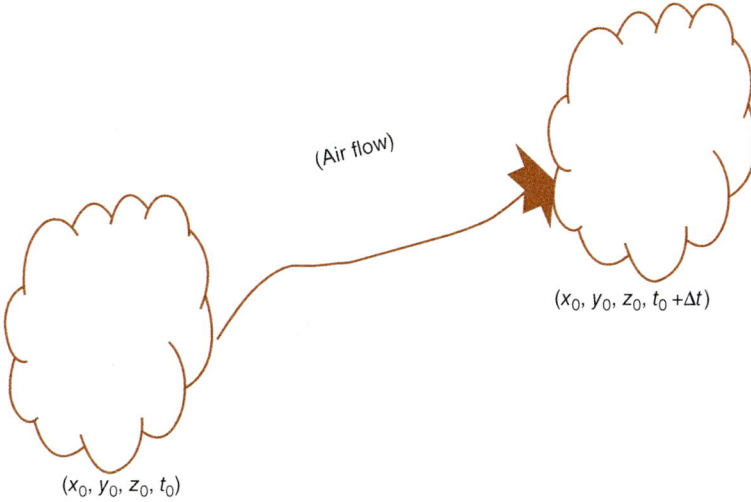

(Air flow)

$(x_0, y_0, z_0, t_0 + \Delta t)$

(x_0, y_0, z_0, t_0)

Fig. 5.7. Movement of a pollutant air parcel (puff) in an air flow trajectory.

A puff further analysed by the mass balance equation is given in Eq. 5.13.

$$\frac{d}{dt}[X] = E + P - L - D \qquad \text{(Eq. 5.13)}$$

Where:

$[X]$ = concentration of X in the puff (molecules cm^{-3}) and sink rate
L_i and D_i = are the sinks of X (molecules cm^{-3} s^{-1}).

The mass balance equation in terms of mixing ratio rather than mass to the size of the puff is assumed to be arbitrary. The puff travels and is diluted in its own travel path, where emissions and production of an atmospheric compound are accounted for as a mass, i.e. mass$_{in}$ is equal to sink and depositions of mass mass$_{out}$ are equal to zero. Whereas, in a box model mass$_{in}$ and mass$_{out}$ are not equal to zero. Exclusion of the transport terms is a foremost advantage of the puff model. Adding to it, the supposition that all points within the puff are transported with the same velocity, where wind gradient gradually expands the puff to the point where it becomes no longer perceptible. One frequent application of the puff model is to follow the chemical evolution of an isolated pollution plume (Fig. 5.8).

In the illustrated example, the puff is allowed to grow with time to simulate dilution with the surrounding air containing a background concentration $[X]_b$. The mass balance equation along the trajectory of the plume is written in Eq. 5.14 which has a substantial advantage over the box model. Where, k_{dil} (s^{-1}) is a dilution rate constant.

$$\frac{d}{dt}[X] = E + P - L - D - K_{dil}\left([X] - [X]_b\right)$$

$$\text{(Eq. 5.14)}$$

A variable puff model is also described through a column model, which follows the chemical evolution in a well-mixed column of air extending vertically from the surface to some mixing depth h and travelling along the surface (Fig. 5.9). Exchange of air mass with the air above h is assumed to be insignificant by a dilution rate constant as in Eq. 5.15. As emission flux E of species X travels along the surface through a column it mixes and varies with time and space.

$$\frac{d}{dt}[X] = \frac{E}{h} + P - L - D \qquad \text{(Eq. 5.15)}$$

The column model can be made more complicated by partitioning the column vertically into well-mixed layers exchanging mass with each other as in a multi-box model. The mixing height h can be made to vary in space and time, with associated entrainment and detrainment of air at the top of the column.

Column models are frequently used to simulate air pollution in cities and in areas downwind. An urban area (Fig. 5.10) of mixing height h ventilated by a steady wind of speed U is assumed. The urban area extends horizontally over a length L in the direction x of the wind. A pollutant X is emitted at a constant and

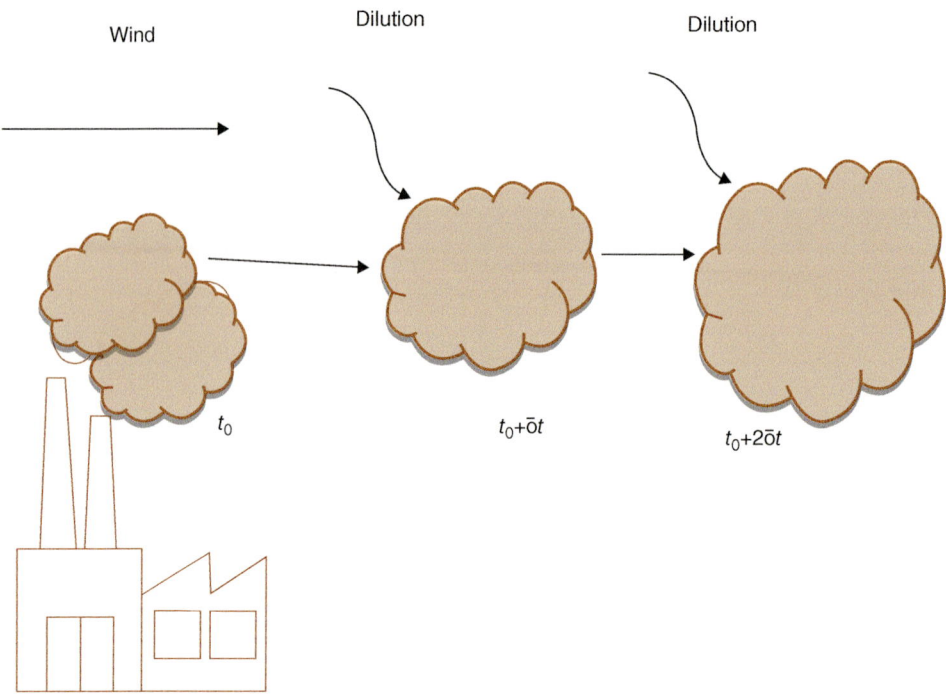

Fig. 5.8. Transportation of pollutants described by the puff model.

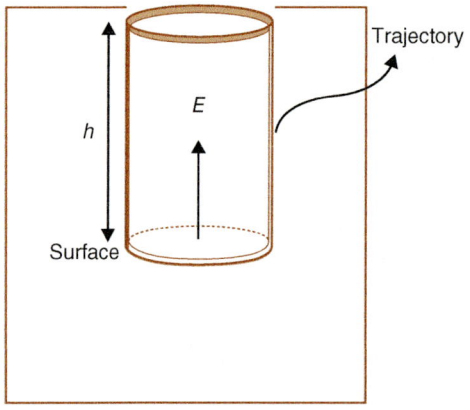

Fig. 5.9. Column model.

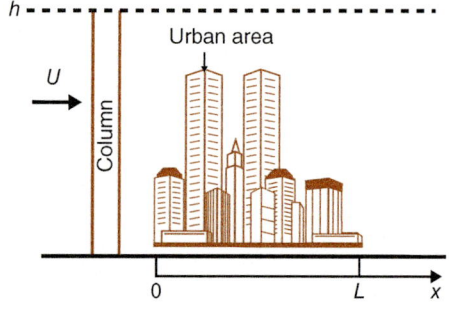

Fig. 5.10. Evolution of pollutant concentrations within and downwind of an urban area in the column model.

uniform rate of E in the urban area and zero outside. Assume no chemical production ($P = 0$) and a first-order loss $L = k[X]$.

The mass balance equation for X in the column is:

$$\frac{d[X]}{dt} = \frac{E}{h} - k[X] \qquad \text{(Eq. 5.16)}$$

which can be re-expressed as a function of x using the chain rule:

$$\frac{d[X]}{dt} = \frac{d[X]}{dx}\frac{dx}{dt} = U\frac{d[X]}{dx} = \frac{E}{h} - k[X] \quad \text{(Eq. 5.17)}$$

$$[X] = 0 \qquad x \leq 0$$

$$[X] = \frac{E}{hkU}\left(1 - e^{-\frac{hx}{U}}\right) \qquad 0 \leq x \leq L$$

$$[X] = [X](L)e^{\frac{h(x-L)}{U}} \qquad x \geq L \qquad \text{(Eq. 5.18)}$$

A plot of the solution is shown in Fig. 5.10. The model captures gradients in pollutant levels across the city and also describes the exponential decay of pollutant levels downwind of the city. If good information on the wind field and the distribution of emissions is available, a column model can offer a considerable advantage over the one-box model at little additional computational complexity.

5.6 Examples of Dispersion Models

There are a number of dispersion models available which are important to the scientific community. USEPA recommended two computer packages for simulation of non-reactive chemicals, discussed below.

5.6.1 AERMOD

The planetary boundary layer (PBL) controls the concentration of the surface level of atmospheric constituents as it is the lowest layer of the troposphere, where surface and atmosphere exchange mass, energy and momentum. After observing the role of PBL in the dispersion of pollutants, the American Meteorological Society (AMS) along with Environmental Protection Agency (EPA) tried to incorporate PBL in the regulatory dispersion models. The model created by the American/ EPA Regulatory Model Improvement Committee (AERMIC) was initially designed to calculate the near field impact caused by industrial sources. From 1991 until 2016, very few changes were introduced in the EPA regulatory models. At that time Industrial Source Complex (ISC3) were the most popular and commonly used dispersion models, used in most of the state implementation plan constructions, new source permit cases, risk assessment cases and in cases of exposure analysis. Therefore, AERMIC decided to overhaul it with the following objectives: (i) adopting ISC3's input/output computer architecture; (ii) updating, where practical, antiquated ISC3 model algorithms with newly developed or current state-of-the-art modelling techniques; and (iii) ensuring that the source and atmospheric processes presently modelled by

ISC3 will continue to be handled by the AERMIC Model (AERMOD), albeit in an improved manner (EPA).

After seven long development processes, the AERMIC was able to find the replacement of the ISC3 model – the AERMOD. After the formulation, to find the scope of further improvement, the model was tested against a variety of field data sets which covered elevated and surface releases, complex and simple terrain, and rural and urban boundary layers. The model underwent many performance evaluations where the adequacy of the model was checked. With this, AERMOD performance was compared against the other regulatory models such as ISC3 (USEPA, 1995), CTDMPLUS (Perry, 1992), RTDM (Paine and Egan, 1987) and HPDM (Hanna and Paine, 1989; Hanna and Chang, 1993). In 2000 EPA become the regulatory model for both complex and simple terrain.

AERMOD is a steady-state plume model; it assumes that the vertical and horizontal distribution of a pollutant is Gaussian under a stable boundary layer. But, in the case of a convective boundary layer, the horizontal distribution is Gaussian while the vertical distribution is a bi-Gaussian probability density function. AERMOD is useful to study the lofting of the plume and also to help track a plume mass which can penetrate the CBL and also can re-enter the boundary layer.

The modelling system of AERMOD is the agglomeration of three main components: (i) AERMOD: a dispersion model; (ii) AERMAP: a pre-terrain processor; and (iii) AERMET: a pre-meteorological data processor. AERMOD as a whole software package is freely downloadable from the USEPA website. Figure 5.11 shows the working system of the model.

5.6.2 AERMET

The role of meteorology in the dispersion of pollutants cannot be overlooked. AERMOD requires two metadata files named as a surface data file and a profile data file. Running AERMET requires generating these two met files, which can be used as a input file by AERMOD. AERMET is a three-stage process and requires three different data sets: (i) hourly surface data; (ii) upper air data; and (iii) on-site data. The data are extracted

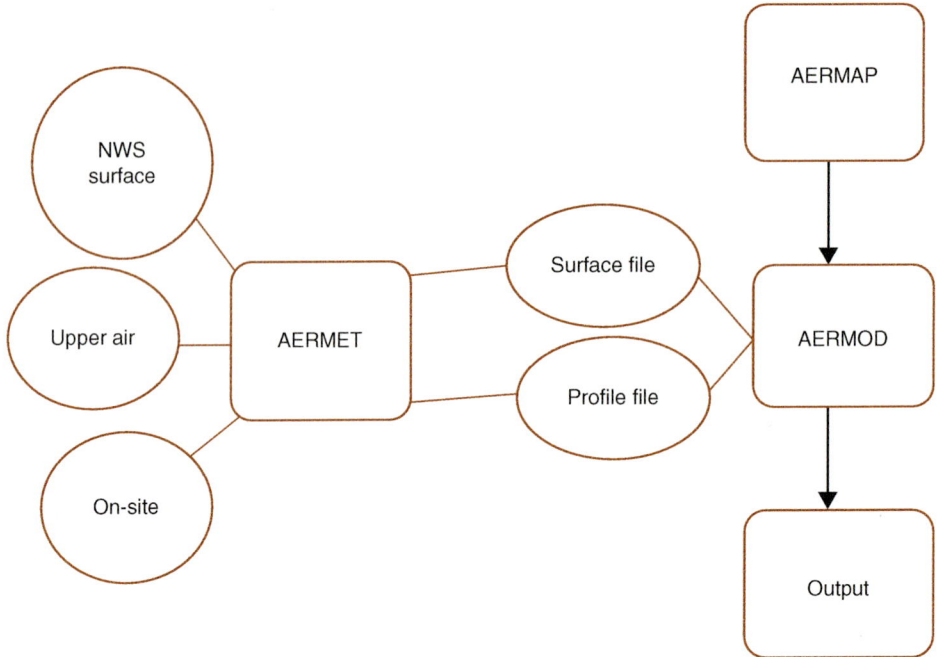

Fig. 5.11. Illustration of the AERMOD modelling system.

and assessed for their quality by the software during the first stage, the data are merged and an intermediate file is written during the second stage, and during the last stage the software reads the merged file and generates a surface data file and profile data file. These work as input for AERMOD.

5.6.3   AERMAP

Using AERMOD, one can assess the concentration of a pollutant over any type of terrain, from flat plains to rigid mountains. As AERMOD can't create its own terrain, AERMAP helps to construct it. AERMAP has several sources from where the user can access the standard map and data files. It first determines a base elevation for each receptor and source. It then helps to find the terrain height and location which have greatest impact over the receptor site, known as hill height. The AERMAP output file can easily be incorporated as an AERMOD input file. To detect the source and receptor position, AERMAP

relies on the Universal Transverse Mercator (UTM) coordinate system.

5.6.4   CALPUFF

As all plume models are steady-state models, they assume that the conditions at the time of emission and the time it reaches the receptor site are the same; therefore, they cannot generate results for the long-range transport of pollutants (Allwine *et al.*, 1998; Scire *et al.*, 2000; Lee, 2016). Puff models are very helpful in these conditions whereas plume models are not, as they do not require a steady state assumption (Lakes Environmental, n.d.; Scire *et al.*, 2000).

An air pollutant puff is considered as a continuous plume having discrete packets of air pollutant species, as shown in Fig. 5.8 (Scire *et al.*, 2000). Most of the puff models follow the snapshot approach to evaluate the effect of a puff over the receptor air pollutant concentrations (Ludwig *et al.*, 1977; van Egmond and Kesseboom, 1983; Petersen, 1986); the pollutants' concentration at the receptor site is not an output of a single puff but a cumulative effect of all nearby puffs. The puff can be frozen at a particular time interval (Scire *et al.*, 2000).

With the aim to conduct the air-quality monitoring of a non-steady state system, an air dispersion model was developed by the Sigma Research Corporation (now part of Earth Tech, Inc.) known as CALPUFF (Lakes Environmental, n.d.; Allwine *et al.*, 1998; Scire *et al.*, 2000, p. 10; Rood, 2014; Lee, 2016). It is a multi-layer non-steady-state puff dispersion model, where multiple species of air pollutant can be studied (Gryning and Batchvarova, 2012). The model can simulate the spatial and temporal variation of meteorological condition and its effect in pollutant transportation, transformation and settling (Gryning and Batchvarova, 2012). The model was developed for a broad perspective; for example, it is useful to study the micro- to macro-scale effects for short-term as well as long-term studies. The model is also very helpful when studying the effects of the wet and dry deposition and chemical transformations of a species. With this, the effect of coastal interaction and the terrain on pollutant concentration can easily be determined using this model. USEPA adopted the CALPUFF model in its Guideline on Air Quality Models as the preferred model for evaluating the long-range transport of pollutants and their impacts (Lakes Environmental, n.d.; Scire *et al.*, 2000; Rood, 2014).

The CALPUFF modelling system has a set of pre-processing and post-processing programmes, and is built using three main programmes: (i) CALMET: a diagnostic three-dimensional meteorological model; (ii) CALPUFF: an air-quality dispersion model; and (iii) CALPOST: a post-processing package. The whole system works on a graphical user interface. There are many other components that help to prepare geographical (terrain) data and meteorological (surface, upper air, precipitation, etc.) data (Lakes Environmental, n.d.; Scire *et al.*, 2000, Gryning and Batchvarova, 2012; Rood, 2014).

To simulate the non-steady-state dispersion puff, it needs three-dimensional meteorological data and that can be generated using the CALMET model. The CALMET model consists of a wind field generator, and boundary layer

modules for over water and over land studies. It can also merge the wind data generated by the CSUMM prognostic wind field model or MM5/MM4 model (Lakes Environmental, n.d.; Scire *et al.*, 2000).

CALPOST is a post-processing programme that helps to compute the time-averaged concentration of pollutants along with their depositional rate. CALPOST works only on the output of CALPUFF (Lakes Environmental, n.d.; Scire *et al.*, 2000).

Cases of stagnation, inversion, recirculation and fumigation can be studied using the CALPUFF model. With this, it is very useful to study the effect of complex terrain, calm wind or light wind conditions. It is one of the best models for the Visibility assessments and Class I area impact studies. It is also very helpful to study secondary pollutant formation and particulate modelling (Lakes Environmental, n.d.; Scire *et al.*, 2000).

5.6.5 Photochemical modelling

Photochemical air-quality models have become widely recognized and utilized tools for regulatory analysis for the control strategies. These photochemical models are large-scale air-quality models that simulate the changes of pollutant concentrations in the atmosphere using a set of mathematical equations, illustrating the chemical and physical processes in the atmosphere. These models are applied at multiple spatial scales from local to global. Two examples of photochemical models follow.

5.6.5.1 Community Multiscale Air Quality (CMAQ)

The primary goals for the modelling system are to improve: (i) the environmental management ability by evaluating the impacts of multiple air pollutants over multiple scales; and (ii) the better understanding of atmospheric processes in the scientific community while simulating chemical and physical interactions in the atmosphere. The newest Models-3/CMAQ version 4.5 is now available for download from CMAS (n.d.).

5.6.5.2 Comprehensive Air quality Model with Extensions (CAMx)

This is a publicly available open-source computer modelling system for the integrated assessment of gaseous and particulate air pollution, built on today's understanding that air-quality issues are complex, interrelated, and reach beyond the urban scale. CAMx is designed to: (i) simulate air quality over many geographic scales; (ii) treat a wide variety of inert and chemically active pollutants such as ozone, inorganic and organic $PM_{2.5}$ and PM_{10}, mercury and other toxic elements; (iii) provide source-receptor sensitivity analyses; and (iv) be computationally efficient and easy to use. USEPA has approved the use of CAMx for numerous ozone and PM State Implementation Plans throughout the US, and has used this model to evaluate regional mitigation strategies (Exponent, n.d.; CAMx, 2016; USEPA, 2016; http://www.src.com/calpuff/calpuff_eula.htm).

5.7 Conclusion

Air pollution in cities is a serious human health issue. Therefore, there is the need for a reliable air-quality management system for reduction of the urban air-pollution problem. There are numerous air-quality models that have been developed using deterministic, statistical and physical approaches for urban air quality. Though models are to a great extent expedient in understanding the complex physicochemical and photochemical processes of the atmosphere, due to the spontaneously changing environment and limitations of proper monitoring, measurements and description of processes of the pollution, the robustness of the models is weakened. Therefore more research is needed to better understand inherent atmospheric processes.

References

Allwine, K.J., Dabberdt, W.F. and Simmons, L.L. (1998) *Peer Review of the CALMET/CALPUFF Modeling System.* EPA Contract No. 68-D-98-092. The KEVRIC Company Inc., Durham, North Carolina. Available at: https://www3.epa.gov/scram001/7thconf/calpuff/calpeer.pdf (accessed 12 September 2018).

Baynton, H.W., Bidwell, J.M. and Befun, W.D. (1965) The association of low level inversions with surface winds and temperature at Point Arguello. *Journal of Applied Meteorology* 4, 509–516.

Boubel, R.W., Fox, D.L., Turner, D.B. and Stern, A.C. (1994) *Fundamentals of Air Pollution,* 3rd edn. Academic Press, San Diego.

Brankovic, C. and Molteni, F. (1997) Sensitivity of the ECMWF model northern winter climate to model formulation. *Climate Dynamics* 13, 75–101.

CAMx (2016) CAMx download. Comprehensive Air Quality Model with Extensions. Available at: http://www.camx.com (accessed 20 July 2018).

CMAS (n.d.) Model Release Calendar. Available at: http://www.cmascenter.org/download/release_calendar.cfm?temp_id=99999 (accessed 20 July 2018).

Exponent (n.d.) CALPUFF end-user license agreement. Exponent Inc. Engineering and Scientific Consulting. Available at: http://www.src.com/calpuff/calpuff_eula.htm (accessed 30 July 2018).

Garcia, R.R., Stordal, F., Solomon, S. and Kiehl, J.T. (1992) A new numerical model of the middle atmosphere 1. Dynamics and transport of tropospheric source gases. *Journal of Geophysical Research,* 97(D12), 12967.

Grimmond, C.S.B., King, T.S., Roth, M. and Oke, T.R. (1998) Aerodynamic roughness of urban areas derived from wind observations. *Boundary-Layer Meteorology* 89, 1–24.

Gryning, S.E. and Batchvarova, E. (2012) *Air Pollution Modeling and Its Application XIII.* Springer Science & Business Media, Boston, MA.

Hanna, S.R. and Chang, J.C. (1993) Hybrid plume dispersion model (HPDM) improvements and testing at three field sites. *Atmospheric Environment. Part A. General Topics* 27(9), 1491–1508.

Hanna, S.R. and Paine, R.J. (1989) Hybrid plume dispersion model (HPDM) development and evaluation. *Journal of Applied Meteorology* 28, 206–224.

Haq, G. and Schwela, D. (2008) *Air Quality Modelling. Foundation Course on Air Quality Management in Asia.* Stockholm Environment Institute, University of York, York, UK.

Jacobson, M.Z. (2005) *Fundamentals of Atmospheric Modeling,* 2nd edn. Cambridge University Press, Cambridge, UK.

Keck, R.-E., Aagaard Madsen, H., Larsen, G.C., Veldkamp, D., Wedel-Heinen, J.J. and Forsberg, J. (2013) *A Consistent Turbulence Formulation for the Dynamic Wake Meandering Model in the Atmospheric Boundary Layer*. DTU Wind Energy. (DTU Wind Energy PhD; No. 0012 (EN)). Available at: http:// orbit.dtu.dk/en/publications/a-consistent-turbulence-formulation-for-the-dynamic-wake-meandering-model-in-the-atmospheric-boundary-layer(0222ac75-fd0b-42e0-9345-aa3dc493bd71).html (accessed 12 September 2018).

Lakes Environmental (n.d.) CALPUFF View brochure. Available at: https://www.weblakes.com/products/calpuff/resources/lakes_calpuff_view_brochure.pdf (accessed 20 September 2017).

Lee, R.F. (2016) Comparison of features and data requirements among the CALPUFF, AERMOD, and ADMS Models. Breeze Software, Dallas, Texas. Available at: https://www.slideshare.net/BREEZESoftware/comparison-of-features-and-data-requirements-among-the-calpuff-aermod-and-adms-models (accessed 20 September 2017).

Liptak, B.G. (1974) *Environmental Engineers' Handbook Volume II, Air Pollution*. Chilton Book Company, Radnor, Pennsylvania.

Ludwig, F.L., Gasiorek, L.S. and Ruff, R.E. (1977) Simplification of a Gaussian puff model for real-time minicomputer use. *Atmospheric Environment* 11(5), 431–436.

Pai, P. and Tsang, T.T.H. (1993) On parallelization of time-dependent three-dimensional transport equations in air pollution modelling. *Atmospheric Environment. Part A. General Topics* 27 (13), 2009–2015.

Paine, R.J. and Egan, B.A. (1987) *User's Guide to the Rough-Terrain Diffusion Model (RTDM) (Rev. 3. 20)* (No. PB-88-171467/XAB; ERT-PD-535-585). Environmental Research and Technology, Inc., Concord, Massachusetts.

Pasquill, F. and Smith, F.B. (1983) *Atmospheric Diffusion*. John Wiley & Sons, Hoboken, New Jersey.

Perry, S.G. (1992) CTDMPLUS: a dispersion model for sources near complex topography. Part I: Technical formulations. *Journal of Applied Meteorology* 31(7), 633–645.

Petersen, W.B. (1986) A demonstration of INPUFF with the MATS data base. *Atmospheric Environment* 20(7), 1341–1346.

Rood, A.S. (2014) Performance evaluation of AERMOD, CALPUFF, and legacy air dispersion models using the Winter Validation Tracer Study dataset. *Atmospheric Environment* 89, 707–720.

Scire, J.S., Strimaitis, D.G. and Yamartino, R.J. (2000) A *User's Guide for the CALPUFF Dispersion Model*. Earth Tech, Inc. Concord, Massachusetts.

Seinfeld, J.H. and Pandis, S.N. (2006) *Atmospheric Chemistry and Physics*. John Wiley & Sons, Inc., Hoboken, New Jersey.

Turner, D.B. (1994) *Workbook of Atmospheric Dispersion Estimates – An Introduction to Dispersion Modelling*, 2nd edn. Lewis Publishers, CRC Press, Boca Raton, Florida.

USEPA (1995) *User's Guide for the Industrial Source Complex (ISC3) Dispersion Models (revised) Volume I – User Instructions*. EPA-454/b-95-003a, US Environmental Protection Agency, Research Triangle Park, North Carolina.

USEPA (2016) Photochemical modeling. Support Center for Regulatory Atmospheric Modeling (SCRAM). Available at: https://www3.epa.gov/scram001/photochemicalindex.htm (accessed 20 July 2018).

van Egmond, N.D. and Kesseboom, H. (1983) Mesoscale air pollution dispersion models – II. Lagrangian puff model and comparison with Eulerian grid model. *Atmospheric Environment* 17(2), 267–274.

6 Indices Used for Assessment of Air Quality

Prashant Rajput,[1,2] Gyanesh Kumar Singh[1] and Tarun Gupta[1]*

[1]*Indian Institute of Technology Kanpur, India;* [2]*University of Surrey, Guildford, UK*

Abstract

Copious evidence suggests that rapid urbanization and widespread anthropogenic activities are the main factors resulting in high levels of ambient air pollutant over the Indo-Gangetic Plain (IGP) in India. Continuous monitoring of levels of air pollutants affecting air quality, human health and the ecosystem is essentially required on a day-to-day basis. The influence of pollutants on the air quality, human health and susceptibility of plants is studied widely with the help of various indices, e.g. air-quality index (AQI), air-quality health index (AQHI), air-pollution index (API), air-pollution tolerance index (APTI) and anticipated performance index. In this chapter, we discuss the important indices: AQI, AQHI, API and APTI. As a case study, we have discussed in detail the AQI. We report on the status of ambient air quality (for the year 2016) over the IGP, utilizing the CPCB (Central Pollution Control Board, India) database. Air quality index (AQI) values have been presented systematically in this chapter to better evaluate the air quality associated with various major and potential atmospheric pollutants (available in CPCB database). Ambient data of air pollutants ($PM_{2.5}$ or PM_{10}, SO_2, NO, NO_2 and CO) have been retrieved from three sites continuously monitored by CPCB: upwind (Panchkula; states of Haryana), central (Lucknow; states of Uttar Pradesh) and downwind locations (Kolkata; West Bengal) in the IGP. Monthly and seasonally averaged data sets have been discussed and compared for the aforementioned three sites. Results suggest that Lucknow (in central IGP) was more polluted than the Kolkata and Panchkula sites (upwind of major polluting sources in the IGP). Relative to Lucknow, low concentrations of air pollutants at Kolkata suggests the influence of sea- and land-breeze wind systems on ventilation coefficient and efficient dispersion of the pollutants. AQI values at Kolkata ranging between 27 and 137 indicate a good to moderately polluted atmospheric scenario. However, AQI at Lucknow ranging between 301 and 400 indicates very poor air quality, particularly during post-monsoon and wintertime, whereas at Panchkula during most of the time, air quality was found to be satisfactory (AQI: 51–100). The present synthesis documents the air quality scenario over the IGP in a versatile and simplified manner that can be also be utilized for public awareness.

6.1 Introduction

Atmospheric chemistry and composition are important factors that govern air pollution (Seinfeld and Pandis, 2006; Rajeev *et al.*, 2018; Rajput *et al.*, 2018; Sorathia *et al.*, 2018). The Indo-Gangetic Plain (IGP) stretches over a vast area from the north-west to the north-east region of India and covers nearly 20% of the Indian subcontinent landmass (Gupta *et al.*, 2004; Ram *et al.*, 2010b; Gupta and Mandaria, 2013; Rajput *et al.*, 2013; Singh *et al.*, 2014a). It is a densely populated region and provides shelter for ~40% of the Indian population (Badarinath *et al.*, 2006).

---

* Corresponding author: tarun@iitk.ac.in

Rapid urbanization and widespread anthropogenic activities, including emissions from industries, agricultural-residue burning, biomass burning and fossil-fuel combustion, etc., are causing high levels of air pollutants over the IGP (Ramanathan *et al.*, 2001; Lawrence and Lelieveld, 2010; Rajput *et al.*, 2011b, 2014b, 2016b, 2017; Rastogi *et al.*, 2014; Chakraborty *et al.*, 2017). Mineral dust (fine and coarse) upliftment due to natural wind systems and anthropogenic activities further elevate the background levels of ambient particulate matter (PM) (Jethva *et al.*, 2005; Lawrence and Lelieveld, 2010; Rajput *et al.*, 2015, 2017; Rajeev *et al.*, 2016). Thus, it has been realized that ambient air quality over the IGP is deteriorating (Rajput and Gupta, 2016). Annual and seasonal practices on biomass burning emission and fossil-fuel combustion in the IGP (Rajput *et al.*, 2016c), with ambient atmospheric conditions, are shown in Fig. 6.1 (Rajput, 2013).

There have been continuous efforts made by several researchers to study the atmospheric chemistry, air pollution, radiative forcing and health risk assessment from various locations within the IGP (Ramanathan *et al.*, 2001; Venkataraman *et al.*, 2005; Ram *et al.*, 2010a; Rajput *et al.*, 2011a; Ramachandran and Srivastava, 2013; Rajput and Sarin, 2014; Rajput *et al.*, 2014a, 2014c, 2016a; Srinivas and Sarin, 2014; Chakraborty *et al.*, 2016; Kumar *et al.*, 2017). Thus, a lot of data are available in the literature from the IGP. In addition, data sets of major parameters (SO_2, CO, NO_2, $PM_{2.5}$, PM_{10}, etc.) from different geographical locations in India influencing the air quality are available from CPCB (Central Pollution Control Board, India) continuous monitoring stations in India. For the present study, we have retrieved data from CPCB monitoring stations to assess the long-term variability pattern in AQI (air-quality index). To assess the impact of air pollutants on air quality, human health and susceptibility of plants, there are various indices of air quality, for example, air-quality index (AQI), air-quality health index (AQHI), air-pollution index (API), air-pollution tolerance index (APTI) and anticipated performance index. The APTI and anticipated performance index are used to infer the susceptibility of plant species to pollution. In the following sections, we discuss in brief important indices: AQI, AQHI, API and APTI. As a case study, we discuss in detail the AQI.

6.2 Important Indices Used for Assessing Air Quality

6.2.1 Air quality index (AQI)

6.2.1.1 *General definition*

To make the impact of air quality accessible to one and all, the National Air Quality Index (AQI) was launched by the Ministry of Environment, Forests and Climate Change (MoEFCC) under Swachh Bharat Mission in September 2014. The air-quality scenario is expressed in terms of AQI (Nigam *et al.*, 2015). The AQI is determined mathematically by involving the weighted values of air pollutants (e.g. concentration) and is represented by a mathematical number (Ott, 1978). Two steps are involved in determining the AQI: (i) formation of sub-indices (corresponding to each pollutant); and (ii) aggregation of sub-indices to get an overall AQI. Sub-indices ($I_1, I_2,....., I_n$) formation for 'n' pollutant variables ($Y_1, Y_2,...., Y_n$) is done using sub-index functions, which are based on the standards of air quality and related health effects.

Mathematically, a relationship between pollutant concentrations and health effects can be represented by each sub-index as given below:

$$I_i = f(Y_i), \text{ where } i = 1, 2, ..., n. \quad \text{(Eq. 6.1)}$$

And the overall index (I) can be obtained by aggregation of sub-indices (I_i) as given below and referred to as AQI:

$$I = F(I_1, I_2, ..., I_n) \quad \text{(Eq. 6.2)}$$

The aggregation function is generally a mathematical operator that involves summation or multiplication.

6.2.1.2 *Applications of AQI*

Six important objectives that can be achieved from AQI have been suggested by a previous study (Ott, 1978) and are listed below.

1. Resource allocation: assisting the administrators in determining priorities and allocating funds for monitoring and improvement of air quality.
2. Location ranking: facilitates comparison of air quality in different cities or locations. Thus, it

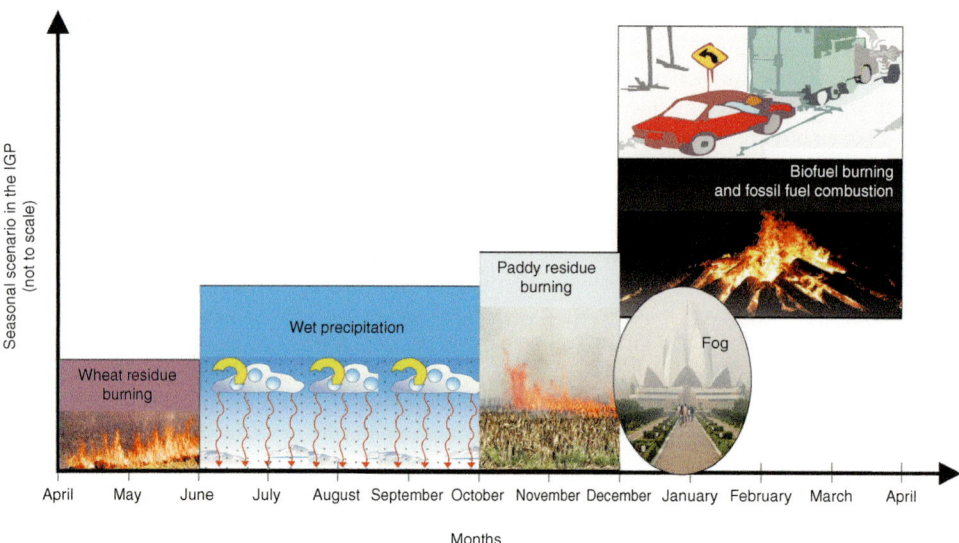

Fig. 6.1. Figure showing typical practices and ambient atmospheric conditions in the Indo-Gangetic Plain. Emissions from bio-fuel burning and fossil-fuel combustion practices and secondary organic aerosol (SOA) formation occur in all seasons but the intensity/amount is highest during winter (as shown above).

could be a helpful entity for determining the areas and frequencies of potential hazards.

3. Standards enforcement: assisting in enforcement of legislative standards and existing criteria. Appropriate standards and adequate monitoring programmes can be identified.

4. Trend analysis: degradation or improvement in air quality over a time period can be assessed quickly and easily through AQI, thus presenting an opportunity for forecasting air quality and planning pollution-control measures accordingly.

5. Public information: creating awareness among people about atmospheric conditions. Thus, it can be beneficial for people who suffer from illness triggered by air pollution, thereby enabling them to alter their daily activities based on the information of high pollution levels.

6. Scientific research: a large set of data can be reduced to a comprehensible form, which can provide better insight for the researchers while studying any atmospheric phenomena.

The contribution of individual pollutants to overall air quality can be identified through the AQI index. Using AQI in conjunction with other data sets, e.g. local emission inventories, could provide detailed and more practical information to understand source-to-receptor impacts.

6.2.2 Indian air-quality index (IND-AQI)

For the protection of human health due to adverse effects of air pollutants, the elimination/reduction of exposure to hazardous air pollutants (HAPS) and providing guidelines for making pollution-control decisions, the development of appropriate air quality standards is essential. Thus, an account of the level of air quality adopted by a regulatory agency is considered as the air-quality standard. These standards provide essential and vital legal structure to mitigate and control the air pollution. In this context, the CPCB has developed a new set of Indian National Air Quality Standards (INAQS) for 12 parameters with regard to carbon monoxide (CO), nitrogen dioxide (NO_2), sulfur dioxide (SO_2), $PM_{2.5}$, PM_{10}, ozone (O_3), lead (Pb), ammonia (NH_3), benzo(a)pyrene (BaP), benzene (C_6H_6), arsenic (As) and nickel (Ni).

One of the outcomes of studying AQI is a quick accountability of air-quality scenarios deduced from major pollutants which have acute and chronic potential health impacts. Most of the pollutants are measured continuously at the CPCB stations and data is shared via an online monitoring network. In the AQI system, the pollutants considered are CO, NO_2, SO_2, $PM_{2.5}$, PM_{10},

O_3, NH_3 and Pb. It is important to mention here that Pb is not involved while estimating real-time AQI, because ambient concentrations of Pb are not measured online. However, the AQI estimation of preceding days with incorporation of Pb helps in analysing the air quality status and its impact.

A single number (index value) labelled with colour and nomenclature represents complex air-quality data of various pollutants. AQI differentiates various air pollution scenarios depending on the concentration of pollutants and the corresponding health effects. AQI is represented by six categories: Good, Satisfactory, Moderately polluted, Poor, Very Poor and Severe (Table 6.1). The sub-indices are determined as a linear function of the measured ambient concentrations of a pollutant. Overall, AQI is determined from the worst sub-index. Various AQI values exhibit the impact of health problems associated with them. For example, a value ranging from 0 to 50 (labelled as Good) indicates minimum air-quality impact on human health, and range 51–100 (Satisfactory) is an indicator of minor breathing discomfort for sensitive people. AQI in the range of 101–200 (Moderate) indicates a scenario of breathing discomfort for people suffering from lung or heart disease, whereas the range 201–300 (Poor) indicates breathing discomfort for human beings under prolonged exposure. Similarly, AQI values ranging from 301–400 (Very Poor) indicate the plausibility of respiratory illness in people due to prolonged exposure and an AQI value greater than 400 (Severe) indicates respiration issues even in healthy people.

The AQI calculation involves the following approach.

1. 24-hourly average concentration (8-hourly in the case of CO and O_3) and health breakpoint concentration range are used to calculate sub-indices for individual pollutants at a monitoring site. AQI is the worst sub-index from that site.
2. All eight pollutants are not necessarily monitored at all the locations. Therefore, a minimum of three pollutants (of which, one should necessarily be either $PM_{2.5}$ or PM_{10}) are required for overall AQI calculation. Otherwise, the data are considered insufficient for AQI calculation.
3. Even if the data are inadequate for determining AQI, the sub-indices for monitored pollutants can be calculated. Air-quality status pertaining to any pollutant can be provided in the form of individual sub-index (i.e. pollutant-wise).
4. AQI on a real-time basis is determined by a web-based system. This automated system captures data from continuous monitoring stations without intervention of human beings and *in situ* determines and exhibits AQI based on running average values.
5. An AQI calculator is developed for manual monitoring stations wherein the measured data can be incorporated manually to obtain the AQI value.

6.2.3 Air-quality health index (AQHI)

Each individual may exhibit different reactions upon exposure (short- versus long-term) to air pollutants. Children, the elderly and those with

Table 6.1. Six categories of air quality index (AQI) and health breakpoint concentration range for eight pollutants. All these data have been retrieved from CPCB site.

AQI Category	AQI	Concentration range*							
		PM_{10}	$PM_{2.5}$	NO_2	O_3	CO	SO_2	NH_3	Pb
Good	0–50	0–50	0–30	0–40	0–50	0–1.0	0–40	0–200	0–0.5
Satisfactory	51–100	51–100	31–60	41–80	51–100	1.1–2.0	41–80	201–400	0.5–1.0
Moderately polluted	101–200	101–250	61–90	81–180	101–168	2.1–10	81–380	401–800	1.1–2.0
Poor	201–300	251–350	91–120	181–280	169–208	10–17	381–800	801–1200	2.1–3.0
Very poor	301–400	351–430	121–250	281–400	209–748*	17–34	801–1600	1200–1800	3.1–3.5
Severe	401–500	>430	>250	>400	>748*	>34	>1600	>1800	>3.5

*CO is in mg/m³ and other pollutants are in µg/m³; 2-hour average values for PM_{10}, $PM_{2.5}$, NO_2, SO_2, NH_3 and Pb whereas 8-hour average values for CO and O_3.

diabetes, heart or lung disease are most sensitive to the adverse health effects of air pollutants. Senior people are at higher risk because of weakening of the heart, lungs and immune system and increased likelihood of health problems such as heart and lung disease. Children are also more vulnerable to air pollution: they have less-developed respiratory and defence systems. Because of their size, they inhale more air per kilogram of body weight than adults. Children generally also spend more time outdoors being physically active, which can increase their exposure to air pollutants. People participating in sports or strenuous work outdoors breathe more deeply and rapidly, allowing more quantum of air pollutants to enter their respiratory tracts. They may experience symptoms like eye, nose or throat irritation, coughing or difficulty breathing when air pollution levels are high. Negative health effects increase as air pollution worsens. Studies have shown that even a modest increase in air pollution can cause small but measurable increases in emergency room visits, hospital admissions and mortality. Small increases in air pollution over a short period of time can increase symptoms of pre-existing illness among those at risk. Thus, depending on the length of exposure time, the subject's health status, his/her genetic background and the concentration of pollutants, air pollution can have a negative effect on the subject's heart and lungs, e.g. it can cause irritation in lungs and other air tracts, worsen chronic diseases such as heart disease, chronic bronchitis obstructive pulmonary disease (COPD) and asthma, etc.

When the AQHI value rises, people can decide whether they need to:

1. Reduce or reschedule outdoor physical activities.
2. Monitor possible symptoms, such as difficult breathing, coughing or irritation in the eyes.
3. Follow a doctor's advice to manage existing conditions such as heart or lung disease. They can also use the index as a reminder of the need to take action under different air pollution episodes (e.g. Good, Satisfactory, Severe, etc.).

AQHI is a type of scale (Table 6.2) designed to help people understand what the quality of the air around them signifies for their health. It is a new tool developed by health and environmental professionals to communicate the health risks posed by air pollution. The AQHI index provides specific advice to the people who are especially vulnerable to the effects of air pollutants, as well as for the general public. It is designed to help people in making decisions to protect their own health and the environment by:

1. Limiting short-term exposure to air pollutants.
2. Adjusting activity during episodes of increased air pollution and encouraging physical activity on days when the index is lower.
3. Reducing personal contribution to air pollution.

The AQHI is designed as a guide (Table 6.2) to the relative risk posed by criteria air pollutants (CAPs), which are known to harm human health. Three specific pollutants are generally chosen as indicators of the overall mixture. (i) Ground-level ozone (O_3), which is formed by photo-chemical reactions in the atmosphere. (ii) Particulate matter, which is a mixture of tiny airborne particles and liquid droplets that can be inhaled deep into the lungs. These particles can either be emitted directly by post-harvest agricultural-waste burning, biomass burning, vehicles, industrial activities or natural sources like bursting of sea-bubbles (sea-salt spray), upliftment of mineral dust, forest fires, or formed as a result of chemical reactions in ambient atmospheric conditions. Particulate matter levels can have influence from both local air pollution sources or from widespread air pollution episodes. (iii) Nitrogen dioxide (NO_2), which is produced as a result of multiple factors that include emissions from motor vehicles and power plants. Nitrogen dioxide is often elevated in the vicinity of high traffic roadways. All three compounds can have serious combined effects on human health – from illness to hospitalization to premature death – even as a result of short-term exposure. Significantly, all of these pollutants appear to threaten human health, even at low levels of exposure, especially among those with pre-existing health problems. In the development of the AQHI, these three pollutants were found to be the best indicator of the health risk of the combined impact of the major pollutants in the air. It is important to mention here that SO_2 and CO (carbon monoxide) were dropped from the AQHI because they were not associated with additional health risks once the effects of ozone, nitrogen dioxide and particulate matter were taken into account.

6.2.4   Air pollution tolerance index (APTI)

Air pollution has become a serious environmental stress to plants due to increasing

Table 6.2. AQHI index-based specific advice for vulnerable people and the general public (MoECC, 2018).

Health risk	AQHI	Health messages	
		Risk population*	General population
Low risk	1–3	Enjoy usual outdoor activities	Ideal air quality for outdoor activities
Moderate risk	4–6	Consider reducing or rescheduling strenuous activities outdoors on experiencing symptoms	No need to modify usual outdoor activities unless experiencing symptoms such as coughing and throat irritation
High risk	7–10	Reduce/reschedule strenuous activities outdoors. Children and older people should also take it easy	Consider reducing/rescheduling strenuous activities outdoors on experience of symptoms such as coughing and throat irritation
Very high risk	> 10	Avoid strenuous activities outdoors. Children and older people should also avoid outdoor physical exertion	Reduce/reschedule strenuous activities outdoors, especially on experience of symptoms such as coughing and throat irritation

*People with heart or breathing problems.

industrialization and urbanization in the last few decades. Particulate matter and gases can cause serious impacts on the overall physiology of plants. It is an established fact that leaves of plants act as bio-filters by removing pollutants from the air (Chen *et al.*, 2017). Plant leaves provide an enormous area for absorption and removal of air pollutants from the air. In this context, APTI is generally utilized to assess plant species susceptibility level (tolerance) to air pollution. Four physiological and biochemical parameters, i.e. leaf extract pH, ascorbic acid, total chlorophyll and relative water content, are generally put together to yield the APTI of plants. It is important to mention here that APTI monitoring throughout different seasons could suggest ideal planting species characteristics in a particular zone of interest.

The APTI of a plant species can be determined using the following mathematical equation:

$$APTI = \frac{A(T+P)+R}{10}$$

where, A is the ascorbic acid content of leaf in mg g^{-1} (dry basis), T is the total chlorophyll of leaf in mg g^{-1} (dry basis), P represents pH of the leaf extract and R is the percentage of relative water content of leaf tissue. In the above equation, the total sum is scaled down by a factor of 10 to obtain a manageable number (i.e. APTI).

6.3 Case Study of Air Quality Assessment (AQI) over the Indo-Gangetic Plain

6.3.1 Monthly and seasonal variability of air pollutants over the IGP

Air quality is discussed for three locations in the IGP: Panchkula (upwind IGP), Lucknow (central IGP) and Kolkata (downwind IGP). The data pertaining to air quality were adopted from CPCB (n.d.). First, we discuss the monthly variability of parameters such as PM$_{2.5}$ (whereas for Kolkata PM$_{10}$ was available), SO$_2$, NO, CO and NO$_2$ over these locations.

6.3.1.1 Panchkula Site (Haryana, upwind IGP)

Temporal variability of monthly averaged PM$_{2.5}$, SO$_2$, NO, NO$_2$ and CO concentrations over Panchkula for the year 2016 is shown in Fig. 6.2.

Furthermore, seasonally averaged concentrations of these species are provided in Table 6.3. PM$_{2.5}$ mass concentrations were higher during wintertime (Dec–Jan: 70 ± 13 µg m^{-3}) and post-monsoon (Oct–Nov: 68 ± 4 µg m^{-3}) as compared to pre-monsoon (March–May: 43 ± 6 µg m^{-3}) and monsoon season (June–Sept: 44 ± 8 µg m^{-3}). As discussed below, PM levels are usually high (average concentration of PM$_{2.5}$: 100–200 µg m^{-3})

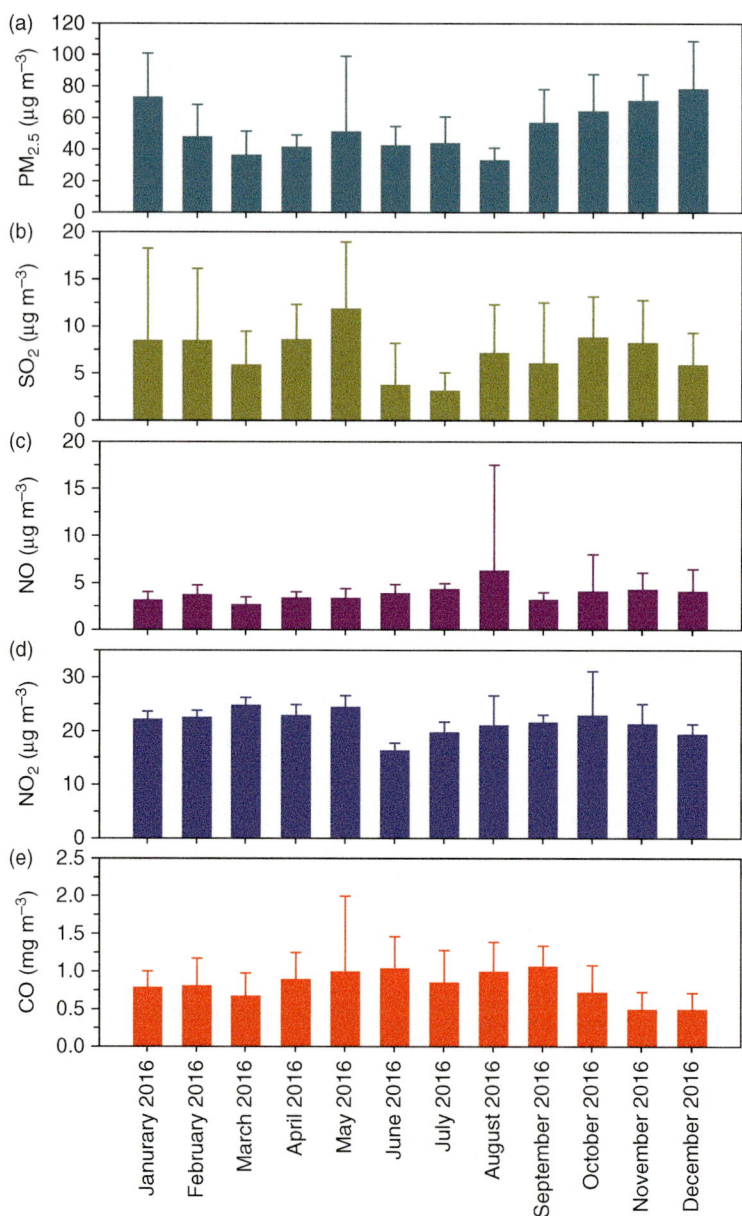

Fig. 6.2. Monthly variability of: (a) $PM_{2.5}$, (b) SO_2, (c) NO, (d) NO_2 and (e) CO at Panchkula (Haryana) in upwind IGP (2016).

during post-monsoon and wintertime in the IGP, unlike lower concentrations observed at the Panchkula site. It is worthwhile mentioning that high-particulate emissions are usually linked with poor combustion practices, e.g. biomass burning (Rajput *et al.*, 2011a,b). Thus, it is evident from these observations that Panchkula is upwind of

major biomass burning practices in the IGP. SO_2 average concentrations were ~8–9 µg m⁻³ in all seasons and exceptionally low during the monsoon period (5 µg m⁻³). During the seasonal-annual cycle at Panchkula, average concentrations of NO varied between 3 and 4 µg m⁻³, whereas NO_2 averaged at 22 ± 2 µg m⁻³. CO average concentrations

Table 6.3. Average concentrations (mean ± 1σ) of atmospheric pollutants during different seasons at Panchkula (upwind IGP; 2016).

Season	Months	$PM_{2.5}$	SO_2	NO	NO_2	CO
		($\mu g\ m^{-3}$)				($mg\ m^{-3}$)
Winter	Dec–Feb	70 ± 13	8 ± 1	4 ± 0.5	21 ± 2	0.7 ± 0.2
Pre-monsoon	March–May	43 ± 6	9 ± 3	3 ± 0.3	24 ± 1	0.8 ± 0.1
Monsoon	June–Sept	44 ± 8	5 ± 2	4.5 ± 1	19 ± 2	1.0 ± 0.1
Post-monsoon	Oct–Nov	68 ± 4	9 ± 0.5	4.2 ± 0.1	22 ± 1	0.6 ± 0.3

were 0.6–0.8 mg m^{-3} from post-monsoon to pre-monsoon and increased to 1 mg m^{-3} during the monsoon period.

Temporal variability of SO_x and NO_x shows large differences in the concentrations, which indicates that atmospheric chemistry and anthropogenic emission sources are highly variable over the region. In the monsoon season, there is an overall decrease in loading of ambient air pollutants. The most plausible reason is a wash-out effect of pollutants due to heavy precipitation (~ 900 mm rainfall) during the monsoon season.

6.3.1.2 Lucknow (Uttar Pradesh, central IGP)

Temporal variability of monthly averaged $PM_{2.5}$, SO_2, NO, NO_2 and CO concentrations over Lucknow in 2016 is shown in Fig. 6.3. Furthermore, seasonally averaged concentrations of these species have been provided in Table 6.4. $PM_{2.5}$ concentrations were higher during wintertime (Dec–Jan: 197 ± 25 μg m^{-3}) and post-monsoon (Oct–Nov: 155 ± 40 μg m^{-3}) when compared to pre-monsoon (March–May: 98 ± 5 μg m^{-3}) and monsoon season (June–Sept: 43 ± 27 μg m^{-3}). However, SO_2 average concentrations were ~9–10 μg m^{-3} in all seasons and exceptionally low during the monsoon season (5 μg m^{-3}). In the central IGP at Lucknow the NO and NO_2 showed pronounced variability. Both of these gaseous species exhibited highest concentrations (μg m^{-3}) during wintertime (NO: 78 ± 44; NO_2: 87 ±12) followed by post-monsoon (NO: 53 ± 33; NO_2: 39 ± 3) and monsoon (NO: 11 ± 2; NO_2: 23 ± 2). Data for these species were not available (on the CPCB website) during the pre-monsoon season and therefore are not discussed herein. CO average concentrations were about 1.2–1.4 mg m^{-3} during the post-monsoon and wintertime and decreased to 0.7–0.8 mg m^{-3} during the pre-monsoon and monsoon period. In general, a higher concentration of pollutants during post-monsoon and wintertime

is attributable to impact from large-scale post-harvest paddy-residue burning and bio-fuel burning, respectively (Rajput et al., 2014b, 2016b, 2017). Long-term data on chemical characteristics and composition of ambient aerosols and rain water from a nearby location at Kanpur in the central IGP suggest that post-monsoon and wintertime are the periods marked with heavy air pollution due to massive emissions from anthropogenic activities, and quite significant contribution (>40%) of secondary aerosols formation resulting from heterogeneous- and aqueous-phase reactions (Gupta et al., 2010; Chakraborty and Gupta, 2010; Kaul et al., 2011; Gupta and Mandaria, 2013; Gupta and Chauhan, 2014; Singh et al., 2014b; Rajeev et al., 2016; Rai et al., 2016; Suryawanshi et al., 2016; Singh and Gupta, 2016a,b; Rajput et al., 2017).

6.3.1.3 Kolkata (West Bengal, downwind IGP)

Temporal variability of monthly averaged $PM_{2.5}$, SO_2, NO, NO_2 and CO concentrations over Kolkata is shown in Fig. 6.4 (2016). Furthermore, seasonally averaged concentrations of these species have been provided in Table 6.5. $PM_{2.5}$ mass concentrations were higher during wintertime (Dec–Jan: 139 ± 13 μg m^{-3}) and post-monsoon (Oct–Nov: 109 ± 34 μg m^{-3}) as compared to the monsoon season (June–Sept: 24 ± 5 μg m^{-3}). The parameter data were not available (NA, at CPCB website) during the pre-monsoon (March–May) and therefore have not been discussed herein. SO_2 average concentrations were ~5–6 μg m^{-3} in all seasons. Thus, there was no significant variability in SO_2 concentration during the study period. This plausibly indicates that nearly a constant impact of major emission sources (stationary and mobile) was observed round the year in Kolkata. In Kolkata, the NO and NO_2 exhibited significant variability. Both of these gaseous

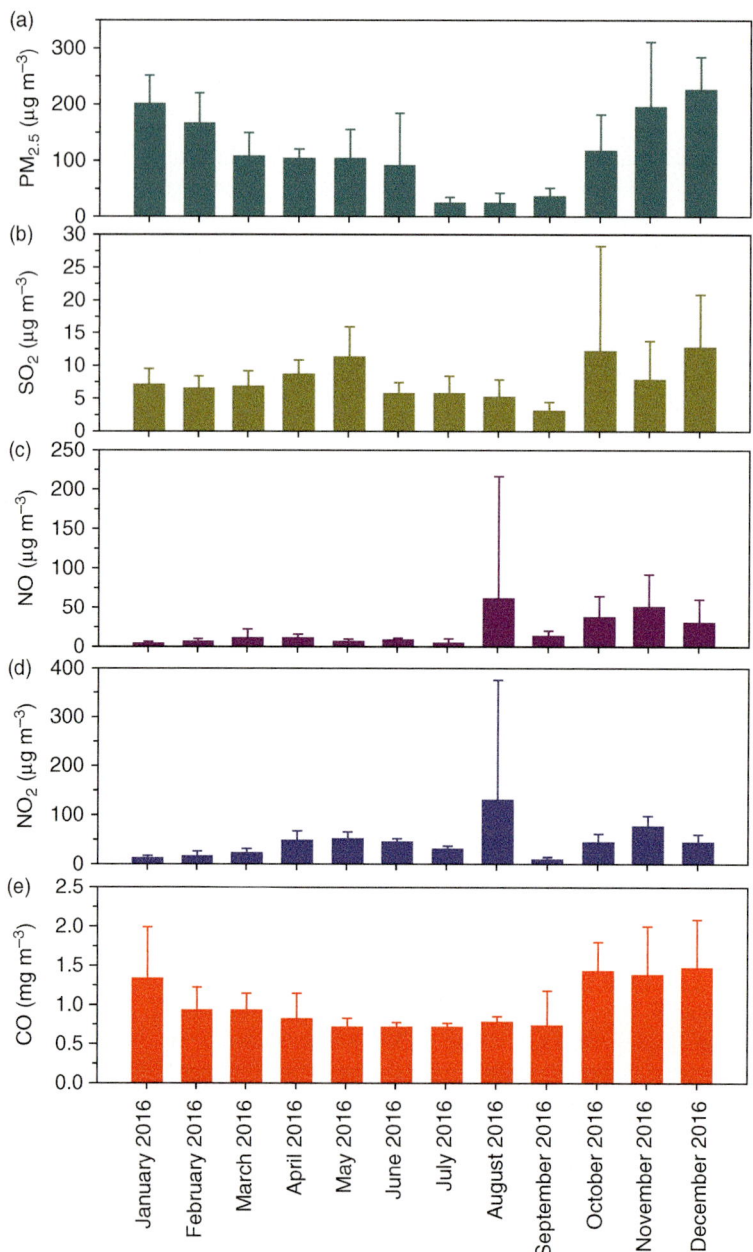

Fig. 6.3. Monthly variability of: (a) PM$_{2.5}$, (b) SO$_2$, (c) NO, (d) NO$_2$ and (e) CO at Lucknow (Central School site; Uttar Pradesh) in central IGP (2016).

species showed higher concentrations (µg m^{-3}) during post-monsoon (NO: 43 ± 7; NO$_2$: 60 ± 15) and monsoon (NO: 22 ± 20; NO$_2$: 52 ± 40). During the wintertime (NO: 13 ± 12; NO$_2$: 25 ± 15) and pre-monsoon (NO: 8 ± 2; NO$_2$: 40 ± 14) these species were relatively low in downwind

IGP. CO average concentrations were 1.9, 1.1 and 0.7 mg m^{-3} during post-monsoon, winter-time and monsoon season, respectively. Thus, the elevated concentrations of CO during wintertime (1.1 mg m^{-3}) and post-monsoon (1.9 mg m^{-3}) can be attributable to poor combustion efficiency

Table 6.4. Average concentrations (mean ± 1σ) of atmospheric pollutants during different seasons at Lucknow (central IGP, 2016).

Season	Months	$PM_{2.5}$	SO_2	NO	NO_2	CO
		($\mu g\ m^{-3}$)				($mg\ m^{-3}$)
Winter	Dec–Feb	197 ± 25	9 ± 3	78 ± 44	87 ± 12	1.2 ± 0.2
Pre-monsoon	March–May	98 ± 5	9 ± 2	NA	NA	0.8 ± 0.1
Monsoon	June–Sept	43 ± 27	5 ± 1	11 ± 2	23 ± 2	0.7 ± 0.1
Post-monsoon	Oct–Nov	155 ± 40	10 ± 2	53 ± 33	39 ± 3	1.4 ± 0.1

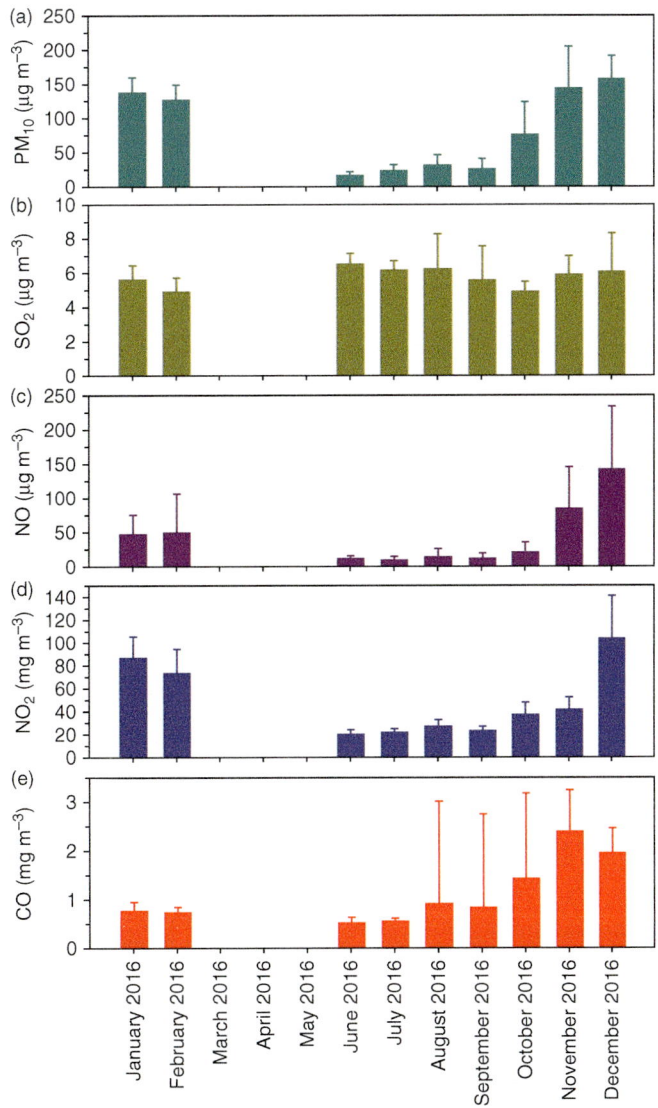

Fig. 6.4. Monthly variability of: (a) PM_{10}, (b) SO_2, (c) NO, (d) NO_2 and (e) CO at Kolkata (Victoria Memorial site; West Bengal) in downwind IGP (2016).

Table 6.5. Average concentrations (mean ± 1σ) of atmospheric pollutants during different seasons at Kolkata (downwind IGP; 2016).

Season	Months	$PM_{2.5}$	SO_2	NO	NO_2	CO
		($\mu g\ m^{-3}$)				($mg\ m^{-3}$)
Winter	Dec–Feb	139 ± 13	6 ± 0.5	13 ± 12	25 ± 15	1.1 ± 0.6
Pre-monsoon	March–May	NA	NA	8 ± 2	40 ± 14	NA
Monsoon	June–Sept	24 ± 5	6 ± 0.4	22 ± 20	52 ± 40	0.7 ± 0.2
Post-monsoon	Oct–Nov	109 ± 34	5.4 ± 0.5	43 ± 7	60 ± 15	1.9 ± 0.5

of biomass burning emission. In fact, previous studies from the IGP have reported that the open field crude burning of moist paddy-residues during the post-monsoon period results in massive emissions of air pollutants (Rajput *et al.*, 2011b, 2014b). It is important to mention here that a prevailing north-westerly wind system during post-monsoon and wintertime results in advective transport of air pollutants from upwind to central and downwind locations in the IGP (Ramanathan *et al.*, 2001; Rajput *et al.*, 2013).

6.3.2 Comparison of annual variation of pollutants for the selected cities

Particulate matter loading in central IGP (Lucknow) appeared to be higher than in upwind (Panchkula) and downwind (Kolkata) sites in the IGP (Fig. 6.5). However, $PM_{2.5}$ concentrations are similar at all sites in monsoon months. NO_x (NO_2 and NO) concentrations at Kolkata site are higher than at Lucknow and Panchkula, indicating dominance of vehicular emissions at Kolkata. However, stationary fossil fuel combustion sources seem to be more or less constant at all three sites as evident from the temporal variability pattern of SO_2 concentration (Fig. 6.5). Variation in CO concentration is indistinguishable at all three sites throughout the year except during wintertime, when CO concentration at Panchkula site is lower than the other two sites.

6.3.3 Comparison of air quality index

As aforementioned, the AQI is usually determined on the basis of ambient concentration of pollutants such as $PM_{2.5}$, PM_{10}, NO_2, SO_2, CO, O_3, NH_3 and Pb, for which national standards have been prescribed. In this study, the sub-index has been calculated for pollutants $PM_{2.5}$ or PM_{10}, SO_2 and CO owing to the availability of these parameters

at the study locations. Subsequently, the overall AQI was determined based on the worst sub-index. AQI presented here are based on the monthly average data of pollutants' concentrations.

6.3.3.1 AQI at Panchkula (upwind IGP)

The AQI obtained for the year 2016 for the Panchkula site (upwind IGP) showed in the satisfactory category (i.e. AQI ranged 51–100) from February to September (broadly during pre-monsoon and monsoon) while from October to December and January (post-monsoon and wintertime) exhibited moderately polluted episodes (AQI varied 51–100; Table 6.6). The various colours associated with the monthly AQI values in Table 6.6 represent the category to which they belong, i.e. from Good to Severe. For the AQI colour coding showing different categories of air quality, see Table 6.1.

6.3.3.2 AQI at Lucknow (central IGP)

The AQI observed for the year 2016 for the Lucknow site (central IGP) was quite different from Panchkula (upwind IGP) and showed significantly higher pollution in this region. The AQI values for the post-monsoon (November) and winter months were on the higher side and represented very poor category (i.e. 301–400) shown in red (Table 6.7). It is important to mention here that during post-monsoon and wintertime, besides the intensification in biomass burning practices, the meteorological conditions (stagnant air mass and shallower boundary layer height particularly in winters) and the geographical location (the Himalaya on one side and Deccan on the other side of the IGP) favours efficient trapping of air pollutants over the IGP. The NW-wind system, prevailing during these seasons, transports massive amounts of air pollutants to further downwind locations in the marine atmospheric boundary layer (MABL) over the Bay of Bengal (BoB), while

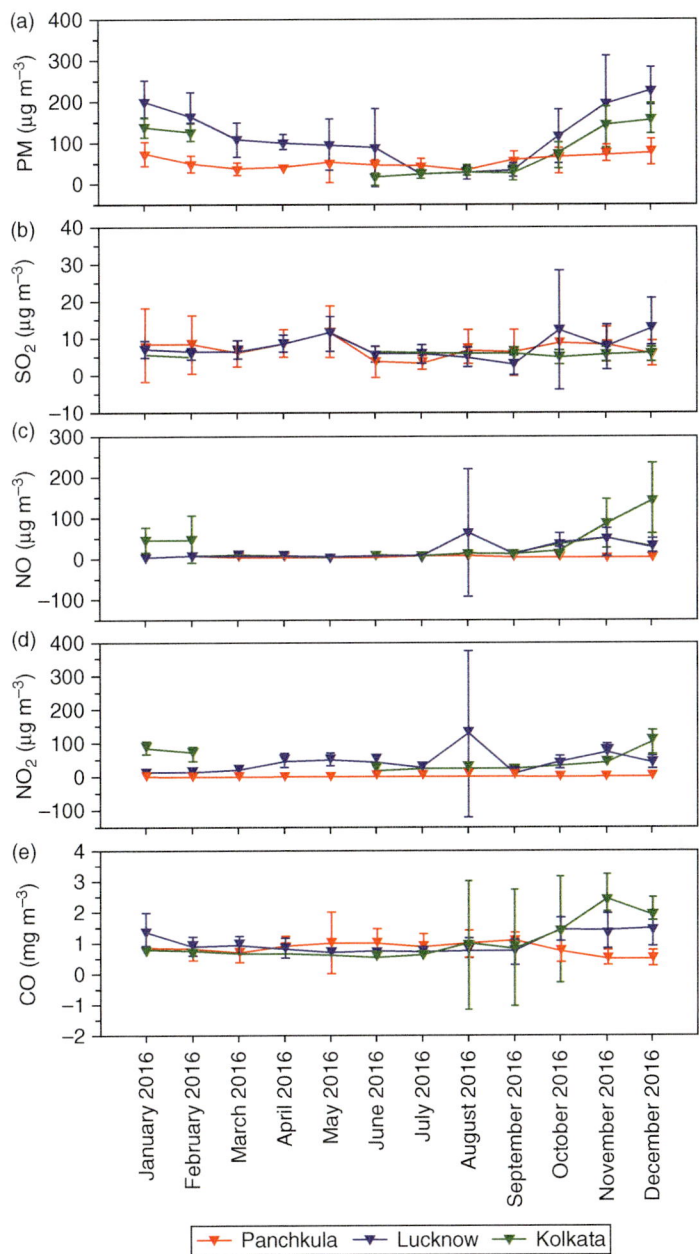

Fig. 6.5. Typical comparison of the variability of PM (PM$_{2.5}$ from Panchkula and Lucknow and PM$_{10}$ from Kolkata), SO$_2$, NO, NO$_2$ and CO in the IGP (2016).

for other periods of the year the AQI showed good to poor category of air quality (Table 6.7). During the monsoon months of July to September the AQI values were relatively low (0–100) representing less air pollution, which can be due to wet scavenging of pollutants under frequent and heavy rainfall.

6.3.3.3   AQI at Kolkata (downwind IGP)

The AQI calculated for the year 2016 for Kolkata (Victoria Memorial site, downwind IGP) was distinctly different from the other two sites discussed above. During the winter months, the obtained AQI values ranged from

Table 6.6. Monthly averaged AQI values (2016) determined from the concentrations of $PM_{2.5}$, SO_2 and CO at Panchkula (Haryana, upwind IGP).

Month	$PM_{2.5}$ ($\mu g\ m^{-3}$)	SO_2	CO ($mg\ m^{-3}$)	Monthly AQI
January	74	8.4	0.8	146
February	49	8.3	0.8	81
March	36	5.9	0.7	61
April	42	8.6	0.9	70
May	52	11.9	1.0	86
June	42	3.8	1.0	70
July	44	3.2	0.9	73
August	33	7.1	1.0	55
September	57	6.1	1.1	95
October	64	8.8	0.7	115
November	71	8.2	0.5	137
December	79	5.9	0.5	162

Table 6.7. Monthly averaged AQI values (2016) determined from the concentrations of $PM_{2.5}$, SO_2 and CO at Lucknow (Central School site; Uttar Pradesh, central IGP).

Month	$PM_{2.5}$ ($\mu g\ m^{-3}$)	SO_2	CO ($mg\ m^{-3}$)	Monthly AQI
January	199	1.3	7.1	361
February	164	6.5	6.5	334
March	108	0.9	6.8	260
April	102	0.8	8.6	239
May	95	0.7	11.2	217
June	89	0.7	5.7	197
July	24	0.7	5.8	41
August	25	0.8	5.1	42
September	35	0.7	3.0	58
October	116	1.4	12.2	286
November	195	1.4	7.8	358
December	226	1.5	12.6	382

Table 6.8. Monthly averaged AQI values (2016) determined from the concentrations of PM_{10}, SO_2 and CO at Kolkata (Victoria Memorial site; West Bengal, downwind IGP).

Month	PM_{10} ($\mu g\ m^{-3}$)	SO_2	CO ($mg\ m^{-3}$)	Monthly AQI
January	136	0.7	5.6	124
February	125	0.7	4.9	117
March	NA	NA	NA	NA
April	NA	NA	NA	NA
May	NA	NA	NA	NA
June	16	0.6	6.5	27
July	23	0.6	6.1	28
August	31	0.9	6.2	47
September	25	0.9	5.5	43
October	75	1.4	4.9	75
November	143	2.4	5.8	129
December	155	1.9	6.0	137

studies should focus on important phenomena pertaining to land- and sea-breeze wind patterns in the downwind IGP, to find some important atmospheric processes and surface ocean–lower atmospheric interactions (Duce and Liss, 2002).

6.3.3.4 Comparison of observed AQI in the IGP

In Fig. 6.6, monthly AQI from three study locations in the IGP has been compared. It can be seen from Fig. 6.6 that for most of the months the AQI values were found to be higher for the central IGP (Lucknow site). Furthermore, for Kolkata site the AQI values were lowest for most of the months. The obtained AQI values suggest that central IGP (Lucknow) is most polluted. It is important to mention here that the local pollution sources are very different in the selected locations: Panchkula is affected mainly by industrial and stationary (power plant) pollution sources whereas Lucknow is affected by industrial and vehicular emissions, brick kilns (Guttikunda *et al.*, 2014) and also by large-scale biomass burning emissions (local vis-à-vis transported from source region in Haryana and Punjab regions located in upwind IGP). In Kolkata (downwind IGP), the major sources contributing to the pollution level include vehicular emissions, thermal power plants, open waste burning and long-range transport (Das *et al.*, 2015). It has been suggested earlier that due to the sea- and land-breeze wind system, owing to proximity to the

101–200, representing a moderately polluted atmosphere (Table 6.8). AQI values could not be calculated for the pre-monsoon season (March, April and May) due to unavailability of the data from the CPCB online monitoring site. For the remaining months, the AQI values exhibit Good and Satisfactory categories, and thus, represent less air pollution during these months in Kolkata. It is important to mention here that a fairly good air quality in Kolkata (downwind IGP) as compared to that over the Lucknow site (central IGP) during post-monsoon and wintertime, when the NW-wind system is prevailing, suggests that future

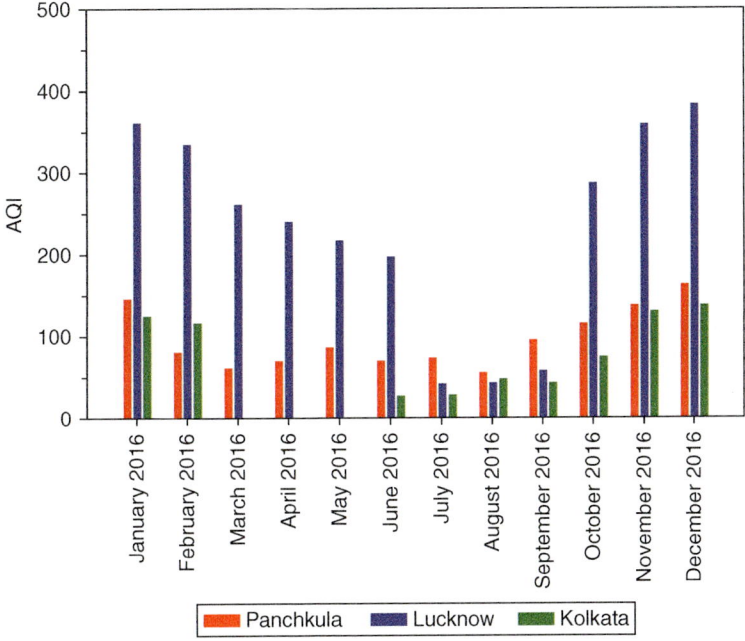

Fig. 6.6. Comparison of monthly averaged AQI values (2016) from different locations in the Indo-Gangetic Plain.

sea, the pollution level in Kolkata is usually lower than other nearby cities (Karar *et al.*, 2006).

6.4 Conclusions

Air pollution has been a cause of scientific concern due to its direct impact on the environment, ecosystem and human health, especially in urban regions. Therefore, a systematic long-term and near-continuous monitoring of air pollutants and air quality is essential. In order to understand the impact of pollutants on air quality, human health and the susceptibility of plants there are various indices, e.g. air-quality index (AQI), air-quality health index (AQHI), air-pollution index (API), air-pollution tolerance index (APTI), and anticipated performance index, etc. The APTI and anticipated performance index are basically used to infer the susceptibility of plant species. In this chapter, we have discussed in brief the important indices: AQI, AQHI, API and APTI. As a case study, we have discussed AQI over the Indo-Gangetic Plain with a set of data adopted from CPCB (Central Pollution Control Board) for three different geographical locations (Panchkula:

upwind IGP, Lucknow: central IGP and Kolkata: downwind IGP) in the IGP. A distinctly different seasonal variability pattern of pollutants' concentration and AQI has been observed over studied locations indicating an inferior air quality during post-monsoon and wintertime over the IGP. For determination of sub-index the pollution parameters CO, SO_2 and $PM_{2.5}$ (PM_{10} in the case of Kolkata) have been opted as per the guidelines and finally, from these data sets, monthly AQI has been estimated. Based on air-pollutant concentration and AQI, it appears that the level of pollution in central IGP (particularly at Lucknow among the studied sites) is relatively high and future studies are urged to focus on assessing the influence of local and long-range transport on the budget and chemistry of pollutants in the airshed of the IGP.

Acknowledgement

The authors express their thanks to the Central Pollution Control Board of India for free access and retrieval of air pollutants' concentration data reported in this study. PR is thankful to the

Council of Scientific & Industrial Research (India) for granting CSIR-SRA fellowship (Pool Scientist # 8934-A/2017). Authors thank the editors Dr. Pallavi Saxena and Dr. Vaishali Naik for providing a wonderful opportunity to contribute this chapter.

References

Badarinath, K.V.S., Chand, T.R.K. and Prasad, V.K. (2006) Agricultural crop residue burning in the Indo-Gangetic Plains – a study using IRS-P6 A WiFS satellite data. *Current Science* 91, 1085–1089.

Chakraborty, A. and Gupta, T. (2010) Chemical characterization and source apportionment of submicron (PM₁) aerosol in Kanpur region. *Aerosol and Air Quality Research* 10, 433–445.

Chakraborty, A., Ervens, B., Gupta, T. and Tripathi, S.N. (2016) Characterization of organic residues of size-resolved fog droplets and their atmospheric implications. *Journal of Geophysical Research: Atmospheres* 121, 4317–4332.

Chakraborty, A., Rajeev, P., Rajput, P. and Gupta, T. (2017) Water soluble organic aerosols in Indo Gangetic Plain (IGP): insights from aerosol mass spectrometry. *Science of the Total Environment* 599–600, 1573–1582.

Chen, L., Liu, C., Zhang, L., Zou, R. and Zhang, Z. (2017) Variation in tree species ability to capture and retain airborne fine particulate matter ($PM_{2.5}$). *Scientific Reports* 7, 3206.

CPCB (n.d.) Average report criteria. Central Pollution Control Board, Ministry of Environment & Forests, Government of India. Available at: http://www.cpcb.gov.in/CAAQM/frmUserAvgReportCriteria.aspx (accessed 23 July 2018).

Das, R., Khezri, B., Srivastava, B., Datta, S., Sikdar, P.K., Webster, R.D. and Wang, X. (2015) Trace element composition of $PM_{2.5}$ and PM_{10} from Kolkata – a heavily polluted Indian metropolis. *Atmospheric Pollution Research* 6, 742–750.

Duce, R.A. and Liss, P.S. (2002) The surface ocean–lower atmosphere study (SOLAS). *Atmospheric Environment* 36, 5119–5120.

Gupta, T. and Chauhan, A. (2014) *Comprehensive Air Sampling Device Application for Indian Patent.* 1474/DEL/2014 filed on 3 June 2014.

Gupta, T. and Mandaria, A. (2013) Sources of submicron aerosol during fog dominated wintertime at Kanpur. *Environmental Science and Pollution Research* 20, 5615–5627. DOI: 10.1007/s11356-013-1580-6.

Gupta, P.K., Sahai, S., Singh, N., Dixit, C.K., Singh, D.P. *et al.* (2004) Residue burning in rice-wheat cropping system: causes and implications. *Current Science* 87, 1713–1717.

Gupta, T., Chakraborty, A. and Ujinwal, K.K. (2010) Development and performance evaluation of an indigenously developed air sampler designed to collect submicron aerosol. *Annals of the Indian National Academy of Engineering (INAE)* 7, 189–193.

Guttikunda, S.K., Goel, R. and Pant, P. (2014) Nature of air pollution, emission sources, and management in the Indian cities. *Atmospheric Environment* 95, 501–510.

Jethva, H., Satheesh, S.K. and Srinivasan, J. (2005) Seasonal variability of aerosols over the Indo-Gangetic basin. *The Journal of Geophysical Research* 110, D21204.

Karar, K., Gupta, A.K., Kumar, A. and Biswas, A.K. (2006) Seasonal variations of PM_{10} and TSP in residential and industrial sites in an urban area of Kolkata, India. *Environmental Monitoring and Assessment* 118, 369–381.

Kaul, D.S., Gupta, T., Tripathi, S.N., Tare, V. and Collett, J.L. (2011) Secondary organic aerosol: a comparison between foggy and nonfoggy days. *Environmental Science & Technology* 45, 7307–7313.

Kumar, V., Goel, A. and Rajput, P. (2017) Compositional and surface characterization of HULIS by UV-Vis, FTIR, NMR and XPS: wintertime study in Northern India. *Atmospheric Environment* 164, 468–475.

Lawrence, M.G. and Lelieveld, J. (2010) Atmospheric pollutant outflow from southern Asia: a review. *Atmospheric Chemistry and Physics* 10, 11017–11096.

MoECC (2018) *Published Plans and Annual Reports 2017–2018.* Ministry of the Environment and Climate Change, Ontario, Canada. Available at: https://www.ontario.ca/page/published-plan-and-annual-report-2017-2018-ministry-environment-and-climate-change (accessed 30 July 2018).

Nigam, S., Rao, B.P.S., Kumar, N. and Mhaisalkar, V.A. (2015) Air quality index – a comparative study for assessing the status of air quality. *Research Journal of Engineering and Technology* 6, 267–274.

Ott, W.R. (1978) *Environmental Indices: Theory and Practice.* Ann Arbor Science, Ann Arbor, MI.

Rai, P., Chakraborty, A., Mandariya, A.K. and Gupta, T. (2016) Composition and source apportionment of PM1 at urban site Kanpur in India using PMF coupled with CBPF. *Atmospheric Research* 178, 506–520.

Rajeev, P., Rajput, P. and Gupta, T. (2016) Chemical characteristics of aerosol and rain water during an El Niño and PDO influenced Indian summer monsoon. *Atmospheric Environment* 145, 192–200.

Rajeev, P., Rajput, P., Singh, D.K., Singh, A.K. and Gupta, T. (2018) Risk assessment of submicron PM-bound hexavalent chromium during wintertime. *Human and Ecological Risk Assessment: An International Journal* 24(6), 1453–1463.

Rajput, P. (2013) Atmospheric polycyclic aromatic hydrocarbons: identification, abundances and spatio-temporal variability. PhD thesis, Physical Research Laboratory, Ahmedabad, India.

Rajput, P. and Gupta, T. (2016) A facile digestion protocol for metal analysis in ambient aerosols: Implications to mineral dust characteristics and human health impact. *Journal of Energy and Environmental Sustainability* 2, 24–29.

Rajput, P. and Sarin, M.M. (2014) Polar and non-polar organic aerosols from large-scale agricultural-waste burning emissions in Northern India: implications to organic mass-to-organic carbon ratio. *Chemosphere* 103, 74–79.

Rajput, P., Sarin, M.M. and Rengarajan, R. (2011a) High-precision GC-MS analysis of atmospheric polycyclic aromatic hydrocarbons (PAHs) and isomer ratios from biomass burning emissions. *Journal of Environmental Protection* 2, 445–453.

Rajput, P., Sarin, M.M., Rengarajan, R. and Singh, D. (2011b) Atmospheric polycyclic aromatic hydrocarbons (PAHs) from post-harvest biomass burning emissions in the Indo-Gangetic Plain: isomer ratios and temporal trends. *Atmospheric Environment* 45, 6732–6740.

Rajput, P., Sarin, M.M. and Kundu, S.S. (2013) Atmospheric particulate matter (PM$_{2.5}$), EC, OC, WSOC and PAHs from NE-Himalaya: abundances and chemical characteristics. *Atmospheric Pollution Research* 4, 214–221.

Rajput, P., Sarin, M.M., Sharma, D. and Singh, D. (2014a) Atmospheric polycyclic aromatic hydrocarbons and isomer ratios as tracers of biomass burning emissions in Northern India. *Environmental Science and Pollution Research* 21, 5724–5729.

Rajput, P., Sarin, M.M., Sharma, D. and Singh, D. (2014b) Characteristics and emission budget of carbonaceous species from post-harvest agricultural-waste burning in source region of the Indo-Gangetic Plain. *Tellus-B* 66(1), 21026. DOI: 10.3402/tellusb.v66.21026.

Rajput, P., Sarin, M.M., Sharma, D. and Singh, D. (2014c) Organic aerosols and inorganic species from post-harvest agricultural-waste burning emissions over northern India: impact on mass absorption efficiency of elemental carbon. *Environmental Science: Processes & Impacts* 16, 2371–2379.

Rajput, P., Mandaria, A., Kachawa, L., Singh, D.K., Singh, A.K. and Gupta, T. (2015) Wintertime source-apportionment of PM$_1$ from Kanpur in the Indo-Gangetic Plain. *Climate Change* 1, 503–507.

Rajput, P., Gupta, T. and Kumar, A. (2016a) The diurnal variability of sulfate and nitrate aerosols during wintertime in the Indo-Gangetic Plain: implications for heterogeneous phase chemistry. *RSC Advances* 6, 89879–89887.

Rajput, P., Mandaria, A., Kachawa, L., Singh, D.K., Singh, A.K. and Gupta, T. (2016b) Chemical characterization and source-apportionment of PM$_1$ during massive loading at an urban location in Indo-Gangetic Plain: impact of local sources and long-range transport. *Tellus-B* 68(1). DOI: 10.3402/tellusb.v68.30659.

Rajput, P., Sarin, M.M., Sharma, D. and Singh, D. (2016c) Characteristics and emission budget of carbonaceous species from post-harvest agricultural-waste burning in source region of the Indo-Gangetic Plain. In: Ragazzi, M. (ed.) *Air Quality: Monitoring, Measuring and Modeling Environmental Hazards*. Apple Academic Press Inc., Oakville, ON, Canada, pp. 243–266. DOI: 10.1201/9781315366074-13.

Rajput, P., Anjum, M.H. and Gupta, T. (2017) One year record of bioaerosols and particles concentration in Indo-Gangetic Plain: implications of biomass burning emissions to high-level of endotoxin exposure. *Environmental Pollution* 224, 98–106.

Rajput, P., Singh, D.K., Singh, A.K. and Gupta, T. (2018) Chemical composition and source-apportionment of sub-micron particles during wintertime over Northern India: new insights on influence of fog-processing. *Environmental Pollution* 233, 81–91.

Ram, K., Sarin, M.M. and Hegde, P. (2010a) Long-term record of aerosol optical properties and chemical composition from a high-altitude site (Manora Peak) in Central Himalaya. *Atmospheric Chemistry and Physics* 10, 11791–11803.

Ram, K., Sarin, M.M. and Tripathi, S.N. (2010b) A 1 year record of carbonaceous aerosols from an urban site in the Indo-Gangetic Plain: characterization, sources, and temporal variability. *Journal of Geophysical Research* 115, D24313.

Ramachandran, S. and Srivastava, R. (2013) Influences of external vs. core-shell mixing on aerosol optical properties at various relative humidities. *Environmental Science: Processes & Impacts* 15, 1070–1077.

Ramanathan, V., Crutzen, P.J., Lelieveld, J., Mitra, A.P., Althausen, D. *et al.* (2001) Indian ocean experiment: an integrated analysis of the climate forcing and effects of the great Indo-Asian haze. *The Journal of Geophysical Research* 106, 28371–28398.

Rastogi, N., Singh, A., Singh, D. and Sarin, M.M. (2014) Chemical characteristics of $PM_{2.5}$ at a source region of biomass burning emissions: evidence for secondary aerosol formation. *Environmental Pollution* 184, 563–569.

Seinfeld, J.H. and Pandis, S.N. (2006) *Atmospheric Chemistry and Physics – From Air Pollution to Climate Change*, 2nd edn. John Wiley & Sons, New York.

Singh, D.K. and Gupta, T. (2016a) Effect through inhalation on human health of PM_1 bound polycyclic aromatic hydrocarbons collected from foggy days in northern part of India. *Journal of Hazardous Materials* 306, 257–268.

Singh, D.K. and Gupta, T. (2016b) Role of transition metals with water soluble organic carbon in the formation of secondary organic aerosol and metallo-organics in PM_1 sampled during post monsoon and pre-winter time. *Journal of Aerosol Science* 94, 56–69.

Singh, A., Rajput, P., Sharma, D., Sarin, M.M. and Singh, D. (2014a) Black carbon and elemental carbon from postharvest agricultural-waste burning emissions in the Indo-Gangetic Plain. *Advances in Meteorology* 2014, 10.

Singh, D.K., Lakshay and Gupta, T. (2014b) Field performance evaluation during fog-dominated wintertime of a newly developed denuder-equipped PM1 sampler. *Environmental Science and Pollution Research* 21, 4551–4564.

Sorathia, F., Rajput, P. and Gupta, T. (2018) Dicarboxylic acids and levoglucosan in aerosols from Indo-Gangetic Plain: inferences from day night variability during wintertime. *Science of the Total Environment* 624, 451–460.

Srinivas, B. and Sarin, M.M. (2014) Brown carbon in atmospheric outflow from the Indo-Gangetic Plain: mass absorption efficiency and temporal variability. *Atmospheric Environment* 89, 835–843.

Suryawanshi, S., Chauhan, A.S., Verma, R. and Gupta, T. (2016) Identification and quantification of indoor air pollutant sources within a residential academic campus. *Science of the Total Environment* 569, 46–52.

Venkataraman, C., Habib, G., Eiguren-Fernandez, A., Miguel, A.H. and Friedlander, S.K. (2005) Residential biofuels in South Asia: carbonaceous aerosol emissions and climate impacts. *Science* 307, 1454–1456.

7 Impact of Air Pollution on the Environment and Economy

Saurabh Sonwani and Vandana Maurya*

Jawaharlal Nehru University, New Delhi, India

It is the predicament of mankind that man can perceive the problematique, yet, despite his considerable knowledge and skills, he does not understand the origins, significance, and interrelationships of its many components and thus is unable to devise effective responses. This failure occurs in large part because we continue to examine single items in the problematique without understanding that the whole is more than the sum of its parts, that change in one element means change in the others.

(Meadows *et al.*, 1972)

Abstract

Air pollution is a growing concern from social, economic and ecological dimensions of society. This chapter focuses on air pollution effects on the environment and on the economy. Several energy-utilizing anthropogenic activities emit large amounts of toxic gases and particulate matter into the environment. Through several atmospheric processes these pollutants create very critical environmental problems such as acid rain, eutrophication, and global climate change. Acid rain has a variety of environmental impacts due to the presence of nitrogen and sulfur in it. It leaches nutrients from soil and adversely affects trees, fish and wildlife, along with building materials. Different ranges of pH may affect a variety of species in the aquatic ecosystem. Haze is one of the significant factors and a result of air pollution, especially in megacities. Haze is responsible for visibility degradation in both developed and developing countries. Apart from visibility reduction, it also creates cloud formation by forming cloud condensation nuclei (CCN). Emissions from thermal power plants and vehicles also contribute to a large amount of nitrogen oxides entering the aquatic ecosystem, which ultimately cause eutrophication in lakes. The depletion of stratospheric ozone allows harmful solar radiation to reach the Earth's surface, which causes a variety of diseases in plants, animals and human beings. In contrast to stratospheric ozone, ground level ozone has harmful effects on living creatures due to its toxic nature. Ground level ozone can have adverse impacts on human health even at very low levels in the atmosphere. Global climate change is also linked to the increasing level of air pollutants, especially greenhouse gases (GHGs). Due to the increasing anthropogenic activities necessary to fulfil today's energy requirements, a large amount of GHGs are emitted into the atmosphere and are causing a several fold rise in global annual temperature. This chapter tries to understand how air pollution is affecting global economic growth and the environment. Pollution may cause direct pressure on economies through increased numbers of deaths due to respiratory disorders or cardiovascular diseases, installation of pollution-control technologies to reduce deaths, management of degraded ecosystems and carving out conservation strategies for pollution threatened species. It can also affect economic

---

* Corresponding author: maurya.vandana09@gmail.com

growth indirectly by increasing morbidities, reducing labour working days and productivity. The World Health Organization reported that 12.6 million deaths per year were found to be linked with environmental pollution (WHO, 2016b). Out of these, 11.6% of deaths are directly linked to air pollution, from indoor and outdoor sources. Another report of the World Health Organization (WHO, 2016c) estimated the expected additional deaths at approximately 250,000 per year from 2030 to 2050 due to malnutrition, malaria, diarrhoea and heat stress. The direct damage cost due to health issues is estimated to be US$ 2–4 billion/year by 2030. Therefore it is crucial to understand that air pollution not only affects ecological systems but also economic systems. Thus, this chapter shows the need for more research in the area of air pollution and its impact on the economy and environment for successful policy making.

7.1   Introduction

The relationship between environmental degradation and economic growth is a hugely debated topic in academic and policy-making circles. 'With the present growth trends of world pollution, resource depletion, industrialization, and population, limits to growth of this planet will be reached in the next hundred years' (Meadows et al., 1972). Air quality is getting worse, especially in Asian countries, as population, traffic, industrialization and energy use increase rapidly, and this is placing a huge burden on economies in terms of deteriorating health, degraded ecosystems, biodiversity loss, and increased mortalities and morbidities. Air pollution may be defined as an 'atmospheric condition in which substances are present at concentrations higher than their normal ambient levels to produce significant effects on humans, animals, vegetation or materials' (Seinfeld, 1986). These substances include gases such as oxides of sulfur and nitrogen, carbon monoxides, hydrocarbons; particulate matter such as smoke, dust, fumes and radioactive materials. Urbanization is a significant factor in the increase in air pollution in many cities in Asia, Africa, the Near East and Latin America (Ashmore, 2005). In 1960, less than 22% of the developing world's population was urban, and the increment reached 34% by 1990. As per extrapolations, population is expected to increase by 50% in urban areas by 2020 (Satterthwaite, 2009). The uncontrolled use of fossil fuels in industries and transport sectors has also become the dominant source of gaseous pollutants such as SO_2, NO_x, VOCs, and also particulate matter. The sources of air pollution can be natural as well as anthropogenic: natural sources are forest fires, emissions from trees, lightning, volcanic eruptions, erosion of surfaces of rocks/minerals/buildings; while anthropogenic sources mainly comprise biomass burning, transportation, vehicular emissions, industrial activities and mining (Chandrappa and Kulshrestha, 2016). Air pollution may affect living (plants, animals and human beings) as well as non-living systems (materials and buildings). In the past, air pollution was considered to be a local problem with a large number of point sources, but due to the application of tall industrial chimneys and long-range transport of pollutants, it has become a regional problem. Remote areas also reported high concentrations of pollutants due to the trans-boundary nature of pollutants. Due to rapid industrialization, population growth and increasing numbers of motor vehicles, there is an adverse effect on the environment, especially in the Asian region. Apart from the traditional pollutants, black carbon (BC), another kind of pollutant found and associated with particulate matter, has the capacity to absorb solar radiation and can also decrease reflection of sunlight on snow/ice (Waliser et al., 2011; Hadley and Kirchstetter, 2012). Air pollution represents the biggest environmental risk to the health of living organisms and poses a huge cost to economies. The costs of air pollution not only includes mortalities but also morbidities and the shortening of life expectancy. Other costs include the impact of air pollution on the built environment, animals and plant health, agriculture, biodiversity and ecosystems. The focus of this chapter is to discuss conventional air pollution and its adverse impacts on the environment and economy.

7.2   Sources of Air Pollution

The sources of air pollution are numerous and can be divided into two categories: (i) natural; and (ii) anthropogenic.

7.2.1 Natural sources

Natural sources include: volcanic eruptions, pollens, forest fires and windblown dust. These release poisonous gases such as sulfur dioxide (SO_2), hydrogen sulfide (H_2S) and carbon monoxide (CO).

7.2.2 Anthropogenic sources

Automobile exhaust, industry and biomass burning are the main anthropogenic sources.

1. Automobiles and thermal power plants: vehicles are the major contributor to total air pollution (i.e. ~60%) (Cadle *et al.*, 1997, 2004). Automobiles, aircraft and locomotives are the main sources of the major air pollutants, which include carbon monoxide, unburned hydrocarbons and nitrogen oxide.

2. Industries: paper and pulp factories, petroleum refineries, fertilizer and steel industries are the major anthropogenic sources of air pollution. They release several toxic gases like carbon monoxide (CO), oxides of sulfur (SO_2 and SO_3), nitric oxide (NO) and hydrocarbons (HC) into the atmosphere. Textile, pesticide and insecticide industries cause serious problems to human health and the environment. The offensive odour released by food processing industries and tanneries also poses a serious threat to the environment. Gases released by various accidents are also responsible for serious threats, e.g. the Bhopal Gas Tragedy, which killed thousands of people due to the leakage of methyl isocyanate (MIC) in Bhopal, India.

3. Burning of fossil fuels: burning wood, fossil fuels and charcoal causes air pollution by releasing carbon dioxide (CO_2), carbon and sulfur dioxide into the atmosphere.

4. Agricultural activities: pesticides and insecticides used in the agriculture sector cause air pollution. When these are inhaled by animals and humans, this can create severe problems.

5. Radioactive fallout: testing of nuclear weapons adds to nuclear pollution. Nuclear pollution is very harmful for flora and fauna.

7.3 Types of Air Pollutants

Air pollutants are mainly classified by category, as listed in Table 7.1. They may be primary or secondary on the basis of their origin; organic or inorganic on the basis of their chemical composition. They can also be classified on the basis of their matter, whether it is natural, particulate or gaseous.

Global emissions of various gases are projected to increase by 2060 (Fig. 7.1). The emissions of nitrogen oxides (NO_x) and ammonia (NH_3) are projected to increase due to a rise in demand for agricultural products and energy. The pollutants such as black carbon (BC), carbon monoxide (CO) and volatile organic compounds (VOCs) are also projected to increase (OECD, 2016).

7.4 Environmental Effects

Industrialization and urbanization are the major causes for the degradation of the environment. Air pollution is a major cause for concern, especially in developing countries. For several decades, authors have explained the relationship between air pollution and ill health effects due to air pollution exposure (Prescott *et al.*, 1998; Pope *et al.*, 2002; Heudorf *et al.*, 2009; Atkinson *et al.*, 2010; Chuang *et al.*, 2011). The most commonly discussed health-related problems were respiratory (asthma and changes in lung function), cardiovascular diseases, and pregnancy outcomes and even deaths. Apart from human health, air pollution is responsible for a variety of environmental effects, such as visibility degradation/haze, eutrophication, acid rain, ozone depletion and global climate change.

7.4.1 Acid rain

Acid rain (acidic deposition) is one of the leading problems of regional air pollution. It is the result of the transformation of atmospheric sulfur and nitrogen emitted from different sources across different locations throughout the world. It is also known for its damaging effects and role in trans-boundary air pollution. Acid rain is a result of the emission of SO_2 (fossil fuel combustion and metal smelter) and NO_x (released from vehicular, industrial and power plant sources) forming sulfuric and nitric acid in precipitation. The acidic

Table 7.1. Classification of pollutants on the basis of their origin, composition and state of matter.

On the basis of origin		On the basis of chemical composition		On the basis of state of matter		
Primary pollutants	Secondary pollutants	Organic pollutants	Inorganic pollutants	Natural contaminants	Particulate matter	Gases and vapours
Directly emitted from identifiable sources into the atmosphere, e.g. particulate matter, SO_2, NO, NO_2, CO and radioactive compounds	Not directly emitted into the atmosphere but formed during atmospheric transformation reactions of different primary pollutants, e.g. ozone, formaldehyde, photochemical smog, peroxyacetyl nitrate (PAN)	This type of pollutant includes carbon and hydrogen, e.g. hydrocarbon, alcohols, aldehyde, ketones and organic sulfur compounds	Inorganic pollutants mostly comprise oxides of carbons (CO and CO_2), carbonates, oxides of sulfur (SO_2 and SO_3), oxides of nitrogen (NO and NO_2), ozone (O_3)	The contaminants produced from natural sources are considered as natural contaminants, e.g. pollen grains are emitted from weeds, grasses and trees	These may be liquid or solid. The particulate matters are identified as aggregates which are larger than 0.002 μm but smaller than 500 μm	Carbon monoxide (CO), oxides of sulfur such as SO, SO_2, SO_3 and SO_4, oxides of nitrogen such as N_2O, NO, NO_2, NO_3, N_2O, and hydrocarbons including both aliphatic hydrocarbons and aromatic hydrocarbons

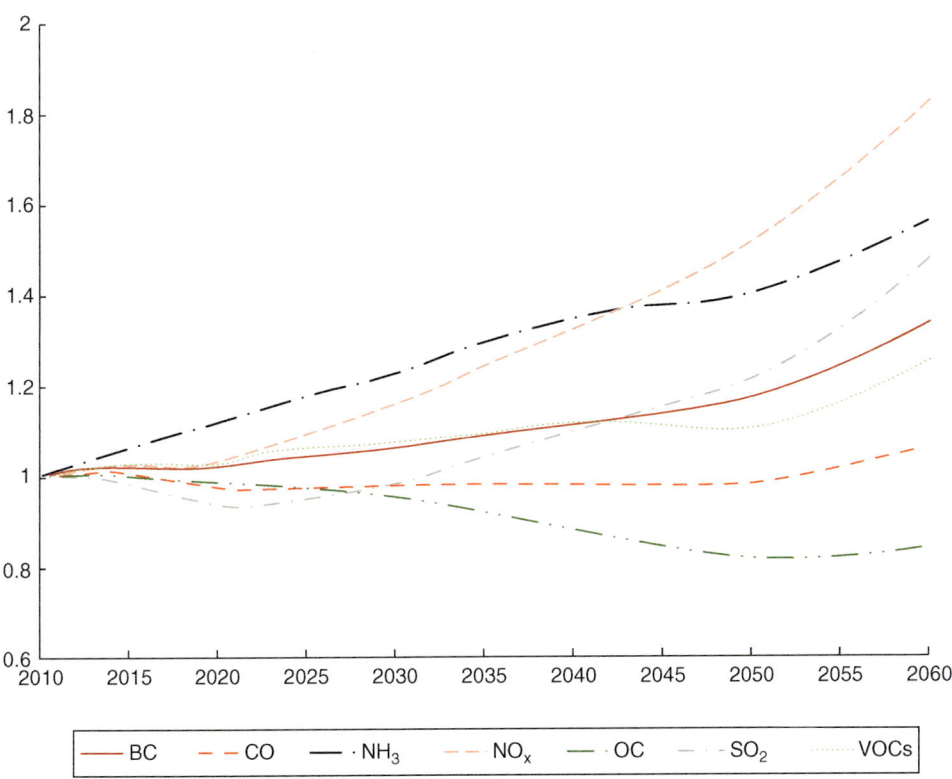

Fig. 7.1. Emission projections of various pollutants (index with respect to 2010). (From OECD, 2016.)

nature of the acid rain (pH range: 4.2 to 4.7) is due to the presence of nitric acids and sulfuric acids. The atmospheric transformation reactions convert emissions of SO_x and NO_x into sulfuric acid and nitric acids (Fig. 7.2).

Dry deposition (gas and particulates) and wet deposition (rain, fog, drizzle or snow) are two important processes by which atmospheric acids are carried to Earth. Atmospheric conditions (stable/unstable) also affect the distribution and transportation of acids into the atmosphere (Taniyasu *et al.*, 2013). When the precursors of the acid deposition are dissolved in water during rain they form various acids. It can be explained by the following reactions:

$$CO_2 + H_2O = H_2CO_3 \text{ (carbonic acid)}$$

$$SO_2 + H_2O = H_2SO_3 \text{ (sulfurous acid)}$$

$$NO_2 + H_2O = HNO_2 \text{ (nitrous acids)}$$
$$+ HNO_3 \text{ (nitric acids)}$$

7.4.1.1   Effects of acid rain on ecosystem

Table 7.2 shows the critical levels of different key organisms at the point at which they may lose their lives, due to increasing acid in their environment. An ecosystem is the interaction of different communities of plants or animals with their environment; a disturbance in any part of the ecosystem can harm the function of other life forms and have a significant impact on everything else.

7.4.1.2   Plants and trees

Acid rain damages the plant foliage and leaves and makes them vulnerable to several bacterial and viral infections. It also makes plants and trees prone to the harmful UV radiation coming from the sun, which ultimately affects their metabolism. Acid rain also removes the essential nutrients and minerals from the soil that plants require to grow, and releases nutrients from

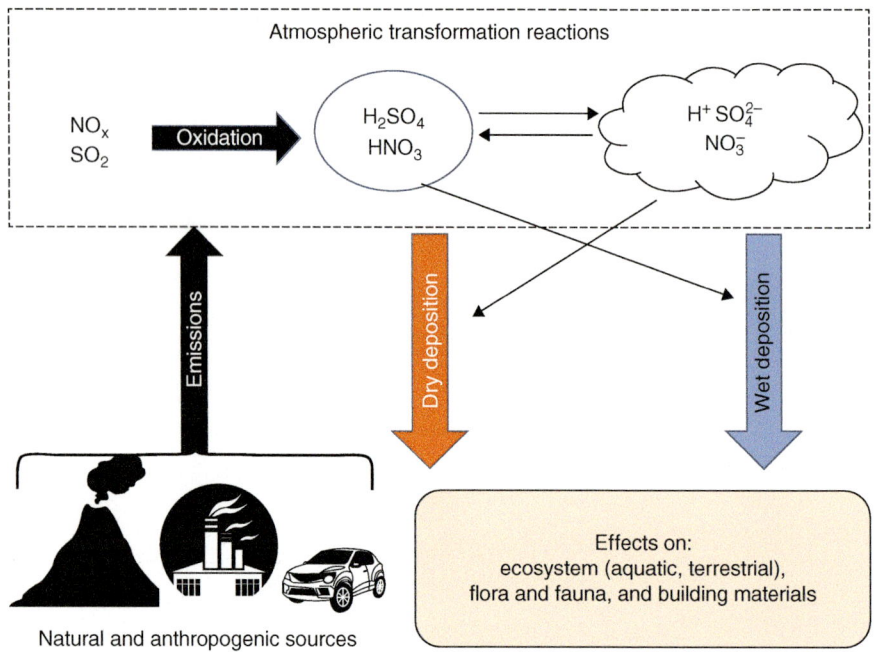

Fig. 7.2. Mechanism involved in acid rain.

plants/trees found at high altitude, which makes them less efficient at absorbing sunlight and ultimately weakens their ability to withstand freezing temperatures.

7.4.1.3   Fish and wildlife

Acid rain significantly alters the pH of the aquatic environment, such as ponds, rivers, lakes and marshy lands, which ultimately harms aquatic life forms. Acid rain causes soil and water bodies to acidify and makes the water inappropriate for life forms. As the acid rain passes through the soil, it leaches the aluminium from the soil/clay particles and dumps it into different water bodies.

The more acid that passes through, the more aluminium is released into lakes, ponds and streams. Some of the plants and animals tolerate the acidic and aluminium-rich water ecosystem, but many of the sensitive life forms lose their life due to the lowering in pH of water bodies. Generally, young and elderly animals are very sensitive to these changes. At pH ≤5, most fish eggs cannot hatch. Most of the non-chordates barely survive at pH 5, while mayflies are more sensitive and may not survive below pH 5.5

(USEPA, n.d.). Some of the areas where the acid rain falls may not be affected by the acid deposition due to the special properties of the soil where rain water passes through. This type of soil has some buffering capacity to neutralize the acidity of the rainwater. This property of soil depends on the thickness, composition of the soil and the type of bedrock underneath it.

7.4.1.4   Episodic acidification

Snow melt and heavy rain can deposit high amounts of acids in to lakes (with much less acidity on normal days), in the absence of soil with buffering capacity. In this short time, high acidity can kill many species present in that lake. Melting snow and heavy rain downpours have resulted in high acidity in lakes of the USA and Canada, due to this episodic acidification.

7.4.1.5   Effects of acid rain on materials

Any form of acid deposition causes the deterioration of building materials (paint and stone), and buildings of historic and cultural importance, such as monuments, statues, sculptures and tombstones. Bronze, limestone, carbon-steel, marble,

Table. 7.2. Critical pH levels of aquatic animals (USEPA, n.d.).

Animal		Critical pH level
Snail		6.0
Clam		6.0
Bass		5.5
Crayfish		5.5
Mayfly		5.5
Trout		5.0
Salamander		5.0
Perch		4.5

paint and some plastics are the most vulnerable to acid deposition. The materials (foundations and pipes) immersed in the acidified water also suffer from corrosion. Calcium carbonate in certain stones dissolves in dilute sulfuric acid to form calcium sulfate:

$$CaCO_3 + H_2SO_4 + H_2O = CaSO_4.2H_2O + CO_2$$

This has two effects. First, it removes the details by breaking down the stones; second, there is a build-up of black skin of gypsum (calcium sulfate) in the sink areas in the buildings. Once the crystal of gypsum forms on the stone, the process may persist for up to 50 years, known as the memory effect. Several studies found links with the impacts of acid deposition on diverse materials. The Taj Trapezium Case (also known as MC Mehta Taj Trapezium Case) is a good example of acid rain effects on building material. The yellowing of the Taj marble was the major issue in the Taj Trapezium Case. A petition was filed against the threat to the deteriorating beauty of the Taj Mahal, to invoke the Air Act 1981, Water Act 1974 and Environmental Protection Act 1986. The purpose behind this petition was to relocate 292 factories to prevent the Taj from emissions released by companies using coke or coal as energy sources. In this case, four National Environmental Engineering Research Institute (NEERI) reports, two Varadharajan reports and several reports by the State Pollution Control Board were presented, which related the pollution emission from the Agra-Mathura region and its impact on the Taj. They statistically explained that by replacing coal with diesel in the railway yards and closing down two thermal power stations, sulfur dioxide emissions could be reduced by 50%. On 11 April 1994, the court, after hearing learned counsel for the parties, passed the order indicating that as a first phase, the industries situated in Agra be relocated out of the Taj Trapezium Zone (TTZ). The decision taken by the court after considering the evidence were the Sustainable Development Principle, the Precautionary Principle and the Polluter Pays Principle. The final judgement in this case was given on 30 December 1996, by Justice Kuldip Singh and Justice Faizan Uddin.

7.4.1.6 Human health

Acid rain does not directly create problems when humans are exposed to it. But the compounds causing acid rain can significantly harm humans with exposure. There are various exposure pathways (dermal, ingestion and inhalation) through which pollutants harm human health. Inhalation of contaminated air (with high levels of particulate matter, metals, sulfate and nitrate) may cause decreased lung function, including difficulty in breathing for people suffering from asthma, cardiovascular problems, otitis media, etc.

7.4.2 Eutrophication

Eutrophication is a condition where a water body has high levels of nutrients, such as nitrogen and phosphorus (Figs 7.3(a) and 7.3(b)). These excessive amounts of nutrient cause algal bloom, which is ultimately responsible for the loss of animal and plant diversity. It is also characterized by excessive plant and algal growth due to the increasing availability of one or more limiting factors (fertilizers, sunlight and carbon dioxide) of photosynthesis (Schindler, 2006). It is a natural ageing process of any lake due to the deposition of sediment into it over centuries (Carpenter, 1981). The N, P and K originating on agricultural land, and fertilizers or animal waste, are the principal nutrients reaching the surface water and involved in eutrophication. Urban and industrial runoff also contribute to eutrophication. Anthropogenic factors also affect eutrophication severely by increasing the rate at which nutrients are added to water bodies. Emissions from thermal power plants and vehicles also allow a large amount of nitrogen oxides to enter aquatic ecosystems.

Basic steps involved in eutrophication are that lakes and streams receiving more fertilizers become more productive. The rich nutrient input triggers the growth of algae to increase its size; this is the condition known as 'population explosion' or 'bloom'.

1. Due to the algal bloom conditions, the penetration of light into the water is diminished, which ultimately decreases the productivity of plants in deeper waters.

2. The water becomes depleted in oxygen. Low oxygen results in more algae dying and also affects the lowering of primary production in the deeper waters.

3. The low levels of oxygen result in the death of large fish which require high amounts of dissolved oxygen (DO), such as trout, salmon and other desirable sport fish.

Essentially, the entire aquatic ecosystem changes with eutrophication.

7.4.3 Haze

Industrialization and urbanization activities around the world have led to an increase in air pollution and a haze problem in developing and developed countries (Fig.7.4(a–f)). When sunlight encounters minute suspended particles in the atmosphere, it reduces visibility, known as haze or smog. Power plants, vehicular traffic, industrial facilities and construction activities play significant roles in the formation of haze by emitting various pollutants, especially fine particulate matters (Watson, 2002). High levels of pollutants trigger more of the haze, due to their absorptive and scattering effects. Haze reduces the clarity and colour of the objects that we see. Some of the pollutants, like sulfate particles, may scatter more light during humid conditions (Li-Jones and Prospero, 1998). Haze mainly originates from cities or crowded areas, and disperses in

Fig. 7.3. (a) Eutrophic condition of lake; (b) discharge of waste water into a reservoir.

Fig. 7.4. Haze problems in different countries around the world from (a) to (f): United Kingdom; United States of America; Singapore; India; China; and Pakistan. Photos are authors' own.

rural and urban areas through wind. Smog can change the weather conditions due to the presence of definite dark particles containing carbon, which play a significant role in altering Earth's radiation budget due to the scattering and absorbing nature of carbon particles. Haze can decrease the quantity of solar energy reaching the Earth's surface by up to 30% (Chameides *et al.*, 1999). Apart from the visibility degradation, air particles are involved in forming cloud condensation nuclei (CCN), which ultimately affect the rainfall pattern. Finer particles (apart from soot particles) were found to be involved in the formation of CCN. It was also found that oxides of sulfur and nitrogen were involved in haze formation. The scavenging processes (dry and wet deposition) were found to be involved in the cleaning of atmospheric pollutants (Gu *et al.*, 2010; Arakaki *et al.*, 2013). Therefore, atmospheric visibility is directly linked to the pollution level in the atmosphere.

7.4.4 Ground level and stratospheric ozone

The gas ozone is ubiquitous throughout the atmosphere with lower levels in the troposphere as compared to the stratosphere. Ground level ozone acts as a pollutant that can harm human health, even at very low concentrations. The elderly, children and people with respiratory diseases are more prone to being affected after exposure. Ozone exposure can cause both short-term and long-term health effects. Wheezing, coughing, painful breathing and irritation and inflammation of the respiratory tract, resulting in decreased lung function, can be caused by short-term ozone exposure (Filippidou and Koukouliata, 2011). Lippmann (1992) noticed the neutrophils in the bronchoalveolar region increase twofold after short-term ozone exposure. Increasing hospital admissions due to ozone exposure were also noticed by the World Health Organization in 2003. In contrast to ground-level ozone, stratospheric ozone is known as good ozone, which acts as a shield and protects the Earth from harmful solar radiation (UV rays). But, this protective covering is gradually weakening day by day, due to the presence of harmful chemicals in the atmosphere, which are the products of anthropogenic activities. Chlorofluorocarbons, hydrochlorofluorocarbons, and halons are important ozone depleting chemical substances. These substances are used for refrigeration, in fire extinguishers, pesticides and as solvents. Thinning of the ozone layer allows harmful UV radiation to reach the Earth's surface, which ultimately causes various types of diseases such as dermal inflammation, skin cancer and cataracts. UV radiation also reduces the yield of sensitive crops, such as soybeans. Apart from crop yields, it also damages forests by affecting tree seedlings, and increases plant susceptibility to diseases, pests and harsh weather.

7.4.5 Global climate change

Since the start of the industrial revolution, the pollution load in the atmosphere has been increasing at a very rapid rate. Large consumption of fossil fuels, and changes in transportation, the agricultural pattern and standards of living are the main reasons for the increase in pollution levels in the atmosphere. At the start of the 21st century, the USA, Canada and the developing nations focused on regional and local air pollutants. The introduction of regulatory measures and policy along with cleaner technologies has resulted in an improvement of air quality (World Bank, 1997). Greenhouse gases (GHGs) emitted from different sources disturb the balance of the naturally present gases in the atmosphere. They also disturb the Earth's radiation budget through various physicochemical transformation processes. Global warming significantly impacts human health, ecology, agriculture, water resources, forests, wildlife and coastal areas. Fig. 7.5 shows anthropogenic global GHG emissions by different groups of gases from 1970 to 2010.

According to the IPCC (2014), GHG emissions are increasing at a very high rate. In the period 2000–2010 they rose with an average growth rate of 2.2% per year, which was very high when compared to the period 1970–2000, where the emission growth rate was only 1.3% per year. The high use of coal has increased the carbon energy content over the last few decades since 1970. The expected global temperature rise is 4–5°C annually as compared to pre-industrial times. The expected rise of CO_2 was 750–1300 ppm of CO_2 equivalent till the end of this century. To explore the most cost-effective way to keep the temperature rise below 2°C, various emission reduction technologies, along with policy implementation, have been taken into consideration to ultimately avoid the unmanageable risks of climate change (IPCC, 2014).

7.5 Air Pollution and Economic Growth

Air pollution is detrimental not only for the environment and its life forms but also to economic growth. Pollution-induced health problems arise from negative market externalities and affect macro-economic performance of an economy. Moreover, the health impacts of pollution are surrounded by uncertainty at the level of individuals and for the aggregate economy (Bretschger and Vinogradoval, 2017). The total market costs due to air pollution include both direct and indirect costs. Direct costs include the

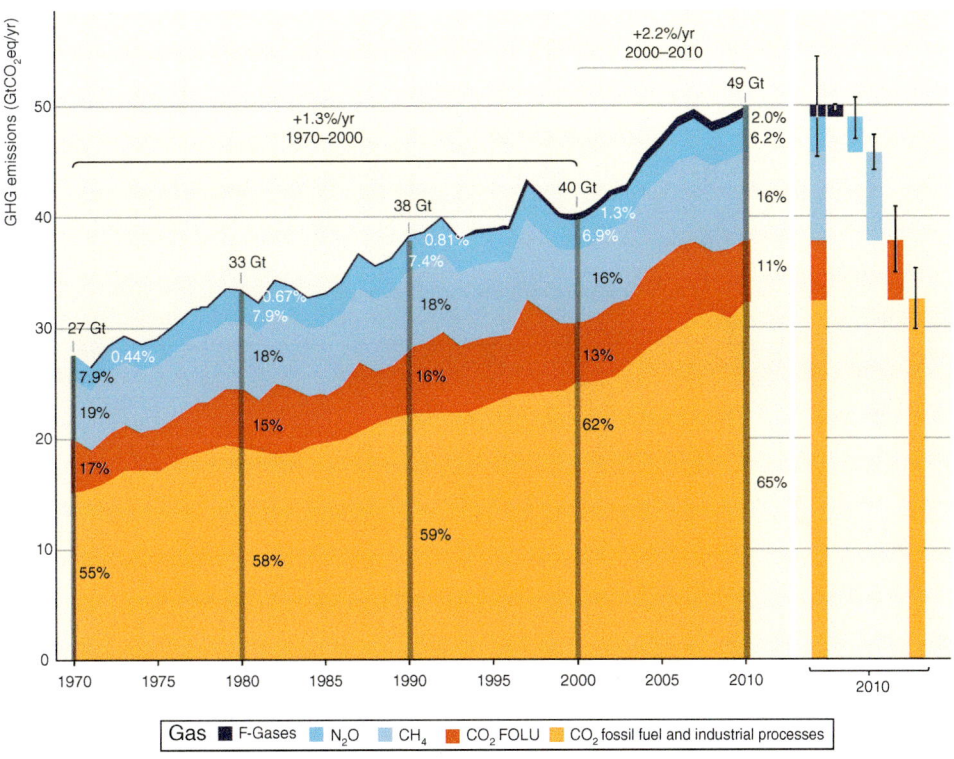

Fig. 7.5. Total annual anthropogenic emissions of greenhouse gases (GHGs) by groups of gases 1970–2010. (From IPCC, 2014.)

changes in value generated in different sectors from changes in labour productivity, increased health expenditures and changes in value generated in agriculture from changes in crop yields. Indirect economic effects include relocation of factors of production, and changes in international trade and savings. The market impacts of air pollution are projected to lead to global economic costs that increase to 1% of global GDP by 2060, and the annual number of lost working days is projected to reach US$3.7 billion at the global level (OECD, 2016).

7.5.1 Debates surrounding the environment and economic growth

The environmental degradation caused by air pollution is affecting world economic growth. The Environment Kuznets Curve (EKC) is one of the most preferred approaches to analyse the relationship between environmental degradation

and economic growth (Álvarez-Herránz, 2017; Özdemir and Özokcu, 2017; Stern and Dijk, 2017). EKC shows an inverted U-shaped curve characterized by an income level (level of GDP per capita) and after this level, environmental degradation decreases with economic growth (Fig. 7.6). It posits that there will be a huge burden on natural resources in the early stages of development and later that this burden will be reduced as the economies grow richer. This relationship is dependent on scale, composition and technology effects.

Grossman and Krueger (1991) applied EKC in a study of the potential influence of the North American Free Trade Agreement (NAFTA) in the US. The model includes SO_2, dark matter and suspended particulate matter (SPM). The main findings support the existence of EKC for both SO_2 and dark matter while there is a negative relationship between GDP and SPM.

They found that environmental degradation can be reduced by economic growth and thus economic growth is not a threat to the environment

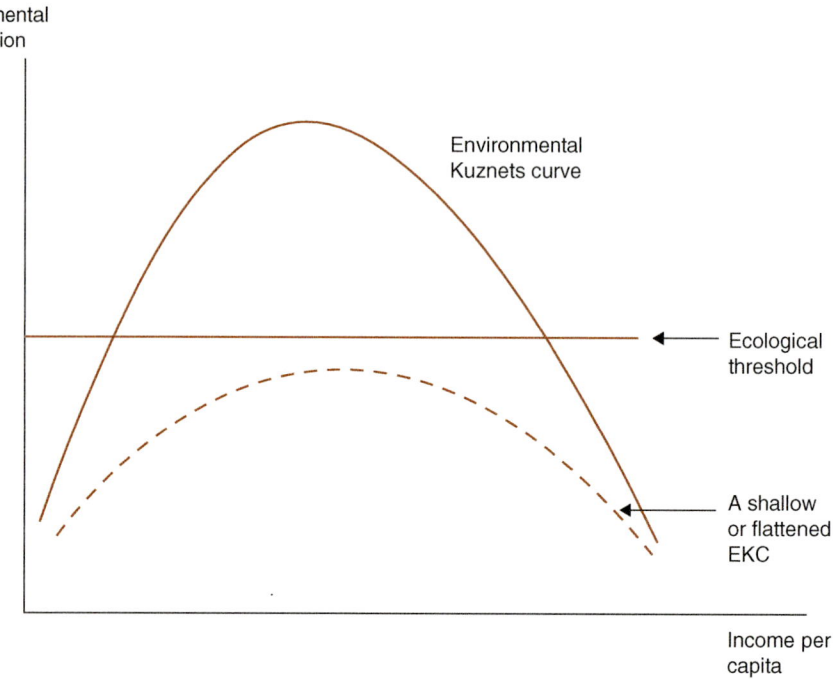

Fig. 7.6. Environmental Kuznets curve. (Adapted from Özdemir and Özokcu, 2017.)

but a remedy for environmental deterioration. Holtz-Eakin and Selden (1995) analysed the relationship between income and carbon dioxide. They found evidence supporting an inverted-U-shaped relationship between per capita carbon emissions and per capita income. And further they discovered that most countries would continue to operate on an upward sloping portion of the curve through 2100 leading to a steady increase in global emissions.

Álvarez-Herránz et al. (2017) used panel data to explore the impact of improvements in Energy Research and Development (ERD) on GHG emissions, using the EKC hypothesis for 28 OECD nations during 1990–2014. They found that energy innovation contributed to the reduction of energy intensity and environmental pollution. Their results indicated that energy innovation measures need time to reach their full effect. Meanwhile, Stern and Dijk (2017) examined the role of income, convergence, time-related factors and spatial effects to explain changes in national level population-weighted $PM_{2.5}$ particulate pollution in various countries between 1990 and 2010, using a model that integrates EKC and convergence approaches. They found

that economic growth has positive but modest effects on the growth in $PM_{2.5}$ concentrations. Convergence effects are small and not statistically significant.

Özdemir and Özokcu (2017) explored the relationship between economic growth and climate change by analysing the relationship of CO_2 emissions and income. They found an inverted N-shaped curve relationship between environment degradation and income in select countries. An N-shaped curve is an indication of insufficiency of environmentally friendly improvements, and asks for urgent policies to mitigate or adapt to climate change. Therefore, the relationship of income and air pollution varies from one nation to another and cannot be generalized globally (Table 7.3).

EKC is one of the most dominant models used by various economists and policy makers to model ambient pollution concentrations and aggregate emissions. But, various studies identified a number of problems with some of the main EKC estimators and their interpretations. Stern et al. (1996) critically examined the concept of EKC and identified econometric problems with the EKC estimators.

Table 7.3. Variations of environmental Kuznets curve (EKC) results. (From Özdemir and Özokcu, 2017.)

Authors and publication year	Environmental indicators	Economic indicators and other variables	Regions and periods	Econometric technique	Results
Shafik and Bandyopa-dhyay (1992)	Deforestation; per capita (cap) CO_2 emissions; water (DO and faecal coliforms)	GDP per cap, a time trend	149 countries, 1961–1986	Panel data, log, quadratic, cubic, fixed effect (FE) model	CO_2; monotonic-ally increasing
Holtz-Eakin and Selden (1992)	Per cap CO_2 emissions	GDP per cap	130 countries, 1951–1986	Panel data, log, quadratic, cubic; FE model	Quadratic inverted-U shape, cubic N-normal
Moomaw and Unruh (1997)	Per cap CO_2 emissions	GDP per cap	16 industrial OECD countries, 1950–1992	Panel data, quadratic and cubic, FE and pooled ordinary least squares (OLS)	Inverted U-shape for quadratic, N-shape for cubic
De Bruyn (1997)	Per cap CO_2 emissions, nitrous oxide (NO_x), sulfur dioxide (SO_2)	GDP per cap, structural changes, technology, population density	Netherlands, West Germany, UK and USA, 1960–1993	Time series, log, decomposition analysis	Monotonically increasing
Galeotti and Lanza (1999)	Per cap CO_2 emissions	GDP per cap	110 countries, 1971–1996	Log, gamma and Weibull functions	Inverted U-shape (all countries, non-OECD and OECD)
Ravallion et al. (2000)	Per cap CO_2	Average per cap GDP, population, time trend, GINI coefficient (income inequality)	42 countries, 1975–1992	Panel data, level and log, quadratic, cubic; FE and pooled OLS	Pooled OLS is better; cubic: insignificant, quadratic: monotonically decrease as income inequality grows
Borghesi (2000)	CO_2 per cap	GDP per cap in PPP, population density, GINI coefficient	126 countries, 1988–1995	Panel data, log and level, linear, quadratic, cubic; FE	Monotonically increase, CO_2 emissions decrease slightly as inequality grows
Dijkgraaf and Vollebergh (2001)	CO_2 per cap	GDP per cap, energy consumption per cap	24 OECD countries, 1960–1997	Panel data, time series, log, cubic; FE and seemingly unrelated regressions (SUR)	N-shape for panel data; N-shape for five countries in time series data

Continued

Table 7.3. Continued.

Authors and publication year	Environmental indicators	Economic indicators and other variables	Regions and periods	Econometric technique	Results
Cole and Neumayer (2004)	CO_2 per cap, 9 more air pollutants and water pollutants	GDP per cap, share of manufacturing in GNP, share of pollution intensive exports and imports in total exports and imports, trade intensity	21 OECD countries, 1980–1987	Panel data, log, cubic (quadratic for some of the equations); generalized least squares (GLS) with random effect (RE) and FE models	Inverted U-shape for CO_2, inverted U-shape and inverted N-shape for other pollutants
Dinda and Coondoo (2006)	Per cap CO_2 emissions	GDP per cap and a time trend	88 countries, 1960–1990	Panel data, log; cointegration test, error correction Test	Bi-directional relationship
Akbostancı et al. (2009)	SO_2, SPM, CO_2	GDP per cap, population density	Turkey, 1968–2003, for time series; 1992–2001, for panel data (provinces)	Time series; Johansen cointegration, panel data; GLS, level and log; cubic	N-shape for SO_2 and SPM; monotonically increasing for CO_2
Dutt (2009)	Per cap CO_2 emissions	GDP per capita, governance, political institutions, socioeconomic conditions, population density, education	124 countries, 1960–2002	Panel data, quadratic; robust OLS, fixed effect model	Linear, 1960–1980; inverted U-shape, 1984–2002
Narayan and Narayan (2010)	Per cap CO_2 emissions	Real GDP	43 developing countries, 1980–2004	Panel data; panel cointegration	Inverted U-shape in Middle Eastern and South Asia panels
Jayanthaku-maran et al. (2012)	Per cap CO_2 emissions	GDP per cap, energy consumption per cap, trade intensity, manufacturing value added	China and India	Time series, log, cointegration and ARDL methodology	Structural breaks are detected

Continued

Table 7.3. Continued.

Authors and publication year	Environmental indicators	Economic indicators and other variables	Regions and periods	Econometric technique	Results
Jobert et al. (2014)	Per cap CO_2 emissions	Real per cap GDP and per cap energy consumption	55 countries, 1970–2008	Bayesian shrinkage estimators	Inverted U-shape is observed in some countries but not all of them
Franklin and Ruth (2012)	Per cap CO_2 emissions	GDP per cap, Gini coefficient, ratio of exports to imports, inflation adjusted energy prices	US, 1800–2000	Time series, level, cubic; OLS, Prais-Winsten AR(1) regression model	Inverted U-shape
Zhang and Zhao (2014)	Per cap CO_2 emissions	GDP per cap, energy intensity, income inequality, urbanization, the share of industry sector in GDP	28 Chinese provinces, 1995–2010	Panel data, log-level, cubic; fixed effect model	N shape
Yang et al. (2015)	Per cap CO_2 emissions, total CO_2, industrial (Ind) dust, Ind gas, Ind smoke, Ind SO_2, Ind waste water	Real GDP, the percentage of exports, imports, domestic trade in GDP, the ratio of entry of FDI over GDP, the population density	29 Chinese provinces, 1995–2010	Panel data; level and log; fixed and random effects models, general sensitivity test	Positive linear. inverted-U and N form
Bölük and Mert (2014)	Per cap CO_2 emissions	GDP per cap in constant USD at 2005 prices, electricity production from renewable sources per cap	Turkey, 1961–2010	Time series; ARDL	Inverted U-shape
Heidari et al. (2015)	Per cap CO_2 emissions	Real GDP per cap in constant USD at 2000 prices, energy consumption per cap	5 Asian countries, 1980–2008	Panel data; panel smooth threshold regression (PSTR) model	Inverted U-shape

Continued

Table 7.3. Continued.

Authors and publication year	Environmental indicators	Economic indicators and other variables	Regions and periods	Econometric technique	Results
Chen et al. (2016)	CO_2 emissions	Real GDP, energy consumption	188 countries, 1993–2010	Panel data; panel cointegration, vector error-correction model	Inverted U-shape
Özdemir and Özokcu (2017)	Per cap CO_2 emissions	Per capita income	26 OECD, 1980–2010; 52 emerging nations	Panel data analysis	Inverted N-shaped curve
Gill et al. (2017)	CO_2 emissions	GDP	Malaysia, 1970–2011	Autoregressive distributed lag (ARDL)	Insignificant relationship between GHG and GDP
Ahmad et al. (2017)	CO_2 emissions	Income	Croatia, 1992–2011	Autoregressive distributed lag (ARDL)	Inverted U-shaped
Atasoy (2017)	CO_2 emissions, population growth rate and energy consumption per capita	GDP per capita	50 US states, 1960–2010	Panel data estimators, augmented mean group (AMG) and common correlated effects mean group estimator (CCEMG)	AMG strongly validates EKC for 30 states and CCEMG shows weak EKC relationship in 10 states

7.5.2 Air pollution and its impact on economic growth

As discussed earlier in the chapter, air pollution affects the economy and environment alike. Its impact on the environment is visible but its impact on the economy is not. It leads to a plethora of burdens on society in the form of health issues, reduced productivity, reduced labour and reduced ecological services, which can pose a huge cost to any nation. In the following sections, the impact of air pollution on various sectors, which affects economic growth, is discussed.

7.5.2.1 Health

WHO (2014) describes air pollution as a major risk factor in several diseases, leading to disabilities and deaths, including cancers, lower respiratory infections, cardiovascular and cerebrovascular diseases. Particulate air pollution causes illness and leads to death from cardiovascular and respiratory diseases by provoking alveolar inflammation causing exacerbation of lung disease and increased blood coagulation (Seaton et al., 1995). Air pollution claims several million premature deaths and also imposes annual costs of trillions of dollars. Global health care costs are projected to increase from US$21 billion (using constant 2010 USD and PPP exchange rates) in 2015 to US$176 billion by 2060 (OECD, 2016).

In 2012, global ambient (outdoor) air pollution caused around 3 million deaths. The western Pacific region and South East Asian Region (SEAR) had the highest burden of 1.1 million and 799,000 deaths per year, respectively.

However, in Europe, America, Western Pacific and the eastern Mediterranean, a reduced number of deaths is observed (Fig. 7.7).

In 2012 household air pollution caused around 4.3 million deaths and these were mainly concentrated in low- and middle-income countries. The highest burden was the South East Asian Region (SEAR) and Western Pacific region, with 1.69 million and 1.62 million deaths, respectively (Fig. 7.8). In these regions,

women have a higher risk of developing health issues as they are more involved in household chores as compared to men. High-income nations experienced the lowest burden with 19,000 deaths for that year (WHO, 2014).

A study by Cohen *et al.* (2017) represented spatial and temporal trends in mortality and burden of disease due to ambient air pollution at country, regional and global levels from 1990 to 2015. They found that exposure to $PM_{2.5}$ caused

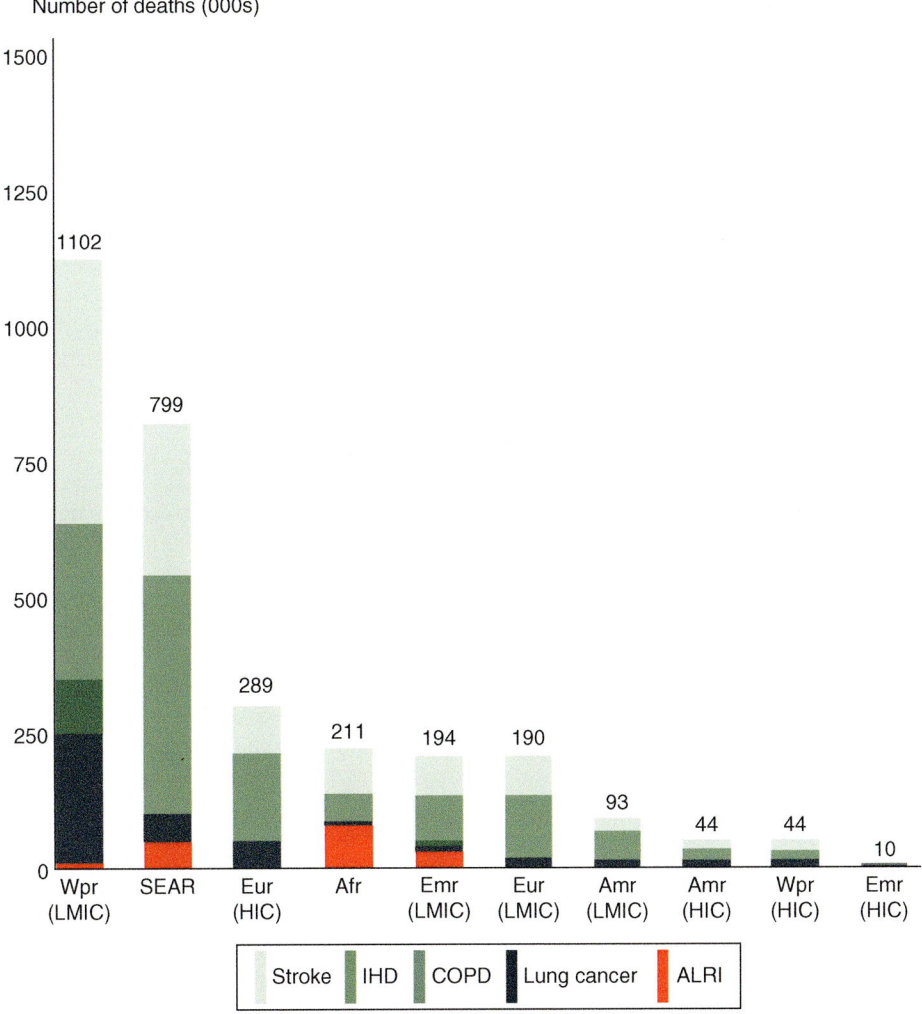

Fig. 7.7. Deaths attributable to ambient air pollution in 2012, by disease and by region (IHD: ischaemic heart disease; COPD: chronic obstructive pulmonary disease; ALRI: acute lower respiratory infections; Wpr: Western Pacific region; LMIC: low- and middle-income countries; SEAR: South East Asian Region; Eur: European Union; HIC: high-income countries; Afr: Africa; Emr: Eastern Mediterranean; Amr: America). (From WHO, 2016a.)

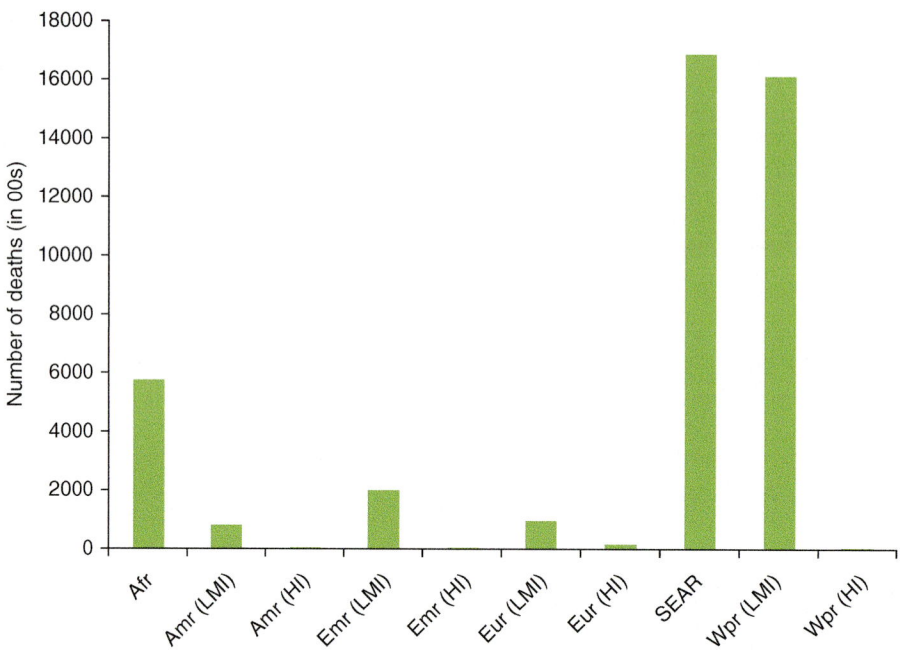

Fig. 7.8. Regional distributions of total deaths due to household air pollution in 2012 (Afr: Africa; Amr: America; LMI: low- and middle-income; HI: high-income; Emr: Eastern Mediterranean; Eur: European Union; SEAR: South East Asian Region; Wpr: Western Pacific region). (From WHO, 2014.)

4.2 million deaths (representing 7.6% of total global mortality) and 103.1 million (representing 4.2% of global burden) disability-adjusted life years (DALYs) in 2015. Globally, $PM_{2.5}$ exposure was responsible for 17.1% of mortality from ischaemic heart disease, 14.2% from stroke, 16.5% from lung cancer, 24.7% from lower respiratory infections (LRIs), and 27.1% from chronic obstructive pulmonary disease (COPD) in 2015. Of all deaths attributable to ambient $PM_{2.5}$ in 2015, deaths from ischaemic heart disease and stroke accounted for 57% (State of Global Air Report, 2017).

Ozone exposure resulted in an additional 254,000 deaths in the same year. The rise in mortality rates and DALYs is observed due to rising levels of pollution and non-communicable diseases in the largest low-income and middle-income countries in East and South Asia, which are experiencing growth in population and ageing. It is also further estimated that the number of premature deaths due to outdoor air pollution will increase to 6–9 million annually by 2060, especially in highly populated regions (OECD, 2016).

7.5.2.2 Ecosystems

Air pollution affects various ecosystems by altering the physical components of the ecosystems. Sulfur oxides, nitrogen oxides and ozone affect their growth and functions. Oxides of sulfur and nitrogen react together to form 'acid rain', which precipitates and increases the acidity of soil. Increased acidity affects the ability of ecosystems to provide various services, i.e. nutrient cycling, productivity and climate regulation. Tropospheric ozone damages cell membranes of the plants, inhibiting important processes required for their growth and development. When these pollutants enter into water bodies, it causes the accumulation of nutrients resulting in eutrophication, which causes loss of life in aquatic ecosystems (UNECE n.d.). Persistent toxic air pollutants are of particular concern in aquatic ecosystems. These toxic compounds may deposit into the aquatic systems and may bio-magnify in the aquatic organisms at the top of the trophic levels. Most of the ecosystems are exposed to multiple air pollutants simultaneously and various species differ in sensitivity to air pollution and its biogeochemical consequences, and this differential sensitivity implies that air pollution will shift the species' composition or will lead to outright loss of sensitive species of various ecosystems (Lovett *et al.*, 2009).

7.5.2.3 Biodiversity

Biodiversity is a vital aspect of our Earth's system and it has been indicated that environmental pollution is the main cause of biodiversity loss on Earth. Air pollution negatively affects biodiversity and can draw species towards the verge of extinction. There are several paths of exposure from which air pollutants can affect wildlife adversely: through direct exposure (inhalation, dermal contact or ingestion) or indirectly (through wet/dry deposition processes soil/water surface) after their exposure to wildlife. Animals can suffer health-related problems after their exposure, with sufficient levels of air pollutants. Air pollutants are responsible for birth defects, reproductive failure, and disease in animals. A study examined the relationship between biodiversity and economic growth using an indicator of species diversity and per capita income (Grossman and Krueger, 1995; Asafu-Adjaye, 2003), which indicated that economic growth has adverse impact on biodiversity. The countries with high agricultural output are found to have comparatively higher biodiversity loss and this can be due to faulty agricultural practices (Asafu-Adjaye, 2003) or other human interventions in ecological systems.

7.6 Conclusion

Rapid urbanization and industrialization increases pollution loads in the ambient atmosphere. Local, regional and transported pollutants are all responsible for the pollution level. The high level of air pollutants may cause environmental problems such as acid rain, eutrophication, haze, ozone depletion and global climate change. Apart from environmental problems, the health of wildlife and human life has also been severely affected by increasing air pollution, which causes various health-related problems, especially respiratory diseases including asthma, bronchitis and diseases related to the heart. GHG emissions from all over the world have severely affected climate change for the last few decades. The US and Canada, along with some developing countries, are taking preventive measures to reduce the global emission budget of GHGs.

Air pollutants affect every economy differently as the sources of each pollutant and demographic factors may vary from place to place. Therefore, no-size-fits-all should be the motto while policy making. Innovation policies can improve the environmental quality. The policies should be addressed to reduce the economic cost of air pollution for the long term.

As also pointed out in Meadows *et al.* (1972), it is possible to stabilize ecological and economic conditions, and global equilibrium can be designed to fulfil the 'needs' of each person, so that each has equal opportunity to realize his/her individual human potential. Moreover, it is also crucial to realize the carrying capacity of our Earth system. It is dependent on preferences, technology, structure of production and consumption and an ever-changing state of interactions between the physical and biotic environment. Thus, it is important to protect the capacity of ecological systems to sustain welfare of all socioeconomic groups (Arrow *et al.*, 1995).

References

Ahmad, N., Du, L., Lu, J., Wang, J., Li, H.-Z. and Hashmi, M.Z. (2016) Modelling the CO_2 emissions and economic growth in Croatia: is there any environmental Kuznets curve? *Energy* 123, 164–172.

Akbostancı, E., Türüt-Aşık, S. and Tunç, G.I. (2009) The relationship between income and environment in Turkey: is there an environmental Kuznets curve? *Energy Policy* 37(3), 861–867.

Álvarez-Herránz, A., Balsalobre, D., Cantos, J.M. and Shahbaz, M. (2017) Energy innovation–GHG emissions nexus: fresh empirical evidence from OECD countries. *Energy* 101, 90–100.

Arakaki, T., Anastasio, C., Kuroki, Y., Nakajima, H., Okada, K. *et al.* (2013) A general scavenging rate constant for reaction of hydroxyl radical with organic carbon in atmospheric waters. *Environmental Science & Technology* 47(15), 8196–8203.

Arrow, K., Bolin, B., Costanza, R., Dasgupta, P., Folke, C. *et al.* (1995) Economic growth, carrying capacity and environment. *Ecological Economics* 15, 91–95.

Asafu-Adjaye, J. (2003) Biodiversity loss and economic growth: a cross country analysis. *Contemporary Economic Policy* 21(2), 173–185.

Ashmore, M.R. (2005) Assessing the future global impacts of ozone on vegetation. *Plant, Cell & Environment* 28, 949–964.

Atasoy, B. (2017) Testing the environmental Kuznets curve hypothesis across the US: evidence from panel mean group estimators. *Renewable and Sustainable Energy Reviews* 77, 731–747.

Atkinson, R.W., Fuller, G.W., Anderson, H.R., Harrison, R.M. and Armstrong, B. (2010) Urban ambient particle metrics and health: a time series analysis. *Epidemiology* 21, 501–511.

Bölük, G. and Mert, M. (2014) Fossil and renewable energy consumption, GHGs (greenhouse gases) and economic growth: evidence from a panel of EU countries. *Energy* 74, 439–446.

Borghesi, S. (2000) The environmental Kuznets curve: a survey of the literature. FEEM Working Paper No. 85–99. Available at SSRN: https://ssrn.com/abstract=200556 (accessed 21 November 2018).

Bretschger, L. and Vinogradoval, A. (2017) Human development at risk: economic growth with pollution-induced health shocks. *Environment and Resource Economics* 66, 481–495.

Cadle, S.H., Gorse Jr, R.J., Belian, T.C. and Lawson, D.R. (1997) Real-world vehicle emissions: a summary of the sixth coordinating research council on-road vehicle emission workshop. *Journal of the Air and Waste Management Association* 47(3), 426–438.

Cadle, S.H., Croes, B.E., Minassian, F., Natarajan, M., Tierney, E.J. and Lawson, D.R. (2004) Real-world vehicle emissions: a summary of the Thirteenth Coordinating Research Council on-road Vehicle Emissions Workshop. *Journal of the Air and Waste Management Association* 54(1), 18–23.

Carpenter, S.R. (1981) Submersed vegetation: an internal factor in lake ecosystem succession. *The American Naturalist* 118 (3), 372–383.

Chameides, W.L., Yu, H., Liu, S.C., Bergin, M., Zhou, X. *et al.* (1999) Case study of the effects of atmospheric aerosols and regional haze on agriculture: an opportunity to enhance crop yields in China through emission controls? *Proceedings of the National Academy of Sciences USA* 96(24), 13626–13633.

Chandrappa, R. and Kulshrestha, U.C. (2016) Sustainable industrial air pollution management. In: *Sustainable Air Pollution Management.* Springer International Publishing, Cham, Switzerland, pp. 207–290.

Cole, M.A. and Neumayer, E. (2005a) Environmental policy and the environmental Kuznets curve: can developing countries escape the detrimental consequences of economic growth? *Handbook of Global Environmental Politics*, 298.

Chuang, K.-J., Yan, Y.-H., Chiu, S.-Y. and Cheng, T.J. (2011) Long-term air pollution exposure and risk factors for cardiovascular diseases among the elderly in Taiwan. *Occupational and Environmental Medicine* 68, 64–68.

Cohen, A.J., Brauer, M., Burnett, R., Vos, T., Murray, C.J.L. and Forouzanfar, M.H. (2017) Estimates and 25 year trends of global burden of disease attributable to ambient air pollution: An analysis of data from the Global burden of diseases study 2015. *The Lancet* 389, 1907–1918.

Cole, M.A. and Neumayer, E. (2005) Environmental policy and the environmental Kuznets curve: can developing countries escape the detrimental consequences of economic growth? *Handbook of Global Environmental Politics* 298.

De Bruyn, S.M. (1997) Explaining the environmental Kuznets curve: structural change and internationalagreements in reducing sulphur emissions. *Environment and Development Economics* 2(4), 485–503.

Dijkgraaf, E. and Vollebergh, H.R. (2001) A note on testing for environmental Kuznets curves with panel data. FEEM Working Paper No. 63.2001. SSRN. Available at: https://ssrn.com/abstract=286692 (accessed 21 November 2018).

Dinda, S. and Coondoo, D. (2006) Income and emission: a panel data-based co-integration analysis. *Ecological Economics* 57(2), 167–181.

Dutt, K. (2009) Governance, institutions and the environment-income relationship: a cross-country study. *Environment, Development and Sustainability* 11(4), 705–723.

Filippidou, E. C. and Koukouliata, A. (2011) Ozone effects on the respiratory system. *Progress in Health Sciences* 1, 144–155.

Franklin, R.S. and Ruth, M. (2012) Growing up and cleaning up: the environmental Kuznets curve redux. *Applied Geography* 32(1), 29–39.

Galeotti, M. and Lanza, A. (1999) Desperately seeking environmental Kuznets. *Environmental Modelling and Software* 20(11), 1379–1388.

Gill, A., Viswanathan, K.K. and Hassan, S. (2017) The environmental Kuznets curve (EKC) and the environmental problem of the day. *Renewable and Sustainable Energy Reviews* 81, 1636–1642.

Grossman, G.M. and Krueger, A.B. (1991) *Environmental Impacts of a North American Free Trade Agreement*. NBER Working Paper No. 3914. National Bureau of Economic Research, Cambridge, MA.

Grossman, G.M. and Krueger, A.B. (1995) Economic growth and the environment. *The Quarterly Journal of Economics* 110(2), 353–377.

Gu, J., Bai, Z., Liu, A., Wu, L., Xie, Y. *et al.* (2010) Characterization of atmospheric organic carbon and element carbon of $PM_{2.5}$ and PM_{10} at Tianjin, China. *Aerosol Air Quality Research* 10, 167–176.

Hadley, O.L. and Kirchstetter, T.W. (2012) Black-carbon reduction of snow albedo. *Nature Climate Change* 2(6), 437.

Heidari, H., Katircioğlu, S.T. and Saeidpour, L. (2015) Economic growth, CO_2 emissions and energy consumption in the five ASEAN countries. *International Journal of Electrical Power and Energy Systems* 64, 785–791.

Heudorf, U., Neitzert, V. and Spark, J. (2009) Particulate matter and carbon dioxide in classrooms – the impact of cleaning and ventilation. *International Journal of Hygiene and Environmental Health* 212, 45–55.

Holtz-Eakin, D. and Selden, T.M. (1992) Stoking the fires? CO_2 emissions and economic growth. *Journal of Public Economics* 57(1), 85–101.

IPCC (2014) Climate change 2014: synthesis report. In: Pachauri, R. and Meyer, L. (eds) *Contribution of Working Groups I, II and III to the Fifth Assessment Report of the Intergovernmental Panel on Climate Change*. IPCC, Geneva, Switzerland.

Jayanthakumaran, K., Verma, R. and Liu, Y. (2012) CO_2 emissions, energy consumption, trade and income: a comparative analysis of China and India. *Energy Policy* 42, 450–460.

Jobert, T., Karanfil, F. and Tykhonenko, A. (2014) Estimating country-specific environmental Kuznets curves from panel data: a Bayesian shrinkage approach. *Applied Economics* 46(13), 1449–1464.

Li-Jones, X. and Prospero, J.M. (1998) Variations in the size distribution of non-sea-salt sulfate aerosol in the marine boundary layer at Barbados: impact of African dust. *Journal of Geophysical Research: Atmospheres* 103(D13), 16073–16084.

Lippmann, M. (1992) A multi-year study of air pollution and respiratory hospital admissions in three New York State metropolitan areas: results for 1988 and 1989 summers. *Journal of Exposure Analysis and Environmental Epidemiology* 2(4), 429–450.

Lovett, G.M., Tear, T.H., Evers, D.C., Cosby, J.B., Findlay, S.E.G. *et al.* (2009) Effect of air pollution on ecosystems in eastern United States. *Annals of New York Academy of Sciences* 1162, 99–129.

Meadows, D.H., Meadows, G., Randers, J. and Behrens III, W.W. (1972) *The Limits to Growth*. Universe Books, New York.

Moomaw, W.R. and Unruh, G.C. (1997) Are environmental Kuznets curves misleading us? The case of CO_2 emissions. *Environment and Development Economics* 2(04), 451–463.

Narayan, P.K. and Narayan, S. (2010) Carbon dioxide emissions and economic growth: panel data evidence from developing countries. *Energy Policy* 38(1), 661–666.

OECD (2016) The economic consequences of outdoor air pollution: policy highlights. OECD. Available at: https://www.oecd.org/environment/indicators-modelling-outlooks/Policy-Highlights-Economic-consequences-of-outdoor-air-pollution-web.pdf (accessed 20 August 2017).

Özdemir, O. and Özokcu, S. (2017) Economic growth, energy and environmental Kuznets curve. *Renewable and Sustainable Energy Reviews* 72, 639-647. DOI: 10.1016/j.rser.2017.01.059.

Pope, C., Burnett, R., Thun, M., Calle, E., Krewski, D. *et al.* (2002) Lung cancer, cardiopulmonary mortality, and long-term exposure to fine particulate air pollution. *JAMA* 287, 1132–1141.

Prescott, G.J., Cohen, G.R., Elton, R.A., Fowkes, F.G. and Agius, R.M. (1998) Urban air pollution and cardiopulmonary ill health: a 14.5 year time series study. *Occupational and Environmental Medicine* 55, 697–704.

Ravallion, M., Heil, M. and Jalan, J. (2000) Carbon emissions and income inequality. *Oxford Economic Papers*, 651–669.

Satterthwaite, D. (2009) *Shaping Urban Environment: Cities Matter*. The World Bank Urban Strategy. The World Bank, Washington, DC.

Schindler, D.W. (2006) Recent advances in the understanding and management of eutrophication. *Limnology and Oceanography* 51(1) (part 2), 356–363.

Seaton, A., Godden, D., MacNee, W. and Donaldson, K. (1995) Particulate air pollution and acute health effects. *The Lancet* 345, 176–178.

Seinfeld, J.H. (1986) *Atmospheric Chemistry and Physics of Air Pollution*. Wiley-Interscience, New York.

Shafik, N. and Bandyopadhyay, S. (1992) Economic growth and environmental quality: time series and cross section evidence. Policy Research Working Paper No. WPS904. The World Bank, Washington, DC, USA.

State of Global Air Report (2017) A special report on global exposure to air pollution and its disease burden, institute for health metrics and evaluation and the Health Effects Institute. Available at: https://www.stateofglobalair.org/sites/default/files/SOGA2017_report.pdf (accessed 17 August 2017).

Stern, D.I. and Dijk, J.V. (2017) Economic growth and global particulate pollution concentrations. *Climate Change* 142, 391–406.

Stern, D.I., Common, M.S. and Barbier, E.B. (1996) Economic growth and environmental degradation: the environmental Kuznets curve and sustainable development. *World Development* 24, 1151–1160.

Taniyasu, S., Yamashita, N., Moon, H.B., Kwok, K.Y., Lam, P.K. *et al.* (2013) Does wet precipitation represent local and regional atmospheric transportation by perfluorinated alkyl substances? *Environment International* 55, 25–32.

UNECE (n.d.) Air pollution, ecosystems and biodiversity. Available at: http://www.unece.org/environmental-policy/conventions/envlrtapwelcome/cross-sectoral-linkages/air-pollution-ecosystems-and-biodiversity.html (accessed 5 September 2017).

USEPA (n.d.) Effects of acid rain – surface waters and aquatic animals. United States Environmental Protection Agency. Available at: https://landuse.alberta.ca/Forms%20and%20Applications/RFR_ACFN%20Reply%20to%20Crown%20Submission%205%20-%20TabD9%20AcidRain_2014-08_PUBLIC.pdf (accessed 20 August 2018).

Waliser, D.E., Li, J.L., L'Ecuyer, T.S. and Chen, W.T. (2011) The impact of precipitating ice and snow on the radiation balance in global climate models. *Geophysical Research Letters* 38(6), 1–6.

Watson, J.G. (2002) Visibility: science and regulation. *Journal of the Air & Waste Management Association* 52(6), 628e713. DOI: 10.1080/10473289.2002.10470813.

WHO (2014) Burden of disease from household air pollution for 2012. Available at: http://www.who.int/airpollution/data/HAP_BoD_results_March2014.pdf (accessed on 20 August 2018).

WHO (2016a) *Ambient Air Pollution: A Global Assessment of Exposure and Burden of Disease*. World Health Organization, Geneva, Switzerland.

WHO (2016b) Public health, environmental and social determinants of health, Issue 88, November. Available at: http://www.who.int/phe/news/eNews-88-nov.pdf?ua=1 (accessed 8 September 2017).

WHO (2016c) Factsheet: climate change and health. Available at: http://www.who.int/mediacentre/factsheets/fs266/en/ (accessed 13 August 2017).

World Bank (1997) *Clear Water, Blue Skies*. World Bank, Washington, DC.

Yang, H., He, J. and Chen, S. (2015) The fragility of the environmental Kuznets curve: revisiting the hypothesis with Chinese data via an 'Extreme Bound Analysis'. *Ecological Economics* 109, 41–58.

Zhang, C. and Zhao, W. (2014) Panel estimation for income inequality and CO_2 emissions: a regional analysis in China. *Applied Energy* 136, 382–392.

8 Effects of Air Pollution on Human Health

Priyanka Kulshreshtha*
University of Delhi, India

Abstract

The drastic rise in urbanization and the growing economy has resulted in an increase in sources of air pollution in the developing and underdeveloped countries. There has been a constant increase in urban sources, i.e. industries, vehicles running on dirty fuels, burning of fossil fuels, waste and stubble burning, coal-based power plants, etc. over the past few decades. With the rising economy, this spurt in industrialization and urbanization has increased the concentration of toxic pollutants such as carbon monoxide (CO), nitrogen dioxide (NO_2), particulate matter (PM), sulfur dioxide (SO_2) and volatile organic compounds (VOCs). This chapter concentrates on the adverse health effects of these air pollutants (both ambient and indoors) on the people living in urban areas especially young children, pregnant ladies, the elderly and people with respiratory and cardiovascular conditions.

8.1 Introduction

With increasing urbanization and infrastructure growth, the world is encountering a dual challenge of protecting the environment as well as advancing the economy. A recent study indicated that about 6.5 million deaths in India are associated with air pollution, with urban areas sharing the majority of its percentage (IHME, 2013).

Air pollution is caused by the presence of certain toxic gases, particulates and other substances in the air, the majority of which are generated by human-made activities in sufficient concentration and duration, which substantially interfere with the health, comfort and safety of humankind and its environment. For centuries, researchers have been trying to understand the dynamics of air pollutants and their effect on health. Air pollution in the present context is not just restricted to ambient air, but has slowly and steadily seeped into our indoor microenvironments. To have a better understanding of the ill effects of air pollution, it is important to know about its sources, both indoors and outdoors. The UN General Assembly has set up an open working group (OWG) on sustainable development goals, which formulated a proposal considering economic, social and environmental dimensions in July 2014. The proposal focused on 17 goals with 169 targets, including poverty, hunger, health, education, sustainable cities, climate change, and protecting oceans and forests. One of the goals is to 'Ensure healthy lives and promote well-being for all at all ages', which aims at securing a healthy life for all. As part of this goal, nine sub-goals

---

* E-mail: priya.kulsh@gmail.com

were defined and one of the sub-goals exclusively targets the reduction of the number of deaths and illnesses from different forms of pollution and contamination by 2030 (TERI, 2015a,b).

Urban air pollution is a result of the combustion of fossil fuels that are used in transportation, power generation, the industrial sector, and other economic activities supplemented by localized sources, such as waste burning, resuspension of road dust, and construction activities. Household air pollution (HAP), also known as indoor air pollution (IAP), is a serious area of concern in rural spaces, as the majority of this population continues to depend on traditional biomass for cooking and space heating and on kerosene or other liquid fuels for lighting, all of which lead to high levels of HAP. More than 70% of the Indian population is dependent on traditional fuels (coal, firewood, agricultural residue, cow dung) for domestic purposes such as cooking, and almost 32% depend on kerosene for lighting their homes. It has been estimated that about 3 billion people in the world rely on traditional biomass for the purpose of cooking, and that 500 million households still rely on kerosene and similar for the purpose of lighting homes (WHO, 2015).

In rural India, as per the records, only 11.4% of the households use LPG for cooking (Census of India, 2011). This chapter discusses in detail the adverse health effects of ambient and indoor air pollution, from various sources in an urban environment.

8.2 Sources of Air Pollution

An increasing percentage of the world's population lives in urban areas. High population density and the concentration of industry exert great pressures on local environments. Air pollution from households, industry power stations and transportation (motor vehicles), are often major problems. As a result, the greatest potential for human exposure to ambient air pollution and subsequent health problems occurs in urban areas. Improving air quality is a significant aspect of promoting sustainable human settlements.

Sources of air pollution can be broadly classified into natural and anthropogenic. The natural sources include volcanoes, forest fires, terrestrial vegetation and pollens. The anthropogenic sources

include everything involving human activities. The other major classifications for outdoor and indoor air pollution are listed in Fig. 8.1.

8.2.1 Major air pollutants

Ambient air pollution has been identified as a national problem, since it is the fifth-largest cause of mortality in India (Atkinson *et al.*, 2012). The Central Pollution Control Board (CPCB) in India implements the National Air Quality Monitoring Programme through a network of 544 operating ambient air-quality stations covering 224 cities and towns in 26 states, and 5 union territories of the country. WHO focuses on four health-related air pollutants, namely, particulate matter (PM), measured as particles with an aerodynamic diameter less than 10 µm (PM_{10}) and less than 2.5 µm ($PM_{2.5}$), nitrogen dioxide (NO_2), sulfur dioxide (SO_2) and carbon monoxide (CO). The above pollutants are primarily monitored for the general state of air quality and are sufficient parameters to assess the impact on the health and well-being of humankind and its environment.

8.2.1.1 Ambient and indoor air pollutants

PARTICULATE MATTER (PM).   The evidence on airborne PM and its public health impact is consistent in showing adverse health effects at exposures that are currently experienced by urban populations in both developed and developing countries. WHO air quality guidelines (AQGs) for PM are represented in Table 8.1.

Airborne particulate matter consists of a group of organic and inorganic substances and mixtures of substances. They are emitted by combustion and metallurgical processes along with other industrial operations (QUARG, 1996). The chemical constituents of non-biological particulate matter include sulfates, nitrates, acids, ammonium ions, metal compounds, water, carbonaceous compounds and crustal materials (formed as a result of accumulation mode particles which include super micrometre particles). They are also formed via natural processes, including suspension of soil dust or sea salt, volcanoes, forest fires and the ingress of micrometeorites into the atmosphere. Further, biological agents like fungal spores, bacteria and viruses also

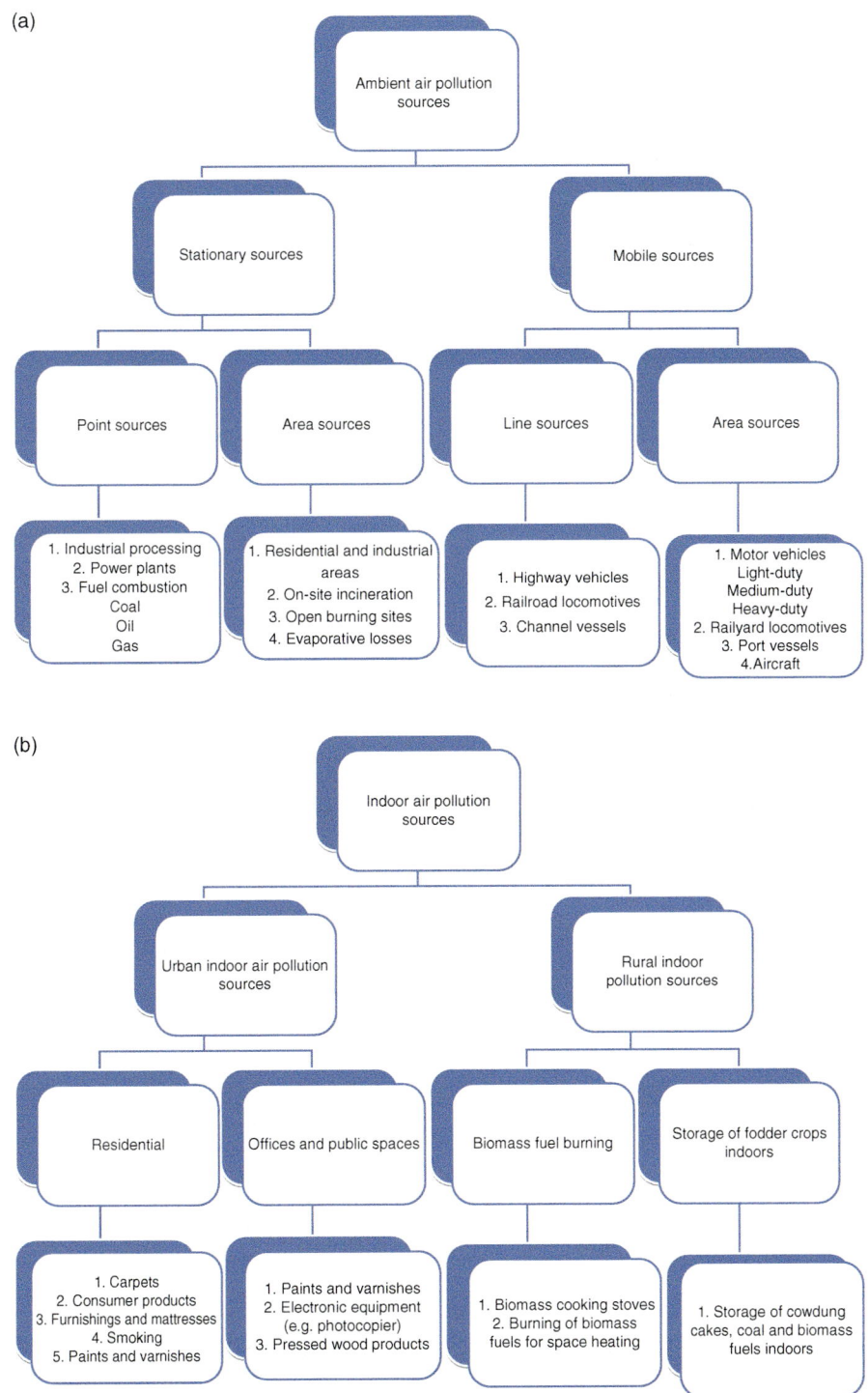

Fig. 8.1. Outdoor and indoor air pollution: (a) major ambient air pollution sources; (b) major indoor air pollution sources.

Table 8.1. WHO air-quality guidelines (2006) for particulate matter.

Particulate matter	Annual mean	24 hour mean (µg/m³)
PM$_{10}$	20	50
PM$_{2.5}$	10	25

contribute to the particulate matter (Tao *et al.*, 2013; Bozlaker *et al.*, 2014).

Since different types of chemicals have been associated with a potential hazard like lung cancer, birth defects, etc., information on their chemical characterization is required for safeguarding the health of citizens (Kimbrough *et al.*, 1977; Boffetta *et al.*, 1993; Boman *et al.*, 2013).

In order to measure the concentrations of particulates, suitable devices have been developed to capture the particles, which have various size fractions. The common measurement parameters are defined below.

1. PM$_{10}$: the particulate matter measured with a sampler that has 50% collection efficiency (cut point) for particles with aerodynamic diameter of 10 µm. These samplers collect particles in the size range equivalent to the International Organization for Standardization (ISO) definition of the term 'thoracic fraction'.
2. PM$_{2.5}$: these samplers measure particulate matter that has a cut point of 2.5-µm aerodynamic particle diameter. This is equivalent to the ISO's term 'high-risk respirable fraction'.
3. RSP (respirable suspended particles): this size range is measured with a sampler having a cut point of 3.5 µm. This is similar to the ISO's term 'respirable fraction'.
4. TSP (total suspended particles): this particulate matter is measured for a cut point of between 25 and 40µm.

When considering the biological activity of PM, the deposition of PM in the human respiratory tract is the main criterion used to define a particle. Hence, ISO has defined the classification of particulate matter in terms of respiratory health as follows.

• The inhalable fraction refers to the mass fraction of total airborne particles that is inhaled through the nose and/or mouth.

• The thoracic fraction refers to the mass fraction of inhaled particles that penetrates beyond the larynx. Generally, the thoracic fraction is represented by a lognormal distribution with a median aerodynamic diameter of 10 ± 1.5 µm.

• The respirable fraction refers to the mass fraction of inhaled particles that penetrate the alveolar region of the lungs where gas exchange occurs. It is represented by means of a cumulative lognormal distribution with a median aerodynamic diameter of 4 ± 1.5 µm.

• The 'high risk' respirable fraction is defined as the respirable fraction for the sick and the infirm, or children. It is represented by means of a cumulative lognormal distribution with a median aerodynamic diameter of 2.5 ± 1.5 µm.

The common sources of outdoor PM are road traffic (including exhaust, non-exhaust emissions, tyre wear and resuspension), power plants, agriculture, industries and domestic heating systems. Anthropogenic sources generally generate primary and secondary fine, ultrafine and nanoscale particulates, while natural sources contribute mainly coarser particles.

In the indoor environment, sources of particles include re-entrainment of existing particles through activities such as sweeping or dusting, emissions from cooking, cigarette smoking, and transport from outdoors by infiltration/leakage through the walls, windows, doors and ventilation systems (Kamens *et al.*, 1991). Additionally, particles of biological origin are also present indoors, occur in a wide range of sizes and can affect health through allergic and toxic mechanisms.

CARBON MONOXIDE (CO). CO is a colourless, odourless gas produced by the incomplete combustion of organic substances and is currently the only air pollutant with a specific and clinically relevant biological marker that is a useful indicator of exposure (Coultas and Lambert, 1991). Prolonged exposure to CO initiates the formation of carboxyhaemoglobin (COHb), which is a stable complex of carbon monoxide and haemoglobin found in blood. Besides the COHb levels in the blood, the concentration of CO in expired breath also provides an indication of CO exposure. Most combustion processes produce some CO, depending upon the availability of oxygen and the efficiency of the process. The main outdoor source of CO is petrol vehicle exhaust fumes. In the indoor environment, relevant combustion sources of CO include gas-fuelled appliances, unvented space heaters, wood stoves and tobacco products. High concentrations of CO are usually associated with the operation of combustion appliances in poorly ventilated rooms or of damaged or badly installed or badly maintained appliances. Exposure to high levels of CO can lead to dizziness and headaches, lethargy, unconsciousness and even death, and there is some evidence that low levels may exacerbate conditions such as angina and have other subtle chronic effects.

As people spend a considerable amount of time indoors, levels of CO inside the home can have a significant impact on personal exposure levels (although particular subgroups, such as commuters and those working in certain occupations, may be more affected by outdoor levels). For example, in the United Kingdom, homes with CO sources such as gas cookers have had recordings of peak concentrations of up to 60 mg/m^3 (52.4 ppm; the WHO air-quality guideline for a 30-minute exposure to CO) (Burr, 1995), and in other cases much higher peak levels have been associated with malfunctioning combustion appliances. However, long-term CO concentrations are generally much lower. In other indoor microenvironments in which internal combustion engines are operated with insufficient ventilation, mean levels of CO can rise to above 115 mg/m^3 (100.4 ppm) for prolonged periods, with much higher short-term values (Burr, 1995).

Emissions of CO in the domestic environment may be classified as accidental or as resulting from the intentional use of combustion devices. Accidental emissions may result from the improper use of combustion appliances and from faulty appliances; other non-intentional sources include the ingress of polluted air from attached garages or from the outdoor environment. Accidental emissions can lead to very high indoor CO levels, which may result in acute and sometimes fatal health effects. Most CO emissions come from the intentional use of partially vented or unvented combustion appliances, such as gas cookers, and other appliances including water

heaters or unvented gas space heaters, along with wood or other solid fuel burning appliances. CO emissions are highly variable between gas cookers as well as between individual burners on the same appliance. Operating a gas cooker with an improperly adjusted flame can lead to very high emission rates (up to and above a fivefold increase, compared with a properly adjusted flame). Emissions from unvented gas space heaters are very variable, but tend to be comparable with gas cooker emissions. Infrared gas space heaters produce higher emissions than convective or catalytic appliances. For unvented kerosene space heaters, radiant appliances produce higher emissions than convective appliances. For these types of sources, the wick setting has a significant effect, with a low setting producing the highest CO emission rates. Among wood and other solid-fuel-burning appliances, the non-airtight wood burning stoves and fireplaces may produce substantial amounts of CO compared with airtight appliances. Tobacco smoking is also a source of indoor CO, with emission rates varying between tobacco brands and with the total number of cigarettes smoked.

The Coburn–Forster–Kane (CFK) equation was the foremost model that was developed to explain the relationship between CO concentration and blood (COHb) levels. Coburn *et al.* (1965) developed a relationship between the concentration of CO in inspired air and the rate of CO production (endogenous), alveolar ventilation, diffusing capacity of the lung, and the mean oxygen tension in the alveolar capillaries. The toxicity of CO is governed by these factors and by exposure duration, respiratory minute volume, cardiac output, tissue oxygen demand and blood haemoglobin concentration (Klausen, 1985). The rate at which arterial blood reaches equilibrium with the inspired concentration of CO is affected by the diffusion capacity of the lungs and alveolar ventilation, and the duration and concentration of exposure. This model, in its non-linear form, may be used to predict CoHb levels at high CO exposures, whereas in linear form, it can be made applicable to typical air pollution situations.

NITROGEN DIOXIDE (NO$_2$). Combustion of atmospheric nitrogen (N$_2$) with oxygen (O$_2$), dissociation of the atmospheric NO$_2$ in the presence of sunlight energy (hυ), and rapid cooling of the vehicular exhausts on dilution in the atmosphere are the three primary mechanisms that yield high outdoor NO concentrations. The major source of atmospheric NO_2 formation is the oxidation of NO as shown by the following reaction.

$$2NO + O_2 \rightarrow 2NO_2$$

The major sources of NO_2 are combustion of fossil fuels in power plants and motor vehicles. According to WHO (2005), NO_2 is a toxic gas with significant health effects. Animal toxicological studies also suggest that long-term exposure to NO_2 at concentrations above current ambient concentrations has adverse effects. Numerous epidemiological studies have used NO_2 as a marker for the cocktail of combustion-related pollutants, in particular, those emitted by road traffic or indoor combustion sources (Dockery et al., 1993; Burnett et al., 2004; Cohen et al., 2004). In these studies, any observed health effects could also have been associated with other combustion products, such as ultrafine particles, nitrous oxide (NO), particulate matter or benzene. Although several studies – both outdoors and indoors – have attempted to focus on the health risks of NO_2, the contributing effects of these other, highly correlated co-pollutants were often difficult to rule out (WHO, 2005).

Almost 50% of NO_2 originates from motor vehicles and is said to be a very difficult pollutant to control. It also plays an important role as a precursor in the formation of tropospheric ozone. Ghosh et al. (2017) have explained the role of NO_2 in the formation of tropospheric ozone at major traffic junctions in New Delhi, India. The study also revealed that the highest annual concentration of tropospheric ozone (56.2 ± 23.5 µg/m³) was reported from Punjabi Bagh among the selected sites of Delhi and was recorded primarily during summer time (March–June 2016). The results indicate that higher temperature, lower humidity and high intensity of solar radiation contribute to the formation of tropospheric ozone in Delhi. Shukla et al. (2010) concluded in their study that the concentration of suspended particulate matter (SPM) and NO_2 was found to be maximum in winter near Delhi due to poor fuel quality, traffic congestion and improperly maintained vehicles. A study done in Bangalore city in India reported that the growing transport sector was responsible

for rising concentration particulate and NO_2 in the city (Thakur, 2017).

There have been numerous studies on the monitoring of indoor concentrations of NO_2. In the absence of emission sources, levels generally correlate with outdoor concentrations (Monn et al., 1998). In homes with gas cooking stoves, Lambert et al. (1993) found that average NO_2 levels were higher in kitchens when compared to bedrooms. Further, the normal use of a gas cooking range adds 25 ppb (47 lg m⁻³) of NO_2 to the background concentration in a home (Samet et al., 1987). In homes with unvented kerosene space heaters, the weekly average concentration exceeds 45 ppb (Leaderer et al., 1986). A similar rise in the weekly average levels has been reported in homes with unvented gas space heaters (Ryan et al., 1988). Dosimetry models (Overton and Miller, 1988) indicate that NO_2 is absorbed along the entire tracheobronchial tree with the highest deposition occurring at the junction of conducting and respiratory airways. Beyond the terminal bronchioles, the dose delivered to the lung epithelium decreases because of the increase in lung surface area. Inhaled NO_2 is combined with water in the lung to form nitric (HNO_3) and nitrous (HNO_2) acids and may react with lipids and proteins present on the epithelial surface to form nitrite anions and hydrogen ions (Postlethwait and Bidani, 1990). While oxidant injury has been identified as the principal mechanism by which NO_2 damages the lung, a lot of uncertainty remains regarding the reactions of tissue with NO_2.

SULFUR DIOXIDE (SO_2). Sulfur dioxide (SO_2) is a colourless gas readily soluble in water. Natural sources such as sulfur bacteria activities, volcanoes and forest fires, contribute to environmental levels of SO_2. Human-made contributions include the use of sulfur-containing fossil fuels for transportation, domestic purposes and coal-fired power generation. Of greater interest, with respect to outdoor air quality are the effects on health of the lower concentration to which human beings may be exposed in the ambient air.

SO_2 levels are generally lower indoors than outdoors, and indoor/outdoor ratios of the concentrations range between 0.1 and 0.6 (Leaderer et al., 1993). As a result of emission reductions, annual mean levels of ambient SO_2 in major cities in Europe and the USA are below

20 ppb (52 µg m^{-3}). However, indoor SO$_2$ concentrations can be higher inside homes that have kerosene heaters and poorly vented coal and gas appliances.

8.2.2 Health impacts of major ambient and indoor air pollutants

Several studies conducted across the world demonstrate that outdoor and indoor air pollution are serious environmental risks that cause or aggravate acute and chronic diseases (Table 8.2). Pairing city-level air pollution measures with child-level data from the National Family Health Survey (2005–2006) for six cities in India shows that an increase in ambient air pollution significantly increases child morbidity (Ghosh and Mukherjee, 2010). The six cities considered in the study were Chennai, Delhi, Hyderabad, Indore, Kolkata and Nagpur. The study found that a rise in ambient air pollution significantly increased the likelihood of a child suffering from coughs and fever in the subsequent week. However, the type of cooking fuel used at home is not

significantly related to child morbidity, after accounting for ambient air pollution and other child- and household-level control variables. Thus, while polluted air is bad for child health, ambient air pollution is a more significant determinant of child health outcomes. There was a significant correlation between the two child morbidity outcomes – fever and cough. Controlling citywide air pollution could significantly lower child morbidity, and should receive greater emphasis in urban planning and infrastructure development.

Amitai *et al.* (1998) found that subjects exposed to CO from residential stoves for up to 2.5 h showed declines in their learning and planning abilities, as well as a drop in their attention and concentration spans. Chronic exposure (at 10–30% carboxyhaemoglobin) often produces symptoms that are easily misdiagnosed or overlooked, such as headache, fatigue, dizziness and nausea (Stewart *et al.*, 1970). There is evidence from animal studies that some foetal damage may occur from maternal exposure to CO at these levels (Longo, 1977). A positive association between air pollution and heart failure hospitalization or heart failure mortality has

Table 8.2. Major air pollutants, their sources and possible health impacts.

Pollutant	Sources	Effect on humans
1. Suspended particulate matter (SPM)	(i) Motor vehicular exhaust (ii) Resuspension of dust (iii) Industrial plants and coal burning (iv) Agricultural processes (v) Construction and earth-moving	(i) Nose and throat irritation (ii) Lung function reduction due to damage of lung tissue (iii) Increased incidence of lung diseases and cancers (iv) Lowers resistance to respiratory infection
2. Carbon monoxide (CO)	(i) Burning of fossil fuels (ii) Incomplete combustion from vehicles and power plants (iii) Forest fires	(i) Reduces oxygen reaching heart and brain due to formation of carboxyhaemoglobin (ii) Slowing of reflexes and impairment of thought process (iii) Vulnerable population becomes prone to cardiovascular diseases (iv) Exposure for long duration leads to drowsiness, unconsciousness and sometimes death
Nitrogen dioxide (NO$_2$)	(i) Combustion of biomass fuels in motor vehicles, power plants and industries	(i) Acute respiratory illness, especially in small children (ii) Irritation of the lungs especially the vulnerable asthmatic population
Sulfur dioxide (SO$_2$)	(i) Coal and oil burning power plants/industries (ii) Oil refineries	(i) Adversely affects breathing (ii) Aggravates existing respiratory and cardiovascular disease and asthmatics

been reported by Shah *et al.* (2013). Among 20 non-smoking men with ischaemic heart disease, Lambert (1994) found that the probability of occurrence of an episode of myocardial ischaemia was 2.1 times higher at CoHb levels of 2% relative to those below 1%. In England and Wales, there are on average around 60 deaths annually associated with accidental CO poisoning (Burr, 1997), and similar rates have been observed in the USA (Cobb and Etzel, 1991). Individuals who survive acute CO poisoning may still exhibit neurological and psychological symptoms many weeks or months after exposure, particularly if a period of unconsciousness has occurred (Choi, 1983).

Honickey *et al.* (1985) examined respiratory symptoms among 62 children in Michigan, USA. They found that 84% of children in homes with wood-burning stoves recounted at least one severe respiratory symptom, compared to only 3% of children in homes without a stove. As part of the Six-City Study, Dockery *et al.* (1993) found an odds ratio of 1.32 (95% CI 0.99–1.76) for respiratory illness in households with wood-burning stoves in comparison with those using other sources of heating. Koenig *et al.* (1993) reported that infants exposed to wood smoke were more likely to recount asthma symptoms, and Abbey *et al.* (1998) observed a reduction in lung function in non-smokers exposed to high concentrations of indoor particles over a period of 20 years. A Gambian study found that girls under the age of five who were carried on their mother's backs during cooking (in smoky cooking huts), had increased risk of developing acute respiratory infection (ARI) by up to six times (Schwela, 1997).

A study conducted in Bangkok looked at the relationship between indoor and outdoor particulate matter concentrations (Chestnut *et al.*, 1998). Indoor PM levels were as high, if not higher, than outdoor levels when there was no air conditioning and some indoor sources present. With no notable indoor sources, and maybe some air conditioning, indoor PM concentrations ranged between 50% and 100% of outdoor concentrations. A Dutch study gave similar findings (Janssen *et al.*, 1998a). A good correlation existed between indoor and outdoor levels, but it decreased with increasing indoor sources, such as environmental tobacco smoke (ETS). These two studies found that in general, outdoor concentrations tended to exceed indoor

concentration, whereas a Korean study found indoor air concentrations to be consistently higher than outdoor, an observation that was magnified during wintertime, possibly due to heating (Baek, 1997). All three studies cited the difference between indoor and outdoor concentrations as attributable in part to human indoor activities, duration of human occupancy, ventilation, type of stove used for cooking and heating, and tobacco smoke (Baek, 1997).

Janssen and colleagues (1998a) found in the Netherlands a greater discrepancy between personal exposure and indoor/outdoor concentrations than between indoor and outdoor concentrations. Personal exposure was found to be hugely underestimated, although the direction and the degree of misrepresentation clearly depended on other factors such as indoor sources, housing type and condition and geographic location (Janssen *et al.*, 1998a,b; Smith, personal communication, 1999). The relationship between indoor and outdoor airborne particles was investigated for 16 residential houses located in a suburban area of Brisbane, Australia (Morawska *et al.*, 2001). Comparison of the time series of indoor to outdoor particle concentrations shows a clear positive relationship existing for many houses under normal ventilation conditions (estimated to be about and above 2 h^{-1}), but not under minimum ventilation conditions (estimated to be about and below 1 h^{-1}). These results suggest that for normal ventilation conditions, outdoor particle concentrations could be used to predict instantaneous indoor particle concentrations but not for minimum ventilation, unless air exchange rate is known, thus allowing for estimation of the 'delay constant'.

Acute lower respiratory infections (ALRI) remain the single most important cause of death globally in children under 5 years and account for around 2 million deaths annually in this age group. There are some 15 studies in less-developed countries (LDCs), which have reported on the association between air pollution exposure and ALRI (Kossove, 1982; Campbell *et al.*, 1989; Pandey *et al.*, 1989a; Cerquiero *et al.*, 1990; Collings *et al.*, 1990; Armstrong and Campbell, 1991; Johnson and Aderele, 1992; Mtango *et al.*, 1992; de Francisco *et al.*, 1993; Shah *et al.*, 1994; Victora *et al.*, 1994; Wesley and Loening, 1996; López-Bravo *et al.*, 1997; Ezzati and Kammen, 2001a,b) and two further studies among Navajo Indians in the USA (Morris et al., 1990; Robin et al., 1996).

There have been numerous studies of people with asthma, chronic obstructive pulmonary disease, or chronic bronchitis, showing that exposure to low levels of nitrogen dioxide can cause small decrements in forced vital capacity and forced expiratory volume in 1 second (FEV1), or increases in airway resistance. Pulmonary function responses have been shown in three studies of asthmatics exposed to 560 µg/m^3 (0.30 ppm) while performing mild to moderate exercise.

Linn et al. (1986) and Linn and Hackney (1984) found no pulmonary function responses at concentrations of 1880–7520 µg/m^3. Even within the same laboratory, results have not been replicated with different groups of asthmatics (Utell et al., 1983; Roger et al., 1990). In concentration–response studies, Von Nieding et al. (1970, 1971, 1973) found that brief exposures to levels of 3000 µg/m^3 (1.6 ppm) increased airway resistance in people with chronic obstructive pulmonary disease. At similar concentrations in mildly exercising subjects exposed for 1 hour, however, no responses were seen (Linn et al., 1985). Longer exposures (4 hours) have caused functional effects in chronic obstructive pulmonary disease patients at lower levels (560 µg/m^3; 0.3 ppm) (Morrow and Utell, 1989). The reasons for these mixed results have not been clarified by further research, although it has been suggested that nitrogen dioxide-induced increases in airway resistance at ambient concentrations may not show the expected monotonic concentration–response relationship (Bylin et al., 1985).

Bauer et al. (1986) found that cold-induced increased airway constriction in asthmatics was potentiated by nitrogen dioxide at a concentration of 560 µg/m^3 (0.3 ppm). However, in a further illustration of the response paradox, Avol et al. (1988) found possible increases in reactivity to cold air at 560 µg/m^3 (0.3 ppm) but not at 1130 µg/m^3 (0.6 ppm), and no such effect was found in a follow-up study (Folinsbee, 1992). No exacerbation of bronchial reactivity to natural allergens was found in asthmatics on exposure to nitrogen dioxide, but this may be because of the low levels used (190 µg/m^3; 0.1 ppm) (Orehek et al., 1981; Ahmed et al., 1983a,b). A meta-analysis (Folinsbee, 1992) of the bronchoconstrictor

studies indicated that of the 105 subjects exposed to <376 µg/m^3 (0.2 ppm), 67 had increased reactivity and 38 had decreased reactivity. For all the nitrogen dioxide studies, the percentage increase in airway responsiveness was 59%, primarily due to the results in subjects exposed at rest.

In relation to outdoor nitrogen dioxide as a potential indicator for traffic-dominated urban air pollution, several epidemiological studies undertaken in Europe and Japan provide suggestive evidence for respiratory effects being related to living near busy roads, presumably in part due to associated higher exposures to traffic-generated nitrogen dioxide. For example, Wjst et al. (1993) reported increased respiratory symptoms (e.g. recurrent wheeze) in children aged 9–11 years as a result of traffic density on main roads through their school districts in Munich, Germany; since there was no use of home addresses and no information was provided on the actual distances of the schools from the indicator roads, however, the results are open to question owing to possible exposure misclassification. Similarly, only limited confidence can be accorded to the findings from a case-control study in the United Kingdom conducted by Edwards et al. (1994), which compared children (aged over 5 years) admitted to hospital with asthma compared to other 'hospital' and 'community' control subjects. After stratification for distance from the road (based on relating the postal code of the home address, good to within 100 m of its true location, to indicator roads), children admitted with asthma were more often found to reside near roads with high traffic density than control children.

In a cross-sectional study performed in 1987 on 4855 children aged 6 years in Stuttgart, Germany, asthma prevalence was investigated in relation to outdoor pollution, based on annual measurements where the children lived (Parkhurst et al., 1988). A significantly elevated relative risk of 2.28 was found if the upper tercile (nitrogen dioxide concentration, 60–70 µg/m^3) was compared with the lower tercile (40–50 µg/m^3). A similarly elevated relative risk was seen for the corresponding comparison with nitric oxide, carbon monoxide or traffic counts on the road, suggesting that nitrogen dioxide was an indicator for the traffic effect in this study. Another study in Duisburg, Germany (Schupp et al., 1994) investigated the influence of living by the side of roads with busy traffic on the airways of 10-year-old children. Nitrogen dioxide (14-day average) was measured indoors (child's bedroom) and outdoors (outside bedroom). Airway responsiveness to cold-air challenge and spirometry were measured. Children exposed to an outdoor concentration of 49 µg/m^3 (90th centile) compared to 25 µg/m^3 (10th centile) showed slightly reduced spirometric parameters. The results became clearer and statistically significant if the outdoor concentrations of a traffic index composed of nitrogen dioxide, toluene and benzene was used in the analysis.

Of the 16 LDC studies, ten are case-control designs (two mortality studies), five cohort studies (all morbidity), and one a case-fatality study. In contrast to the relatively robust definitions of ALRI, the measurement of exposure has relied in almost all studies on proxies, including the type of fuel and stove (Cerqueiro et al., 1990; Collings et al., 1990; Johnson and Aderele, 1992; Mtango et al., 1992; Shah et al., 1994; Victora et al., 1994; Wesley and Loening, 1996; López-Bravo et al., 1997), whether the child stays in the smoke during cooking (Kossove, 1982; Mtango et al., 1992; Victora et al., 1994), reported hours spent near the stove (Kossove, 1982; Pandey et al., 1989b), and whether the child is carried on the mother's back during cooking (Campbell et al., 1989; Armstrong and Campbell, 1991; de Francisco et al., 1993). Apart from Ezzati's study (Ezzati and Kammen, 2002), only one previous study made direct measurements of pollution (particulates) and exposure (carboxyhaemoglobin (COHb) concentrations) in a sub-sample (Collings et al., 1990). In that study, respirable particulates in the kitchens of cases were substantially higher than for controls (1998 mg/m^3 vs. 546 mg/m^3; $p<0.01$), but there was no significant difference in COHb levels. Five studies reported no significant association between ALRI incidence and exposure (Johnson and Aderele, 1992; Shah et al., 1994; Victora et al., 1994; Wesley and Loening, 1996; López-Bravo et al., 1997), but the remainder reported significantly elevated ORs (for incidence or deaths) in the range 2–8. Not all, however, have dealt adequately with confounding factors (Kossove, 1982; Pandey et al., 1989b; Cerquiero et al., 1990; Collings et al., 1990; Johnson and Aderele, 1992), although

accounting for confounding factors in studies of this exposure may in any case be problematic (Armstrong and Campbell, 1991; Bruce et al., 1998). In several studies finding no association, relatively small proportions of the samples were exposed. Thus, in urban Brazil only 6% of children were exposed to indoor smoke (Victora et al., 1994) and in another South American study, 97% of homes used gas for cooking, although 81% used polluting fuels (kerosene, wood, coal) for heating (López-Bravo et al., 1997). This study also excluded neonates with birth weight <2500 g – the group most vulnerable to ALRI. In the study from Durban, only 19% of cases and 14% of controls used wood or coal stoves (Wesley and Loening, 1996). In the study reported by Shah, a so-called 'smokeless chullah' was used as an indicator of lower exposure, but such stoves can be little better than traditional ones (Smith, 1989). The most recent report on this topic, by Ezzati and Kammen (2002), describes a cohort study of 345 rural Kenyan people (of which 93 were aged under 5 years), living in 55 homes on a rural cattle ranch (Ezzati and Kammen, 2001a,b). Households used mainly wood or charcoal, in open fires and improved (chimneyless) stoves. Detailed personal exposure assessment was combined with weekly (initially bi-weekly) health outcomes assessment for adults and children using WHO criteria for ALRI. This is the first study that has reported (and presented) exposure–response relationships for particulates.

Hu and Jiang (2013) reported that PM$_{2.5}$ inhalation is the key factor responsible for different types of lung diseases. Atkinson et al. (2014) found that the increase in rate of lung cancer and other similar diseases increases mortality and morbidity in humans. Several studies have reported associations of acute and chronic health effects of PM exposure in developed countries, but such studies are lacking in developing countries and are therefore highly needed (Dockery et al., 1993; Dockery and Pope, 1994). The quantitative assessment of exposure to PM$_{2.5}$ in both rural and urban areas, especially for women who are exposed to high concentrations during cooking, is missing for Asian environments, and further research is clearly needed in this direction.

From the experimental research, it is evident that exposure to NO$_2$ may increase respiratory infections, and adversely affect lung function (Frampton et al., 1991). In this regard, two important projects have discovered possible associations between the above factors.

One of the studies was conducted in Italy by Viegi *et al.* (1992), which has found the use of bottled gas for cooking and its association with increased reporting of coughs in males. At a later stage, in England, Jarvis *et al.* (1996) found that a sample of females who reported they used mainly gas for cooking were more likely to report respiratory symptoms in the 12 months prior to the survey. Moreover, in the Harvard six cities study, it has been observed that serious illness in infants was more common among infants from homes which rely on gas cooking (Ware *et al.*, 1984).

Moreover, Goldstein *et al.* (1988) examined the relationship between NO_2 concentrations in kitchens and spirometric lung function. They also found that exposure to NO_2 levels was associated with a reduction in lung capacity (measured by FEV) of the order of 10%.

However, another study suggested that risk associated with NO_2 exposure is significantly increased for children aged over 2 years (Hasselblad *et al.*, 1992). In an experiment conducted by Salome *et al.* (1996), it was found that the exposure of 600 ppb of NO_2 over a period of 1 hour was associated with an increase in airway hyper-responsiveness. In a similar study conducted in Australia, Pilotto *et al.* (1997) monitored NO_2 exposures among children aged between 6 and 11 years and found that hourly peak levels were associated with significant increases in the reporting of respiratory illness due to short-term peak exposures. However, studies conducted by Samet *et al.* (1993) and Brunekreef *et al.* (1990) failed to find an association between NO_2 and the pulmonary function of a sample of children.

8.3 Conclusion

The health effects of air pollution are tremendous and are emerging with every passing decade. With greater urbanization, the rising concentration of air pollutants, both outdoors and indoors, poses a significant health risk to humans and needs to be tackled on a priority basis. Currently, the environmental quality standard regarding $PM_{2.5}$ in Asian environments is only based on their mass concentrations. However, these specific limits should be revisited in the future given that particle types and sizes govern the actual harmfulness of the PM, especially of respirable fraction. Due to sudden population growth, an increase in solid fuel consumption, greater private vehicle or transportation movement and direct exposure to $PM_{2.5}$ during household activities, the risk of exposure to $PM_{2.5}$ in Asian environments is increasing. The sustainable shift from fossil fuels used in vehicle transportation to electric or renewable sources is a must to bring down the concentration of harmful oxides of nitrogen and carbon monoxide. The policy makers have a tough job in hand, to keep amending the existing policies in order to keep this upsurge in air pollution under control. It is imperative to let the coming generations breathe and live in clean and healthy air.

References

Abbey, D.E., Burchette, R.J., Knutsen, S.F., McDonnell, W.F., Lebowitz, M.D. and Enright, P.L. (1998) Long-term particulate and other air pollutants and lung function in non smokers. *American Journal of Respiratory Critical Care Medicine* 158 (1), 289–298.

Ahmed, T., Dougherty, R. and Sackner, M.A. (1983a) *Effect of NO_2 Exposure on Specific Bronchial Reactivity in Subjects with Allergic Bronchial Asthma. Final Report.* General Motors Research Laboratories, Warren, MI.

Ahmed, T., Dougherty, R. and Sackner, M.A. (1983b) *Effect of 0.1 ppm NO_2 on Pulmonary Functions and Non-Specific Bronchial Reactivity of Normals and Asthmatics Final Report.* General Motors Research Laboratories, Warren, MI.

Amitai, Y., Zlotogorski, Z., Golan-Katzav, V., Wexler, A. and Gross, D. (1998) Neuropsychological impairment from acute low level exposure to carbon monoxide. *Archives of Neurology* 55(6), 845–848.

Armstrong, J.R. and Campbell, H. (1991) Indoor air pollution exposure and lower respiratory infections in young Gambian children. *International Journal of Epidemiology* 20(2), 424–429.

Atkinson, R.W., Cohen, A., Mehta, S. and Anderson, H.R. (2012) Systematic review and meta-analysis of epidemiological time-series studies on outdoor air pollution and health in Asia. *Air Quality, Atmosphere, and Health* 5(4), 383–391.

Atkinson, R.W., Kang, S., Anderson, H.R., Mills, I.C. and Walton, H.A. (2014) Epidemiological time series studies of $PM_{2.5}$ and daily mortality and hospital admissions: a systematic review and meta-analysis. *Thorax*. DOI: 10.1136/thoraxjnl-2013-204492.

Avol, E.L. *et al.* (1988) Laboratory study of asthmatic volunteers exposed to nitrogen dioxide and to ambient air pollution. *American Industrial Hygiene Association Journal* 49, 143–149.

Baek, S., Kim, Y.S. and Perry, R. (1997) Indoor air quality in homes, offices, and restaurants in Korean urban areas – indoor/outdoor relationships. *Atmospheric Environment* 31(4), 529–544.

Bauer, M.A. Utell M.J., Morrow P.E., Speers D.M. and Gibb F.R. (1986) Inhalation of 0.30 ppm nitrogen dioxide potentiates exercise-induced bronchospasm in asthmatics. *American Review of Respiratory Disease* 134, 1203–1208.

Boffetta, P., Merler, E. and Vainio, H. (1993) Carcinogenicity of mercury and mercury compounds. *Scandinavian Journal of Work, Environment Health* 19, 1–7.

Boman, J., Shaltout, A.A., Abozied, A.M. and Hassan, S.K. (2013) On the elemental composition of $PM_{2.5}$ in central Cairo, Egypt. *X-Ray Spectrometry* 42, 276–283.

Bozlaker, A., Spada, N.J., Fraser, M.P. and Chellam, S. (2014) Elemental characterization of $PM_{2.5}$ and PM_{10} emitted from light duty vehicles in the washburn tunnel of Houston, Texas: release of rhodium, palladium, and platinum. *Environmental Science Technology* 48, 54–62.

Bruce, N., Neufeld, L., Boy, E. and West, C. (1998) Indoor biofuel air pollution and respiratory health: the role of confounding factors among women in highland Guatemala. *International Journal of Epidemiology* 27, 454–458.

Brunekreef, B., Houthuijs, D., Dijkstra, L. and Boleij, J.S. (1990) Indoor nitrogen dioxide exposure and children's pulmonary function. *Journal of the Air and Waste Management Association* 40(9), 1252–1256.

Burnett, R.T., Stieb, D., Brook, J.R., Cakmak, S., Dales, R. *et al.* (2004) Associations between short-term changes in nitrogen dioxide and mortality in Canadian cities. *Archives of Environmental Health* 59, 228–236.

Burr, M. (1995) Carbon monoxide. In: Raw, G.J. and Hamilton, R.M. (eds) *Building Regulation and Health*. Construction Research Communication Ltd, London, pp. 26–28.

Burr, M.L. (1997) Health effects of indoor combustion products. *Journal of the Royal Society of Health* 117(6), 348–350.

Bylin, G., Lindvall, T., Rehn, T. and Sundin, B. (1985) Effects of short-term exposure to ambient nitrogen dioxide concentrations on human bronchial reactivity and lung function. *European Journal of Respiratory Disease* 66, 205–217.

Campbell, H., Armstrong, J.R. and Byass, P. (1989) Indoor air pollution in developing countries and acute respiratory infection in children [letter]. *The Lancet* 1(8645), 1012.

Census of India (2011) *Census Info India 2011: Houses, Household Amenities and Assets (Version 2)*. Ministry of Home Affairs, Government of India, New Delhi.

Cerqueiro, M.C., Murtagh, P., Halac, A., Avila, M. and Weissenbacher, M. (1990) Epidemiologic risk factors for children with acute lower respiratory tract infection in Buenos Aires, Argentina: a matched case-control study. *Reviews of Infectious Diseases* 12, Suppl 8, S1021–1028.

Chestnut, L.G., Vichit-Vadakan, N., Ostro, B., Smith, K.R. and Tsai, F.C. (1998) *Executive Summary: Health Effects of Particulate Matter Air Pollution in Bangkok*. World Bank, Washington, DC.

Choi, I.S. (1983) Delayed neurological sequelae in carbon monoxide intoxication. *Archives of Neurology* 40(7), 433–435.

Cobb, N. and Etzel, R.A. (1991) Unintentional carbon monoxide-related deaths in the United States, 1979 through 1988. *Journal of the American Medical Association* 266(5), 659–663.

Coburn, R.F., Forster, R.E. and Kane, P.B. (1965) Considerations of the physiological variables that determine the blood carboxyhemoglobin concentration in man. *Journal of Clinical Investigation* 44, 1899–1910.

Cohen, A.J., Anderson, H.R., Ostro, B., Pandey, K.D., Krzyzanowski, M. *et al.* (2004) Mortality impacts of urban air pollution. In: Ezzati, M., Lopez, A.D., Rodgers, A. and Murray, C.J.L. (eds) *Comparative Quantification of Health Risks: Global and Regional Burden of Disease Attributable to Selected Major Risk Factors*. World Health Organization, Geneva, Switzerland, pp. 1353–1434.

Collings, D.A., Sithole, S.D. and Marti, K.S. (1990) Indoor wood smoke pollution causing lower respiratory disease in children. *Tropical Doctor* 20, 151–155.

Coultas, D.B. and Lambert, W.E. (1991) Carbon monoxide. In: Samet, J.M. and Spengler, J.D. (eds) *Indoor Air Pollution: A Health Perspective*. Johns Hopkins University Press, Baltimore, MD, pp. 187–208.

de Francisco, A., Morris, J., Hall, A.J., Armstrong Schellenberg, J.R. and Greenwood, B.M. (1993) Risk factors for mortality from acute lower respiratory tract infections in young Gambian children. *International Journal of Epidemiology* 22(6), 1174–1182.

Dockery, D.W. and Pope, C.A. (1994) Acute respiratory effects of particulate air pollution. *Annual Review of Public Health* 15, 107–132.

Dockery, D.W., Pope, C.A., Xu, X., Spengler, J.D., Ware, J.H. *et al.* (1993) An association between air pollution and mortality in six United States cities. *New England Journal of Medicine* 329, 1753–1759.

Edwards, J., Walters, S. and Griffiths, R.K. (1994) Hospital admissions for asthma in preschool children: relationship to major roads in Birmingham, United Kingdom. *Archives of Environmental Health* 49, 223–227.

Ezzati, M. and Kammen, D. (2001a) Quantifying the effects of exposure to indoor air pollution from biomass combustion on acute respiratory infections in developing countries. *Environmental Health Perspectives* 109(5),481-488.

Ezzati, M. and Kammen, D.M. (2001b) Indoor air pollution from biomass combustion and acute respiratory infections in Kenya: an exposure response study. *The Lancet* 358(9282), 619-624.

Ezzati, M. and Kammen, D.M. (2002) The health impacts of exposure to indoor air pollution from solid fuels in developing countries: knowledge, gaps, and data needs. *Environmental Health Perspectives*, 110(11), 1057–1068.

Folinsbee, L.J. (1992) Does nitrogen dioxide exposure increase airways responsiveness? *Toxicology and Industrial Health* 8, 273–283.

Frampton, M.W., Morrow, P.E., Cox, C., Gibb, F.R., Speers, D.M. and Utell, M.J. (1991) Effects of nitrogen dioxide exposure on pulmonary function and airway reactivity in normal humans. *American Review of Respiratory Disorders* 143(3), 522–527.

Ghosh, A. and Mukherjee, A. (2010) *Air Pollution and Child Health in Urban India*. Indian Statistical Institute, New Delhi.

Ghosh, R., Gupta, P., Khare, M. and Kulshreshtha, P. (2017) Seasonal variation of tropospheric ozone and its association with the chemical and meteorological precursors in Delhi, India. *Sustainability in Environment* 2(2), 223–257.

Goldstein, I.F., Andrews, L.R. and Hartel, D. (1988) Assessment of human exposure to nitrogen dioxide, carbon monoxide, and respirable particles in New York inner-city residences. *Atmospheric Environment* 22(10), 2127–2139.

Hasselblad, V., Eddy, D.M. and Kotchmar, D.J. (1992) Synthesis of environmental evidence: nitrogen dioxide epidemiology studies. *Journal of the Air and Waste Management Association* 42(5), 662–671.

Honickey, R.E., Osborne, J.S. and Akpom, C.A. (1985) Symptoms of respiratory illness in young children and the use of wood-burning stoves for indoor heating. *Pediatrics* 75(3), 587–593.

Hu, D. and Jiang, J. (2013) A study of smog issues and $PM_{2.5}$ pollutant control strategies in China. *Journal of Environmental Protection* 4, 746–752.

IHME (2013) *The Global Burden of Disease: Generating Evidence, Guiding Policy*. Institute for Health Metrics and Evaluation, Seattle, WA.

Janssen, N.A.H., Hoek, G., Brunekreef, B., Harssema, H., Mensink, I. and Zuidhof, A. (1998a) Personal sampling of particles in adults: relation among personal, indoor and outdoor air concentrations. *American Journal of Epidemiology* 147(6), 537–547.

Janssen, N.A.H., Hoek, G., Harssema, H. and Brunekreef, B. (1998b) Personal sampling of airborne particles: method performance and data quality. *Journal of Exposure Analysis and Environmental Epidemiology* 8, 37–49.

Jarvis, D., Chinn, S., Luczynska, C. and Burney, P. (1996) Association of respiratory symptoms and lung function in young adults with the use of domestic gas appliances. *Lancet* 1(8999), 426–431.

Johnson, A.W. and Aderele, W.I. (1992) The association of household pollutants and socio-economic risk factors with the short-term outcome of acute lower respiratory infections in hospitalized preschool Nigerian children. *Annals of Tropical Paediatrics* 12(4), 421–432.

Kamens, R., Lee, C.T, Wiener, R. and Leith, D. (1991) A study to characterize indoor particles in three non-smoking homes. *Atmospheric Environment Part A General Topics* 25, 939–948.

Kimbrough, R.D., Carter, C.D., Liddle, J.A. and Cline, R.E. (1977) Epidemiology and pathology of a tetrachlorodibenzodioxin poisoning episode. *Archives of Environmental Health* 32, 77–86.

Klausen, K., Andersen, C. and Nandrup, S. (1985) Acute effects of cigarette smoking and inhalation of carbon monoxide during maximal exercise. *European Journal of Applied Physiology and Occupational Physiology* 51, 371–379.

Koenig, J.Q., Larson, T.V., Hamley, Q.S., Rebolledo, V., Dumler, K. *et al.* (1993) Pulmonary lung function in children associated with fine particulate matter. *Environmental Research* 63(1), 26–38.

Kossove, D. (1982) Smoke-filled rooms and lower respiratory disease in infants. *South African Medical Journal* 61(17), 622–624.

Lambert, W.E. (1994) Urban exposures to carbon monoxide and myocardial ischemia in men with ischemic heart disease. PhD thesis, University of California.

Lambert, W.E., Samet, J.M., Hunt, W.C., Skipper, B.J., Schwab, M. and Spengler, J.D. (1993) Nitrogen dioxide and respiratory illness in children: Part II: Assessment of exposure to nitrogen dioxide. *Research Report (Health Effects Institute)* 58, 33–50. Available at: https://www.healtheffects.org/system/files/Research-Report-58-Parts-1-and-2.pdf (accessed 24 August 2018).

Leaderer, B.P., Zagraniski, R.T., Berwick, M. and Stolwijk, J.A.J. (1986) Assessment of exposure to indoor air contaminants from combustion sources: methodology and application. *American Journal of Epidemiology* 124(2), 275–289.

Leaderer, B.P., Stowe, M., Li, R., Sullivan, J., Koutrakis, P. *et al.* (1993) Residential levels of particle and vapor phase acid associated with combustion sources. In: Jantunen, M., Kalliokoski, P., Kukkonen, E., Saarela, K., Seppänen, A. *et al.* (eds) *Proceedings of the Sixth International Conference on Indoor Air Quality and Climate*. Helsinki, Finland, pp. 147–152.

Linn, W.S. and Hackney, J.D. (1984) *Short-Term Human Respiratory Effects of Nitrogen Dioxide: Determination of Quantitative Dose–Response Profiles, Phase II. Exposure of Asthmatic Volunteers to 4 ppm NO$_2$*. Report No. CRC-CAPM-48-83-02. Coordinating Research Council, Inc., Atlanta, GA.

Linn, W.S., Solomon, J.C., Trim, S.C., Spier, C.E., Shamoo, D.A. *et al.* (1985) Controlled exposure of volunteers with chronic obstructive pulmonary disease to nitrogen dioxide. *Archives of Environmental Health* 40, 313–317.

Linn, W.S., Avol, E.L., Shamoo, D.A., Venet, T.G. and Anderson, K.R. (1986) Dose–response study of asthmatic volunteers exposed to nitrogen dioxide during intermittent exercise. *Archives of Environmental Health* 41, 292–296.

Longo, L.D. (1977) The biological effects of carbon monoxide on the pregnant woman, fetus, and new-born infant. *American Journal of Obstetrics and Gynaecology* 129(1), 69–103.

López-Bravo, I.M., Sepúlveda, H. and Valdés, I. (1997) Acute respiratory illness in the first 18 months of life. *Pan American Journal of Public Health* 1, 9–17.

Monn, C., Schindler, C., Brändli, O., Ackermann-Liebrich, U. and Leuenberger, P. (1998) Personal exposure to nitrogen dioxide in Switzerland. SAPALDIA team. Swiss study on air pollution and lung diseases in adults. *Science of the Total Environment* 215, 243–251.

Morawska, L., He, C., Hitchins, J., Gilbert, D. and Parappukkaran, S. (2001) The relationship between indoor and outdoor airborne particles in the residential environment. *Atmospheric Environment* 35(20), 3463–3473.

Morris, K., Morganlander, M., Coulehan, J.L., Gahagen, S. and Arena, V.C. (1990) Wood-burning stoves and lower respiratory tract infection in American Indian children. *American Journal of Diseases of Children* 144(1), 105–108.

Morrow, P.E. and Utell, M.J. (1989) *Responses of Susceptible Subpopulations to Nitrogen Dioxide*. Research Report No. 23. Health Effects Institute, Cambridge, MA.

Mtango, F.D., Neuvians, D., Broome, C.V., Hightower, A.W. and Pio, A. (1992) Risk factors for deaths in children under 5 years old in Bagamoyo district, Tanzania. *Tropical Medicine and Parasitology* 43(4), 229–233.

Orehek, J., Massari, J.P., Gayrard, P., Grimand, C. and Charpin, J. (1981) Reponse bronchique aux allergènes après exposition controlée au dioxide d'azote [Bronchial response to allergens after controlled NO$_2$ exposure]. *Bulletin Européen de Physiopathologie Respiratoire* 17, 911–915.

Overton, J.H. and Miller, F.J. (1988) Dosimetry modeling of inhaled toxic reactive gases. In: Watson, A.Y., Bates, R.R. and Kennedy, D. (eds) *Air Pollution, the Automobile and Public Health*. National Academy Press, Washington, DC, pp. 367–385.

Pandey, M.R., Neupane, R.P., Gautam, A., Shrestha, I.B. (1989a) Domestic smoke pollution and acute respiratory infections in a rural community of the hill region of Nepal. *Environment International* 15(1–6), 337–340.

Pandey, M.R., Smith, K.R. and Boleij, J.S.M. (1989b) Indoor air pollution in developing countries and acute respiratory infection in children. *The Lancet* 1(8635), 427–429.

Parkhurst, W.J., Humphreys, M.P., Harper, J.P. and Spengler, J.D. (1988) Influence of indoor combustion sources on indoor air quality. *Environmental Progress and Sustainable Energy* 7(4), 257–261.

Pilotto, L.S., Douglas, R.M., Attewell, R.G. and Wilson, S.R. (1997) Respiratory effects associated with indoor nitrogen dioxide exposure in children. *International Journal of Epidemiology* 26(4), 788–796.

Postlethwait, E.M. and Bidani, A. (1990) Reactive uptake governs the pulmonary air space removal of inhaled nitrogen dioxide. *Journal for Applied Physiology*, 68, 594–603.

QUARG (1996) *Airborne Particulate Matter in the United Kingdom (Third Report)*. Institute of Public and Environmental Health, University of Birmingham, Birmingham, UK.

Robin, L.F., Lees, P.S.J., Winget, M., Steinhoff, M., Moulton, L.H. *et al.* (1996) Wood-burning stoves and lower respiratory illnesses in Navajo children. *Pediatric Infections Diseases Journal* 15, 859–865.

Roger, L.J., Horstman, D.H., McDonnell, W., Kehrl, H., Ives, P.J. *et al.* (1990) Pulmonary function, airway responsiveness, and respiratory symptoms in asthmatics following exercise in NO_2. *Toxicology and Industrial Health* 6, 155–171.

Ryan, P.B., Soczek, M.L., Spengler, J.D. and Billick, I.H. (1988) The Boston residential NO_2 characterization study: I. Preliminary evaluation of survey methodology. *Journal of the Air Pollution Control Association* 38, 22–27.

Salome, C.M., Brown, N.J., Marks, G.B., Woolcock, A.J., Johnson, G.M. *et al.* (1996) Effect of nitrogen dioxide and other combustion products on asthmatic subjects in a home-like environment. *European Respiratory Journal* 9(5), 910–918.

Samet, J.M., Mercury, M.C. and Spengler, J.D. (1987) Health effects and sources of indoor air pollution: Part 1. *American Review of Respiratory Disease* 136, 1486–1508.

Samet, J.M., Lambert, W.E., Skipper, B.J., Cushing, A.H., Hunt, W.C. *et al.* (1993) Nitrogen dioxide and respiratory illnesses in infants. *American Review of Respiratory Disease* 148(8), 1258–1265.

Schupp, A., Kaaden, R., Islam, M.S., Kreienbrock, L. Porstmann, F. *et al.* (1994) Relation of traffic-related air pollution and airway responsiveness in children from Duisburg. *Allergologie* 17(12), 591–597.

Schwela, D.H. (1997) Cooking smoke: a silent killer. *People and the Planet* 6(3), 1–9.

Shah, N., Ramankutty, V., Premila, P.G. and Sathy, N. (1994) Risk factors for severe pneumonia in children in south Kerala: a hospital-based case-control study. *Journal of Tropical Pediatrics* 40(4), 201–206.

Shah, A.S.V., Langrish, J.P., Nair, H., McAllister, D.A., Hunter, A.L. *et al.* (2013) Global association of air pollution and heart failure: a systematic review and meta-analysis. *The Lancet* 382, 1039–1048.

Shukla, V., Dalal, P. and Chaudhary, D. (2010) Impact of vehicular exhaust on ambient air quality of Rohtak city, India. *Journal of Environmental Biology* 31(6), 929–932.

Smith, K.R. (1989) Dialectics of improved stoves. *Economic and Political Weekly* 11, 517–522.

Smith, K.R. (1999) *Indoor Air Pollution. Pollution Management in Focus*. World Bank, Washington, DC.

Stewart, R.D., Peterson, J.E., Baretta, E.D., Bachand, R.T., Hosko, M.J. *et al.* (1970) Experimental human exposure to carbon monoxide. *Archives of Environmental Health* 21(2), 154–164.

Tao, J., Cheng, T., Zhang, R., Cao, J., Zhu, L. *et al.* (2013) Chemical composition of $PM_{2.5}$ at an urban site of Chengdu in southwestern China. *Advances in Atmospheric Sciences* 30, 1070–1084.

TERI (2015a) *Air Pollution and Health*. Discussion paper. The Energy and Resources Institute, New Delhi.

TERI (2015b) *TERI Energy and Environment Data Directory and Yearbook 2014/15*. The Energy and Resources Institute, New Delhi.

Thakur, A. (2017) Study of ambient air quality trends and analysis of contributing factors in Bangalore, India. *Oriental Journal of Chemistry* 33(2), 1051–1056.

Utell, M.J., Morrow, P.E., Speers, D.M., Darling, J. and Hyde, R.W. (1983) Airway responses to sulfate and sulfuric acid aerosols in asthmatics: an exposure–response relationship. *American Review of Respiratory Disease* 128, 444–450.

Victora, C.G., Fuchs, S.C., Flores, J.A., Fonseca, W. and Kirkwood, B. (1994) Risk factors for pneumonia among children in a Brazilian metropolitan area. *Pediatrics* 93(6), 977–985.

Viegi, G., Carrozzi, L., Paoletti, P., Vellutini, M., DiViggiano, E. *et al.* (1992) Effects of the home environment on respiratory symptoms of a general population sample in middle Italy. *Archives of Environmental Health* 47(1), 64–70.

Von Nieding, G., Wagner, H.M., Krekeler, H., Smidt, U. and Muysers, K. (1970) Absorption of NO_2 in low concentrations in the respiratory tract and its acute effects on lung function and circulation. Paper presented at the Second International Clean Air Congress, Washington, DC.

Von Nieding, G., Wagner, M., Krekeler, H., Smidt, U. and Muysers, K. (1971) Minimum concentrations of NO_2 causing acute effects on the respiratory gas exchange and airway resistance in patients with chronic bronchitis]. *Internationale Archiv für Arbeitsmedizin* 27, 338–348.

Von Nieding, G., Krekeler, H., Fuchs, R., Wagner, M. and Koppenhagen, K. (1973) Studies of the acute effects of NO$_2$ on lung function: influence on diffusion, perfusion and ventilation in the lungs. *Internationale Archiv für Arbeitsmedizin* 31, 61–72.

Ware, J.H., Dockery, D.W., Spiro, A., Speizer, F.E. and Ferris, B.G. (1984) Passive smoking, gas cooking, and respiratory health of children living in six cities. *American Review of Respiratory Disease* 129(3), 366–374.

Wesley, A.G. and Loening, W.E. (1996) Assessment and 2-year follow-up of some factors associated with severity of respiratory infections in early childhood. *South African Medical Journal* 64, 365–368.

WHO (2005) *Air Quality Guidelines for Particulate Matter, Ozone, Nitrogen Dioxide and Sulfur Dioxide. Global Update 2005 Summary of Risk Assessment*. World Health Organization, Geneva, Switzerland.

WHO (2006) *Air Quality Guidelines Global Update*. WHO Regional Office for Europe, Copenhagen, Denmark.

WHO (2015) World Health Statistics 2015. World Health Organization. Available at: http://apps.who.int/iris/bitstream/handle/10665/170250/9789240694439_eng.pdf;jsessionid=F5CF3084F0FEF5D1DC50936FB6EDB828?sequence=1 (accessed 24 August 2018).

Wjst, M., Reitmeir, P., Dold, S. Wulff, A., Nicolai, T. *et al.* (1993) Road traffic and adverse effects on respiratory health in children. *British Medical Journal* 307, 596–600.

9 Megacities of Developing Countries

Arti Choudhary, Manisha Gaur and Anuradha Shukla*

Central Road Research Institute, New Delhi, India

Abstract

More than half of the world's population now lives in urban areas of the developing world. It is predicted that by 2030 there will be 41 megacities in the world and most of these megacities will be located in developing countries. Urban air pollution in most of the megacities (with a population of >10 million), such as Delhi, Beijing and Jakarta, has worsened due to the cumulative effects of industrialization, population growth, and increased use of automobiles. The sources that are responsible for higher emission loads in the megacities can be grouped into several sectors such as domestic, commercial, transport and industrial activities for anthropogenic sources and biogenic sources. This chapter provides a brief introduction to megacity air pollution. We also bring together recent comprehensive reviews from particular megacities of the developing world. In developing countries, megacities are suffering from high particulate matter (PM) loads, which are associated with increased mortality rates. In contrast, reduced emissions of SO_2 have been reported for megacities in developed countries due to use of cleaner, low sulfur fuel, such as natural gas. Various control measures have been implemented to improve air quality in developing countries but there is a need to focus on non-exhaust emissions as well, including solid waste and biomass burning. A mix of different ideas on policy measures will be needed to improve the air quality of megacities rather than working on a single strategy.

9.1 Introduction

Urban populations keep growing due to increasing options for livelihood, particularly in urban areas of developing countries. Due to the cumulative effects of population growth, industrialization and increased vehicle use, several problems have arisen; for example, urban air pollution, global warming and several health problems associated with the heart and lungs. By the year 2030, there will be 41 megacities in the world and most of them will be located in developing countries (UN, 2014). The Indian megacities such as

Delhi, Mumbai and Kolkata collectively have a population of more than 46 million and are facing several issues such as poor air quality, climate change, global warming and numerous health problems. The air pollution consists of higher concentration pollutants such as particulate matter (PM), SO_2, NO_x, ozone (O_3) and greenhouse gases (GHGs), which affect the atmosphere on local, regional and global scales. The rising concentration of air pollution in megacities leads to a high risk of respiratory diseases for human beings. Other associated large-scale problems are acidification and global climate change – which

---

* Corresponding author: anuradha.crri@gmail.com

© CAB International 2019. *Air Pollution: Sources, Impacts and Controls*
(eds P. Saxena and V. Naik)

have considerable impacts on agricultural yield – reduced biodiversity of flora and fauna, and damage to buildings and cultural monuments. Fresh and clean air is a basic requirement for the health and welfare of humankind, and also a necessity for sustainable economic development.

Molina and Molina (2002) defined a megacity as a city having a population of more than 10 million; however, the boundaries of megacities are not defined. In 2017 there were 31 megacities in existence compared to just 10 in 1990, and another 10 cities are projected to reach megacity status by 2030 (UN, 2016). Most of the megacities are located in developing countries such as Asia, Latin America and Africa, and several studies reported that this number will increase by 2030 (UN, 2016; UN Habitat, 2016). Currently, the largest cities among these are Tokyo and Jakarta, having populations of over 30 million. The world's megacities and allied populations are listed in Table 9.1. This clustering of inhabitants increases the consumption of resources and releases extensive pollution in different forms, which cause deterioration of the environment and also cause several health issues for human beings.

There are numerous sources of air pollution, which can be categorized into transport, residential, commercial and industrial activities for anthropogenic sources and biogenic sources. The transport sector includes mainly motor vehicles, trains and aircraft; industrial and residential activities include burning of wood, coal and gas. Burning of vegetation and soils is a source of biogenic (natural) emissions (Guenther *et al.*, 2006). The severity of air pollution is reported in megacities of developing countries, where most of the air pollutants exceed World Health Organization (WHO) guidelines. Although London has fairly favourable pollutant levels compared to the world's other megacities, excesses of NO_x, O_3 and PM over the prescribed limit occur most readily in London (Akimoto, 2003). According to the survey by UNEP (2016), it was found that seven megacities – Mexico City, Beijing, Cairo, Jakarta, Los Angeles, São Paulo and Moscow – exceeded the WHO defined limit of pollutant levels for more than three pollutants. Mexico City has a serious problem with pollutants SO_2, PM, CO and O_3, and moderate-to-heavy pollution with lead (Pb) and nitrogen dioxide (NO_2). This higher PM concentration for Mexico City is due to its preferable conditions (high altitude and climate) and also due to higher use of old and poorly maintained vehicles. Developing countries' megacities are suffering from the highest PM load, which is associated with increasing mortality rates. Many of the megacities also reported reduced emissions of SO_2 due to use of cleaner, low sulfur fuel, such as natural gas (Gurjar *et al.*, 2016). These scenarios indicate the need for implementation of control measures in megacities around the world to improve air quality and protect public health. This chapter offers a brief introduction to the air pollution in megacities of developing countries. It is an extensive topic but here the chapter only comprises the key reasons behind the emergence of megacities, the effect of excessive pollution in megacities of developing countries, and air-quality assessment tools and mitigation.

9.2 The Reason Behind the Emergence of Megacities

Megacities have very large urban sprawls that are associated with numerous environmental challenges, such as increasing air-pollution emissions (Molina and Molina, 2004; Gurjar and Lelieveld, 2005). The global population is expected to rise up to 9.1 billion by 2050 (UN Habitat, 2016), and among this most of the people will be living in urban areas, with megacity populations expected to increase from 3.4 to 6.3 billion during the period 2009 to 2050 (Kim, 2007). According to the UN Report the world population in urban areas has reached 50%. In 2010, about 75% of the inhabitants of the more developed regions lived in urban areas and about 45% in the less-developed regions. These developments have changed the correlation of physical, social and economic processes in cities. For example, uncontrolled urban sprawl leads to increases in environmental problems due to high traffic volume, irregular industry, low-quality housing, etc.

Rapid urbanization has resulted in a significantly increased number of road vehicles everywhere but especially in megacities, which is making people more susceptible to air pollution and its health effects. This rapid development process is associated with several problems, therefore a sustainable environment has become a challenge around the world. In today's developing countries, megacities are well known

Table 9.1. The world's megacities and allied populations. (From UN, 2016.)

Rank	City, country	Population in 2016 (thousands)	City, country	Projected population in 2030 (thousands)
1	Tokyo, Japan	38,140	Tokyo, Japan	37,190
2	Delhi, India	26,454	Delhi, India	36,060
3	Shanghai, China	24,484	Shanghai, China	30,751
4	Mumbai (Bombay), India	21,357	Mumbai (Bombay), India	27,797
5	São Paulo, Brazil	21,297	Beijing, China	27,706
6	Beijing, China	21,240	Dhaka, Bangladesh	27,374
7	Ciudad de México (Mexico City), Mexico	21,157	Karachi, Pakistan	24,838
8	Kinki M.M.A. (Osaka), Japan	20,337	Al-Qahirah (Cairo), Egypt	24,502
9	Al-Qahirah (Cairo), Egypt	19,128	Lagos, Nigeria	24,239
10	New York-Newark, USA	18,604	Ciudad de México (Mexico City), Mexico	23,865
11	Dhaka, Bangladesh	18,237	São Paulo, Brazil	23,444
12	Karachi, Pakistan	17,121	Kinshasa, Democratic Republic of the Congo	19,996
13	Buenos Aires, Argentina	15,334	Kinki M.M.A. (Osaka), Japan	19,976
14	Kolkata (Calcutta), India	14,980	New York-Newark, USA	19,885
15	Istanbul, Turkey	14,365	Kolkata (Calcutta) India	19,092
16	Chongqing, China	13,744	Guangzhou, Guangdong, China	17,574
17	Lagos, Nigeria	13,661	Chongqing, China	17,380
18	Manila, Philippines	13,131	Buenos Aires, Argentina	16,956
19	Guangzhou, Guangdong, China	13,070	Manila, Philippines	16,756
20	Rio de Janeiro, Brazil	12,981	Istanbul, Turkey	16,694
21	Los Angeles-Long Beach-Santa Ana, USA	12,317	Bangalore, India	14,762
22	Moskva (Moscow), Russian Federation	12,260	Tianjin, China	14,655
23	Kinshasa, Democratic Republic of the Congo	12,071	Rio de Janeiro, Brazil	14,174
24	Tianjin, China	11,558	Chennai (Madras), India	13,921
25	Paris, France	10,925	Jakarta, Indonesia	13,812
26	Shenzhen, China	10,828	Los Angeles-Long Beach-Santa Ana, USA	13,257
27	Jakarta, Indonesia	10,483	Lahore, Pakistan	13,033
28	Bangalore, India	10,456	Hyderabad, India	12,774
29	London, United Kingdom	10,434	Shenzhen, China	12,673
30	Chennai (Madras), India	10,163	Lima, Peru	12,221
31	Lima, Peru	10,072	Moskva (Moscow), Russian Federation	12,200
32			Bogota, Colombia	11,966
33			Paris, France	11,803
34			Johannesburg, South Africa	11,573
35			Krung Thep (Bangkok), Thailand	11,528
36			London, United Kingdom	11,467
37			Dar es Salaam, United Republic of Tanzania	10,760
38			Ahmadabad, India	10,527

Continued

Table 9.1. Continued.

Rank	City, country	Population in 2016 (thousands)	City, country	Projected population in 2030 (thousands)
39			Luanda, Angola	10,429
40			Thanh Pho Ho Chi Minh (Ho Chi Minh City), Vietnam	10,200
41			Chengdu, China	10,204

for having the highest levels of pollution. Therefore, it is increasingly important to reduce pollution levels from different sources, particularly from high traffic volumes, industrial activities and domestic heating emissions.

9.3   Effect of Excessive Pollution in Megacities

9.3.1   Adverse health impacts

Megacities have become hotspots of air pollution, making people susceptible to respiratory and heart diseases (Molina and Molina, 2004; Gurjar *et al.*, 2008). The Asian megacities such as Dhaka (Bangladesh) and Karachi (Pakistan) were ranked as the poorest megacities in the world in terms of air quality (Gurjar *et al.*, 2008). This poorest air quality raises mortality rates (Gurjar *et al.*, 2010). According to the World Health Organization, around 7 million people died prematurely due to indoor and/or outdoor air pollution. This indicates the severity of the environmental health risk (WHO, 2014; Ravindra *et al.*, 2015). According to the Global Burden of Disease report, India accounts for 19% of the world's premature deaths and of these, around 628,000 premature deaths are linked with air pollution diseases annually (GBD, 2013). Numerous studies have been conducted on air pollution emissions and their implications in megacities (e.g. Mage *et al.*, 1996; Madronich, 2006; Lawrence *et al.*, 2007; Butler *et al.*, 2008; Gurjar *et al.*, 2008), and different agencies and governments have been notified and encouraged to take immediate steps to combat the problems by joining hands at national and international levels.

A major threat to clean air is vehicular pollution, especially in the swiftly motorizing megacities

of East and Southeast Asia (Kojima *et al.*, 2000; Krupnick and Harrington, 2000). A variety of pollutants such as SO_2, CO, NO_x, volatile organic compounds (VOCs) and particulates are present in the atmosphere, due to partial burning of petrol, diesel and coal vehicles, industrial boilers and power plants. These pollutants impact on urban and regional air quality, and cause significant damage in the cities (Davis *et al.*, 2000). The higher the magnitude of pollutant concentration, the higher the number of people suffering from respiratory diseases, leading to deaths and serious health hazards. For a sustainable and healthier environment, special efforts should be made for educating the general public and local leaders about the adverse effects of urban pollution, how causes lead to consequences, and mitigation strategy. The range of adverse health effects linked with air pollution is very wide, but most commonly it affects the respiratory and cardiovascular systems. Human beings of all ages (children, women, men and older adults) are susceptible to these adverse health effects (WHO, 2005; Peel *et al.*, 2006; Tecer *et al.*, 2008). The health risks increase with the intensity of exposure. According to a study, most developing countries are suffering from respiratory diseases, of which 42% can be attributed to air pollution (Prúss-Utsun and Corvalán, 2006).

9.3.2   Climate change

Climate change is the most critical global challenge in present times as well as in the future. It is linked with several problems such as extreme weather, heat waves, pollution episodes, increased pollen season, forest fires, desertification, indirect costs, etc. More vulnerable and significant impacts are reported on human health, visibility level and the ecosystem. The increasing fine particles and photochemical oxidants are present in the urban centres, which have a substantial impact on local and regional climates (Berntsen *et al.*, 1996; Mayer *et al.*, 2000; Rotstayn *et al.*, 2000; Yienger *et al.*, 2000). The fine particles include black carbon, organic carbon, soot, smoke and dust, and are emitted from industry, transport and residential sources. Black carbon warms the atmosphere at a higher altitude, which causes the melting of glaciers, such as the Hindu Kush-Himalaya and Tibetan glaciers (Ramanathan and Feng, 2009). On the other hand, greenhouse gases (GHGs) warm the surface and the atmosphere, which significantly interferes with rainfall, the retreat of glaciers and sea ice, and sea level, among other factors. Climate change increases human mortality through changes in air pollution.

9.4 Megacities of Developing Countries

This section will discuss the level of air pollution in different megacities of developing countries, such as India and China. In India, the total population was 1.23 billion in 2011, of which 31.17% lived in urban areas, and it is expected that the urban population of the country will reach 590 million (38%) by 2030 (UN, 2014). As of 2014, there were 28 megacities all over the world, such as Tokyo, Guangzhou, Seoul, Delhi, Mumbai, Mexico, New York, São Paulo

and Istanbul, with a total of 453 million people (UN, 2014) compared to just 14 million in 1995. Mumbai had the highest population, followed by Delhi and Kolkata (Fig. 9.1). In 2013 Delhi had the maximum number of vehicles, i.e. 8.827 million, followed by Mumbai (2.5 million) and Kolkata (440,000) (Fig. 9.2).

Poor air quality is one of the most serious environmental problems faced by people living in urban areas of developing countries. The increase in the deterioration of air quality in Mumbai, Delhi and Kolkata is much more significant than that observed in megacities of developed countries. This may be due to high-rise buildings in Indian megacities, which restrict the self-cleaning capabilities to reduce the level of air pollutants (Kumar *et al.*, 2013). With worsening air pollution and its adverse health impacts, there has been an increase in the number of air-quality studies in India (Sood, 2012; Chatterjee *et al.*, 2013; Guttikunda and Goel, 2013; Rizwan *et al.*, 2013; Sandeep *et al.*, 2013; Guttikunda *et al.*, 2014; Pant *et al.*, 2015; Gaur *et al.*, 2016). The most common sources, responsible for increased emission loads, have been identified as the transport, domestic, commercial and industrial sectors.

9.4.1 Beijing

Beijing, the capital of China, is a rapidly developing megacity with a population of 20 million, as of 2014. The city is located on the northern side of the Great China Plain, which links with the south

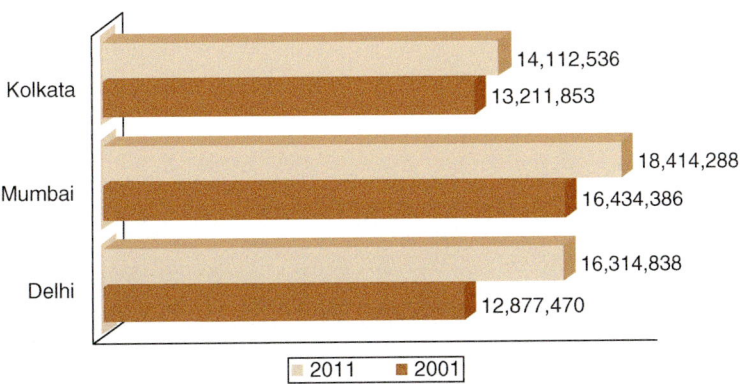

Fig. 9.1. Population of megacities in India. (From DES, 2013.)

of Beijing and is surrounded by mountainous ranges on three sides – the west, north and northeast. Thus, the city behaves like a dustpan that can accumulate air pollutants. Beijing has a semi-humid climate with hot and humid summers and dry winters (Molina and Molina, 2004). In 1998 Beijing was highly dependent on coal for its energy needs and this, coupled with the use of more than 1 million vehicles, led to a change from coal-dominated to a mix of coal- and vehicle-based air pollution. Due to unplanned urbanization and a growing population, the increase in the use of private vehicles

and traffic congestion led to high emissions of harmful pollutants in the city (Deng and Huang, 2004; Zhao *et al.*, 2005). Yu *et al.* (2013) have reported 27% secondary sulfur, 17% vehicle exhaust and 24% of fossil fuel and biofuels burning as major contributors to $PM_{2.5}$ emissions. Beijing was found to be one of the largest emitters of VOC pollution in China (Bo *et al.*, 2008). However, after the implementation of air-pollution control strategies and mitigation after 1998, CO and SO_2 levels showed decreasing trends and are now below the limits set by the National Ambient Air Quality Standard (NAAQS) of China, of 4 mg/m^3 and 60 µg/m^3, respectively. Concentrations of NO_2 and PM_{10} are meeting China NAAQS at 40 µg/m^3 and 70 µg/m^3 for NO_2 and PM_{10}, respectively. Switching to the use of liquefied petroleum gas (LPG) and compressed natural gas (CNG), and phasing out the high emissions in sectors such as power plants, metallurgy and chemical industries, has resulted in the decreasing trend of various pollutants such as CO, NO_x, SO_2 and PM (Fig. 9.3). Despite control strategies, the issues of residential coal burning, construction activities and dust storms remain the primary cause of PM_{10} in Beijing city. A higher level of ozone has also been found at various sites,

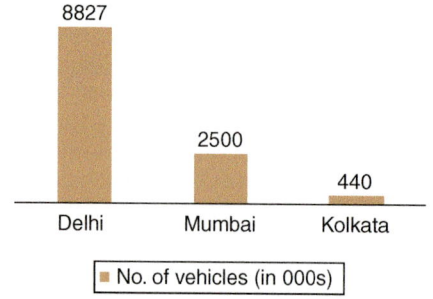

Fig. 9.2. Numbers of vehicles in all the three megacities of India. (From DES, 2013.)

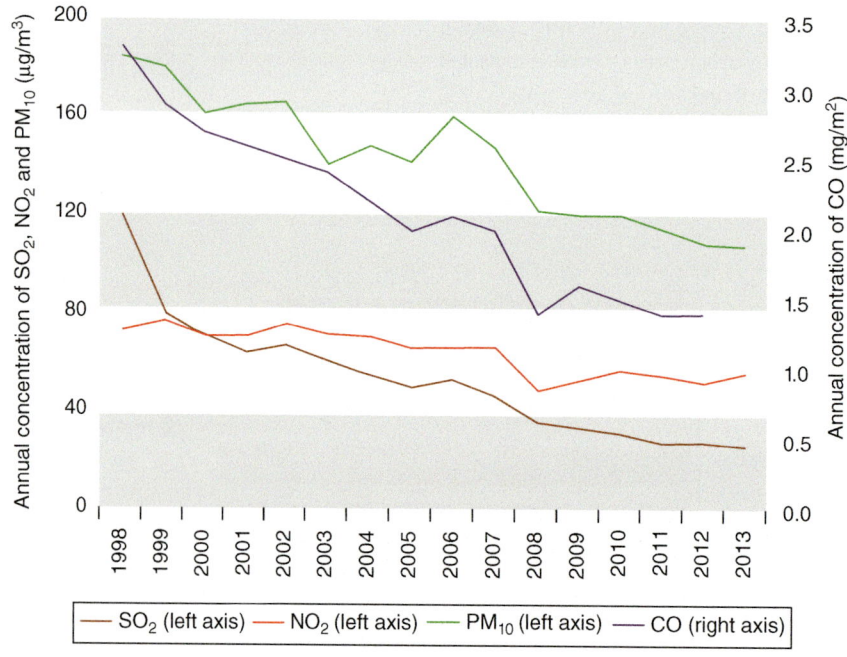

Fig. 9.3. Annual concentration of air pollutants in Beijing, 1998–2013. (From UNEP, 2016.)

which plays a significant role in smog pollution (Shao *et al.*, 2006; Wang *et al.*, 2006).

9.4.2   Delhi

Delhi (28°24'17"N to 28°53'N, 76°20'37"E to 77°20'37"E) lies in the Gangetic plains with the Thar Desert of Rajasthan state to the west, central hot plains to the south, and hilly regions to the north and east. Delhi is one of the largest megacities of South Asia (Gurjar *et al.*, 2016) and is one of the most polluted cities in the world, where air pollutant concentrations are often found to exceed the NAAQS. Delhi has 29 industrial areas and factory complexes with a different range of industries, including food and beverages, metal and alloys, paper and leather products (DES, 2013; Pant *et al.*, 2015). It has a semi-arid climate with long summers (April to October), followed by a monsoon season and winter (October to mid-February) with heavy fog (Gurjar *et al.*, 2016). According to the 2011 census (Statistical Abstract of Delhi 2012, 2012), Delhi had a population of 16.7 million; about 97.5% of the population lived in urban areas and the remaining 2.5% in rural areas. Significant sources of urban air pollution in Delhi are fuel combustion from power plants, industries and dense vehicular population.

In 2014 Delhi was the second-most-populated megacity in the world after Tokyo, with a population of 25 million (UN, 2014). The combination of various sources of pollution such as industries, power plants, brick kilns in surrounding areas, vehicular emission, road dust and mineral dust through long-range transport from the Thar desert are responsible for Delhi's pollution (Gurjar *et al.*, 2004; Kulshrestha, 2009). The vehicular population, including private vehicles, buses, jeeps, two wheelers and taxis, has been continuously rising over the past few decades. Power plants and some industries in the city emit high levels of particulates SO_x and NO_x, and the changing land-use pattern of this rapidly expanding city also causes high dust emission (Kumar *et al.*, 2014). In Delhi, the transport sector is one of the major sources of CO emission (Ravindra *et al.*, 2006) and was estimated to contribute 86% during 1990–2000 (Gurjar *et al.*, 2004; Gurjar *et al.*, 2016). In terms of particulate matter, Delhi appears to be the most

polluted city in the world. According to some studies, 80% of Delhi's PM_{10} emissions result from industrial sources and power plants and about 15% are from traffic emissions (Reddy and Venkataraman, 2002; Gurjar *et al.*, 2004). According to Guttikunda *et al.* (2011), anthropogenic sources such as transport, road dust, power plants and municipal solid waste (MSW) burning share the highest percentage at 8%, 59%, 12% and 8%, respectively, in 2010.

According to Kulshrestha (2015), Delhi needs a new feasible policy through an interstate task force in order to reduce carbon soot emissions from the transport sector, dust emissions from construction, reactive nitrogen (NH_3, NO_2, N_2O) emissions from transport and domestic sectors, and trans-boundary pollution. Studies have shown that every year about 40,000 people are dying prematurely due to air pollution, including 7500 in Delhi (Brandon and Homman, 1995). Health risks caused by air pollution are 12 times higher in Delhi in comparison to the national average. A report published by the World Bank (2003) found a 10% reduction in PM_{10} concentration might lead to 1000 fewer deaths every year in Delhi. A study carried out by CPCB (2012) has reported that respiratory diseases are 1.7 times more prevalent in Delhi when compared to a control environment. Findings also suggested that reduction in lung function is 20% more prevalent in residents of Delhi when compared to a control environment. Another study has shown that children in Delhi are more prone to lung function impairment, particularly in winters (CPCB, 2012).

Delhi has undertaken a major intervention by shifting to the metro rail system from the motor transport pattern. A study revealed that transport from existing metro lines can reduce criteria pollutants by up to 7% and this can reach up to 20–25% if the full length starts functioning (Zhu *et al.*, 2012). The metro railway system has also made significant improvements to pollution levels in various other megacities around the world, such as Shanghai, Beijing, Hong Kong, Bangkok and Mumbai. For traffic management, various measures have been taken by the Delhi government such as road infrastructure development, bus-only lanes, parking area demarcation, and the construction of flyovers and multi-lane roads. Interstate trucks that are not destined for Delhi are also prohibited from entering the city.

Recently in 2015, the Delhi high court has taken strict actions against vehicles older than 10 years, and industries are also advised to switch to the use of cleaner fuel and to install pollution control devices. For energy, natural gas replaced coal to reduce pollution during the Commonwealth Games in 2010. Only stringent action and effective policies can manage the air quality of the Delhi megacity to meet the requirement of its growing population.

9.4.3   Kolkata

Kolkata, meaning 'City of Joy', had a population of around 14.1 million in 2011, making it the third-most-populous megacity in India (Bhaduri, 2013). It is located between 22°32'N latitude and 88°20'E longitude in north-eastern India in the Ganges Delta near the Bay of Bengal, at an average elevation of 1.5–9 m above mean sea level (MSL). Kolkata has a humid tropical climate with hot and dry summers (February–April), monsoon season (May–October) and moderate winters (November–January).

According to Chakraborty and Bhattacharya (2004), in Kolkata, 50% of the total suspended particulate matter (SPM) is emitted from the transport sector and 48% comes from industries. In Kolkata, PM per capita emission was 0.28 kg in 1990 and increased to 0.33 kg in 1995, resulting in 19% growth (Sharma *et al.*, 2002). Emissions from vehicles (50%) and industries (48%) are also the main contributor of the higher SPM levels. Due to these high levels of SPM and PM_{10}, Kolkata is among the most polluted cities of the world (Chakraborty and Bhattacharya, 2004). Nagpure *et al.* (2010) reported that in 2010, the transport sector was responsible for about 8 Gg of PM emissions and 111 Gg of CO emissions in Kolkata, and similar to Delhi, heavy commercial vehicles were the biggest source (29%) of PM emissions, followed by two- and three-wheeled vehicles. Karar and Gupta (2006) also suggested that vehicle exhaust and clusters of industries were the primary causes of high levels of SPM. Levels of particulate matter, volatile organic compounds (VOCs), ozone and oxides of nitrogen are polluting the air of the Kolkata city at an alarming rate and exceeding the standard limits for a larger number of days (CSE, 2011).

Due to chronic exposure, Kolkata showed a positive relationship between air pollution and a number of diseases such as cardiovascular and respiratory disease and cancer (Ghose, 2009). The study also proved that an average citizen of Kolkata is seven times more prone to respiratory illness in comparison to the rural citizen. A higher pollution level also reduces visibility and can damage the infrastructure. In order to improve the air quality of Kolkata city, the West Bengal Pollution Control Board (WBPCB) took strict action against fireworks in 2004. In partnership with India, the Canada Environment Facility, WBPCB, has taken steps to convert coal-fired boilers to oil-fired to improve air quality. This action has reduced PM emissions from boilers by about 90% (Chakraborty and Bhattacharya, 2004). In July 2008, the high court issued an order to phase out old commercial vehicles and two-stroke autos by December 2008 from the city streets. In Delhi, all diesel vehicles that are more than 15 years old and are BS-I, BS-II shall be scrapped and no NOC (no objection certificate) for transfer of such vehicles will be issued by the government. Furthermore, public awareness, proper management of traffic and promotion of public transport will certainly lead to successful results.

9.4.4   Mumbai

According to the World Health Organization (WHO, 2016), Mumbai has been declared as the fifth-most-polluted megacity in the world in terms of PM_{10} levels. It is located between 18°56'N latitude and 72°51'E longitude at the Arabian Sea, with an average elevation from 10 m to 15 m above MSL. The city exhibits a moderately hot climate with high humid conditions, due to its coastal location. The post-monsoon and winter months in Mumbai have calm winds with stagnant conditions leading to air pollution build-up (Gurjar *et al.*, 2016). Rapid commercial and industrial growth in the city led to an increase in the population and so the need for better infrastructure and more consumption of energy. A study done by CPCB (2010) suggests that road dust suspension due to vehicular activities (30%), power plants (21%), solid waste burning (14%) and construction activities (9%) are the major contributors for most of the PM emissions in the city. SO_2 emissions have been found to be highest in Mumbai, followed by Delhi and Kolkata. The prime sources are industries and power plants in all three Indian megacities. Interestingly, emissions from the transport sector have been reduced by four to five times after the implementation of various policies and strategies (Gurjar *et al.*, 2016).

9.5 Air Quality Index (AQI)

Along with the measurement of pollution, AQI focuses on its health effects, depending on the level of exposure from a few hours to several days. AQI ranges from 0 to 500 are shown in Fig. 9.4. The different numerical value or scale assigned indicates the level of pollution and hazard to human health. The higher the AQI value, the higher the air pollution severity and therefore the health effects. For example, an AQI value of 50 represents good air quality with little potential to affect public health, while an AQI value over 300 represents hazardous air quality. The Environment Protection Agency (EPA) calculates the AQI for five major air pollutants and has established NAAQS (National Ambient Air Quality Standards) to protect public health. The pollutants are carbon monoxide, ground-level ozone, particle pollution (also known as particulate matter), NO_x and sulfur dioxide. The EPA has also assigned a specific colour to each AQI category so that it will be easier for people to understand quickly whether air pollution in their communities reaches unhealthy levels. For example, orange means that conditions are 'unhealthy for sensitive groups', while red means that conditions may be 'unhealthy for everyone', and so on.

In India, NAAQS are set by the Central Pollution Control Board (CPCB) and are applicable nationwide. The CPCB has been conferred this power by the Air Act (Prevention and Control of Pollution), 1981. In India the AQI initially covered ten cities – Delhi, Faridabad, Agra, Kanpur, Lucknow, Ahemdabad, Bangalore, Chennai, Varanasi and Hyderabad, with each having monitoring stations displaying AQI. The AQI was initiated on 5 April 2015 for monitoring air quality in the main urban centres across the country based on real-time monitoring of the concerned pollutants and enhancing public awareness for taking mitigative action. There are six AQI categories (Table 9.2), namely Good, Satisfactory, Moderately polluted, Poor, Very

Air quality index levels of health concern	Numerical value	Meaning
Good	0 to 50	Air quality is considered satisfactory, and air pollution poses little or no risk.
Moderate	51 to 100	Air quality is acceptable; however, for some pollutants there may be a moderate health concern for a very small number of people who are ununsually sensitive to air pollution.
Unhealthy for sensitive groups	101 to 150	Members of sensitive groups may experience health effects. The general public is not likely to be affected.
Unhealthy	151 to 200	Everone may begin to experience health effects; members of sensitive groups may experience more serious health effects.
Very unhealthy	201 to 300	Health alert: everyone may experience more serious health effects.
Hazardous	301 to 500	Health warnings of emergency conditions. The entire population is more likely to be affected.

Fig. 9.4. AQI and related health concerns. (From USEPA, 2014.)

Table 9.2. AQI category, pollutants and health break points.

AQI	Associated health impacts	Colour code
Good (0–50)	Minimal impact	
Satisfactory (51–100)	May cause minor breathing discomfort to sensitive people	
Moderately polluted (101–200)	May cause breathing discomfort to people with lung disease such as asthma, and discomfort to people with heart disease, children and older adults	
Poor (201–300)	May cause breathing discomfort to people with prolonged exposure, and discomfort to people with heart disease	
Very poor (301–400)	May cause respiratory illness to people with prolonged exposure. Effect may be more pronounced in people with lung and heart diseases	
Severe (401–500)	May cause respiratory impact even on healthy people, and serious health impacts on people with lung/heart disease. The health impacts may be experienced even during light physical activity	

Table 9.3. National Air Quality Index.

AQI category (range)	PM_{10} (24 hr)	$PM_{2.5}$ (24 hr)	NO_2 (24 hr)	O_3 (24 hr)	CO (8 hr)	SO_2 (24 hr)	NH_3 (24 hr)	Pb (24 hr)
Good (0–50)	0–50	0–30	0–40	0–50	0–1.0	0–40	0–200	0–0.5
Satisfactory (51–100)	51–100	31–60	41–80	51–100	1.1–2.0	41–80	201–400	0.5–1.0
Moderately polluted (101–200)	101–250	61–90	81–180	101–168	2.1–10	81–380	401–800	1.1–2.0
Poor (201–300)	251–350	91–120	181–280	169–208	10–17	381–800	801–1200	2.1–3.0
Very poor (301–400)	351–430	121–250	281–400	209–748	17–34	801–1600	1200–1800	3.1–3.5
Severe (401–500)	430+	250+	400+	748+	34+	1600+	1800+	3.5+

poor and Severe, and a total of eight pollutants (PM_{10}, $PM_{2.5}$, NO_2, SO_2, CO, O_3, NH_3 and Pb) for which short-term (up to 24-hourly averaging period) National Ambient Air Quality Standards are prescribed (MoEF, 2009), shown in Table 9.3.

9.6   Discussion and Conclusion

There is a need for accurate indicators and metrics for air pollution control, and management for natural resources, which have been brought into light by the United Nations Millennium Development Goals (MDGs). Indeed, the fight against climate change is still often criticized for being insufficiently defined and inadequately measured. It is necessary to analyse all the aspects of pollution, such as background concentrations from anthropogenic emissions. So, the creation of new ways of measuring pollution would constitute a large step in the preservation of the environment and the reduction of pollution. Many studies have reported the local- to global-scale air pollution of megacities (Gurjar *et al.*, 2008; Zhang *et al.*, 2008).

It has been observed that the transport sector is a dominant source of air pollution in American and European megacities, whereas mineral dust, local material and fuel used for domestic purposes are the main sources of pollution in Asian and African megacities, along with transportation and industrial activities. Though there are many social, economic and political barriers to using new technologies, cities like Los Angeles and Mexico City have proved the promotion of comprehensive emission reduction measures can improve air quality. Using cleaner energies such as natural gas, efficient fuel burning and increasing reliance on renewable sources of energy (solar, hydro, wind and geothermal) are some of the best ways to control and reduce air pollution without limiting economic growth.

References

Akimoto, H. (2003) Global air quality and pollution. *Science* 302(5651), 1716–1719.

Berntsen, T., Isaksen, I.S., Wang, W.C. and Liang, X.Z. (1996) Impacts of increased anthropogenic emissions in Asia on tropospheric ozone and climate. *Tellus B* 48(1), 13–32.

Bhaduri, S. (2013) Vehicular growth and air quality at major traffic intersection points in Kolkata City: an efficient intervention strategy. *The SIJ Transactions on Advances in Space Research & Earth Exploration (ASREE)* 1(1), 19–25.

Bo, Y., Cai, H. and Xie, S.D. (2008) Spatial and temporal variation of historical anthropogenic NMVOCs emission inventories in China. *Atmospheric Chemistry and Physics* 8(23), 7297–7316.

Brandon, C. and Homman, K. (1995) The cost of inaction: valuing the economy-wide cost of environmental degradation in India. Asia Environment Division, World Bank. October memo.

Butler, T.M., Lawrence, M.G., Gurjar, B.R., Van Aardenne, J., Schultz, M. and Lelieveld, J. (2008) The representation of emissions from megacities in global emission inventories. *Atmospheric Environment* 42(4), 703–719.

Chakraborty, D. and Bhattacharya, P. (2004) Air quality management strategy for Kolkata city under India Canada environment facility project – a case study. West Bengal Pollution Control Board (WBPCB). Available at: http://www.cleanairnet.org/baq2004/1527/articles-59332_bhattacharya.pdf (accessed 18 February 2010).

Chatterjee, A., Sarkar, C., Adak, A., Mukherjee, U., Ghosh, S.K. and Raha, S. (2013) Ambient air quality during Diwali festival over Kolkata – a mega-city in India. *Aerosol and Air Quality Research* 13(13), 1133–1144.

CPCB (2010) *Air Quality Monitoring, Emission Inventory and Source Apportionment Study for Indian Cities.* Central Pollution Control Board, New Delhi. Available at: http://www.moef.nic.in/downloads/public-information/Rpt-air-monitoring-17-01-2011.pdf (accessed 1 August 2018).

CPCB (2012) *Study on Ambient Air Quality, Respiratory Symptoms and Lung Function of Children in Delhi.* Environment Health Management Series: EHMS/01/2012. Central Pollution Control Board, New Delhi.

CSE (2011) *Citizen's Report: Air Quality and Mobility in Kolkata.* Centre for Science and Environment, New Delhi. Available at: http://www.indiaenvironmentportal.org.in/files/file/Kolkata%20Report.pdf (accessed 1 August 2018).

Davis, D.L., Krupnick, A. and McGlynn, G. (2000) *Ancillary Benefits and Costs of Greenhouse Gas Mitigation: Proceedings of an IPCC Co-Sponsored Workshop.* Organisation for Economic Co-operation and Development, Paris, France.

Deng, F.F. and Huang, Y. (2004) Uneven land reform and urban sprawl in China: the case of Beijing. *Progress in Planning* 61(3), 211–236.

DES (2013) *Delhi Statistical Handbook 2013.* Directorate of Economics & Statistics, Government of National Capital Territory of Delhi. Available at: http://www.indiaenvironmentportal.org.in/files/file/delhi%20statistical%20handbook%202013.pdf (accessed 1 August 2018).

Gaur, M., Singh, R. and Shukla, A. (2016) Variability in the levels of BTEX at a pollution hotspot in New Delhi, India. *Journal of Environmental Protection* 7(10), 1245.

GBD (2013) Global Burden of Disease. World Health Organization, Health Statistics and Information Systems. Available at: http://www.who.int/healthinfo/global_burden_disease/about/en (accessed 1 August 2018).

Ghose, M.K. (2009) Air pollution in the city of Kolkata: health effects due to chronic exposure. *Environmental Quality Management* 19(2), 53–70.

Guenther, C.C., Karl, T., Harley, P., Wiedinmyer, C., Palmer, P.I. and Geron, C. (2006) Estimates of global terrestrial isoprene emissions using MEGAN (Model of Emissions of Gases and Aerosols from Nature). *Atmospheric Chemistry and Physics* 6, 3181–3210.

Gurjar, B.R. and Lelieveld, J. (2005) New directions: megacities and global change. *Atmospheric Environment* 39(2), 391–393.

Gurjar, B.R., Butler, T.M., Lawrence, M.G. and Lelieveld, J. (2008) Evaluation of emissions and air quality in megacities. *Atmospheric Environment* 42(7), 1593–1606.

Gurjar, B.R., Van Aardenne, J.A., Lelieveld, J. and Mohan, M. (2004) Emission estimates and trends (1990–2000) for megacity Delhi and implications. *Atmospheric Environment* 38(33), 5663–5681.

Gurjar, B.R., Jain, A., Sharma, A., Agarwal, A., Gupta, P., Nagpure, A.S. and Lelieveld, J. (2010) Human health risks in megacities due to air pollution. *Atmospheric Environment* 44(36), 4606–4613.

Gurjar, B.R., Ravindra, K. and Nagpure, A.S. (2016) Air pollution trends over Indian megacities and their local-to-global implications. *Atmospheric Environment* 142, 475–495.

Guttikunda, S.K. and Goel, R. (2013) Health impacts of particulate pollution in a megacity – Delhi, India. *Environmental Development* 6, 8–20.

Guttikunda, S., Calori, G., Velay-Lasry, F. and Ngo, R. (2011) Air quality forecasting system for cities: modeling architecture for Delhi. *Simple Interactive Models for Better Air Quality, SIM-air Working Paper Series* 36, p. 2011.

Guttikunda, S.K., Goel, R. and Pant, P. (2014) Nature of air pollution, emission sources, and management in the Indian cities. *Atmospheric Environment* 95, 501–510.

Karar, K. and Gupta, A.K. (2006) Seasonal variations and chemical characterization of ambient PM_{10} at residential and industrial sites of an urban region of Kolkata (Calcutta), India. *Atmospheric Research* 81(1), 36–53.

Kim, S. (2007) *Immigration, Industrial Revolution and Urban Growth in the United States, 1820–1920: Factor Endowments, Technology and Geography* (No. w12900). National Bureau of Economic Research, Cambridge, MA.

Kojima, M., Brandon, C. and Shah, J.J. (2000) *Improving Urban Air Quality in South Asia by Reducing Emissions from Two-Stroke Engine Vehicles* (No. 21911). World Bank, Washington, DC.

Krupnick, A. and Harrington, W. (2000) Energy, transportation, and environment: policy options for environmental improvement. *ESMAP Paper* (224).

Kulshrestha, U.C. (2009) Atmospheric dust in India – a natural geo-engineering tool to combat climate change. *ENVIS Newsletter SES JNU* 14(3), 2–5. (School of Environmental Studies and Jawaharlal Nehru University.)

Kulshrestha, U. (2015) Some facts about recent air pollution problem in Delhi. *Journal of Indian Geophysical Union* 19(3), 351–352.

Kumar, P., Jain, S., Gurjar, B.R., Sharma, P., Khare, M. *et al.* (2013) New directions: can a 'blue sky' return to Indian megacities? *Atmospheric Environment* 71, 198–201.

Kumar, B., Verma, K. and Kulshrestha, U. (2014) Deposition and mineralogical characteristics of atmospheric dust in relation to land use and land cover change in Delhi (India). *Geography Journal* 2014, Article ID 325612, 1–11.

Lawrence, M.G., Butler, T.M., Steinkamp, J., Gurjar, B.R. and Lelieveld, J. (2007) Regional pollution potentials of megacities and other major population centers. *Atmospheric Chemistry and Physics* 7(14), 3969–3987.

Madronich, S. (2006) Chemical evolution of gaseous air pollutants down-wind of tropical megacities: Mexico City case study. *Atmospheric Environment* 40(31), 6012–6018.

Mage, D., Ozolins, G., Peterson, P., Webster, A., Orthofer, R. *et al.* (1996) Urban air pollution in megacities of the world. *Atmospheric Environment* 30(5), 681–686.

Mayer, M., Wang, C., Webster, M. and Prinn, R.G. (2000) Linking local air pollution to global chemistry and climate. *Journal of Geophysical Research: Atmospheres* 105(D18), 22869–22896.

MoEF (2009) Ministry of Environment and Forests notification. *The Gazette of India*, 16 November. Available at http://www.moef.nic.in/legis/ep/826.pdf (accessed 1 August 2018).

Molina, L. and Molina, M.J. (eds) (2002) *Air Quality in the Mexico Megacity: An Integrated Assessment*, vol. 2. Springer Science+Business Media, Berlin, Germany.

Molina, M.J. and Molina, L.T. (2004) Megacities and atmospheric pollution. *Journal of the Air & Waste Management Association* 54(6), 644–680.

Nagpure, A.K., Gurjar, B.R., Sahni, N. and Kumar, P. (2010) Pollutant emissions from road vehicles in megacity Kolkata, India: past and present trends. *Indian Journal of Air Pollution Control*, 10(2), 18–30.

Pant, P., Shukla, A., Kohl, S.D., Chow, J.C., Watson, J.G. and Harrison, R.M. (2015) Characterization of ambient $PM_{2.5}$ at a pollution hotspot in New Delhi, India and inference of sources. *Atmospheric Environment* 109, 178–189.

Peel, J.L., Metzger, K.B., Klein, M., Flanders, W.D., Mulholland, J.A. and Tolbert, P.E. (2006) Ambient air pollution and cardiovascular emergency department visits in potentially sensitive groups. *American Journal of Epidemiology* 165(6), 625–633.

Prúss-Utsun, A. and Corvalán, C. (2006) *Preventing Disease Through Healthy Environments: Towards an Estimate of the Environmental Burden of Disease.* World Health Organization, Geneva. Available at: http://www.who.int/quantifying_ehimpacts/publications/preventingdisease.pdf (accessed 28 August 2018).

Ramanathan, V. and Feng, Y. (2009) Air pollution, greenhouse gases and climate change: global and regional perspectives. *Atmospheric Environment* 43(1), 37–50.

Ravindra, K., Wauters, E., Tyagi, S.K., Mor, S. and Van Grieken, R. (2006) Assessment of air quality after the implementation of compressed natural gas (CNG) as fuel in public transport in Delhi, India. *Environmental Monitoring and Assessment* 115(1–3), 405–417.

Ravindra, K., Sidhu, M.K., Mor, S., John, S. and Pyne, S. (2015) Air pollution in India: bridging the gap between science and policy. *Journal of Hazardous, Toxic, and Radioactive Waste* 20(4), A4015003.

Reddy, M.S. and Venkataraman, C. (2002) Inventory of aerosol and sulphur dioxide emissions from India: I – Fossil fuel combustion. *Atmospheric Environment* 36(4), 677–697.

Rizwan, S.A., Nongkynrih, B. and Gupta, S.K. (2013) Air pollution in Delhi: its magnitude and effects on health. *Indian Journal of Community Medicine: Official Publication of Indian Association of Preventive & Social Medicine* 38(1), 4.

Rotstayn, L.D., Ryan, B.F. and Penner, J.E. (2000) Precipitation changes in a GCM resulting from the indirect effects of anthropogenic aerosols. *Geophysical Research Letters* 27(19), 3045–3048.

Sandeep, P., Saradhi, I.V. and Pandit, G.G. (2013) Seasonal variation of black carbon in fine particulate matter ($PM_{2.5}$) at the tropical coastal city of Mumbai, India. *Bulletin of Environmental Contamination and Toxicology* 91(5), 605–610.

Shao, M., Tang, X., Zhang, Y. and Li, W. (2006) City clusters in China: air and surface water pollution. *Frontiers in Ecology and the Environment* 4(7), 353–361.

Sharma, C., Dasgupta, A. and Mitra, A.P. (2002) Future scenarios of inventories of GHGs and urban pollutants from Delhi and Calcutta. *Population (Million)* 839(1001), 1164.

Sood, P.R. (2012) Air pollution through vehicular emissions in urban India and preventive measures. *International Conference on Environment, Energy and Biotechnology. IPCBEE* 33.

Statistical Abstract of Delhi 2012 (2012) Directorate of Economics & Statistics, Government of NCT of Delhi, Delhi, India. Available at: http://delhi.gov.in/DoIT/DES/Publication/abstract/SA2012.pdf (accessed 19 November 2018).

Tecer, L.H., Alagha, O., Karaca, F., Tuncel, G. and Eldes, N. (2008) Particulate matter ($PM_{2.5}$, $PM_{10-2.5}$, and PM_{10}) and children's hospital admissions for asthma and respiratory diseases: a bidirectional case-crossover study. *Journal of Toxicology and Environmental Health, Part A* 71(8), 512–520.

UN (2014) *World Urbanization Prospects: The 2014 Revision, Highlights (ST/ESA/SER.A/352).* Department of Economic and Social Affairs, Population Division, United Nations, New York. Available at: https://esa.un.org/Unpd/Wup/Publications/Files/WUP2014-Highlights.pdf (accessed 1 August 2018).

UN (2016) *The World's Cities in 2016 – Data Booklet (ST/ESA/ SER.A/392).* Department of Economic and Social Affairs, Population Division, United Nations.

UNEP (2016) *A Review of Air Pollution Control in Beijing: 1998–2013.* United Nations Environment Programme, Nairobi, Kenya.

UN Habitat (2016) *Urbanization and Development: Emerging Futures; World Cities Report 2016.* UN Habitat, Nairobi, Kenya.

USEPA (2014) Air quality index: a guide to air quality and your health. Available at https://www3.epa.gov/airnow/aqi_brochure_02_14.pdf (accessed 1 August 2018).

Wang, T., Ding, A., Gao, J. and Wu, W.S. (2006) Strong ozone production in urban plumes from Beijing, China. *Geophysical Research Letters* 33(21).

WHO (2005) *Air Quality Guidelines Global Update, Reporting on a Working Group Meeting.* World Health Organization, Bonn, Germany.

WHO (2014) 7 million premature deaths annually linked to air pollution. News release, 25 March. Available at: http://www.who.int/mediacentre/news/releases/2014/air-pollution/en.

WHO (2016) *Ambient air pollution: A global assessment of exposure and burden of disease.* World Health Organization, Geneva, Switzerland. Available at: http://apps.who.int/iris/bitstream/handle/10665/250141/9789241511353-eng.pdf (accessed 19 November 2018).

World Bank (2003) *Urban Air Pollution: Economic Valuation of the Health Benefits of Reduction in Air Pollution*. South Asia urban air quality management briefing note no. 12. World Bank, Washington, DC.

Yienger, J.J., Galanter, M., Holloway, T.A., Phadnis, M.J., Guttikunda, S.K., Carmichael, G.R., Moxim, W.J. and Levy, H. (2000) The episodic nature of air pollution transport from Asia to North America. *Journal of Geophysical Research: Atmospheres* 105(D22), 26931–26945.

Yu, L., Wang, G., Zhang, R., Zhang, L., Song, Y. *et al.* (2013) Characterization and source apportionment of $PM_{2.5}$ in an urban environment in Beijing. *Aerosol and Air Quality Research* 13(2), 574–583.

Zhang, Y.H., Min Hu, L.J., Zhong, A., Wiedensohler, S.C., Liu, M.O. *et al.* (2008) Regional integrated experiments on air quality over Pearl River Delta 2004 (PRIDE-PRD2004): overview. *Atmospheric Environment* 42(25), 6157–6173.

Zhao, P.S., Feng, Y.C., Zhu, T. and Wu, J.H. (2006) Characterizations of resuspended dust in six cities of North China. *Atmospheric Environment* 40, 5807–5814.

Zhu, T., Melamed, M., Parrish, D., Gauss, M., Klenner, L.G. *et al.* (2012) *Impacts of Megacities on Air Pollution and Climate*. World Meteorological Organization, Geneva, Switzerland.

10 Cost-Effective Technologies Used to Curb Air Pollution

Ravi Prakash Singh[1]* and Saumya Singh[2]
[1]*National Remote Sensing Centre, Telangana, India;*
[2]*Jawaharlal Nehru University, Delhi, India*

Abstract

Attention to the air-pollution problem is being given by the scientific community and overall global community because of the scale of its impact on human health, the living environment and climate change. Efforts to reduce air pollution have stepped up in various parts of the world. In general, technology plays a significant role in dealing with air pollution and other environmental problems faced by every section of society. In this chapter, various available technologies to curb air pollution are discussed with a focus on some cost-effective technologies. The use of air-pollutant control technologies depends upon regulatory compliance and cost-effectiveness. In the present scenario, most of the commercially available technologies, such as electrostatic precipitator (ESP), scrubs and cyclone, are efficient, but the economical operation is still questionable because the available effective pollution control technologies are not always economically viable and environmentally sound. The cost-effectiveness of any technology also varies with the level of operation, the nature of the pollutant and achievable regulatory compliance, and requires an assessment of its potential today and in the future. Some emerging cost-effective and economically viable technologies are being processed and some are at a developmental stage, such as biofiltration, phytoremediation and carbon capture technologies, which have shown potential to curb atmospheric air pollutants. In this chapter, some recent and cost-effective technologies will be discussed, although some technologies are still in the infancy stage and have limited efficiency which requires further research and development to commercially exploit these emerging technologies.

10.1 Introduction

Increasing population and economic growth are the driving forces in increasing emissions of air pollutants in many developing regions of the world. The increase in the concentration of air pollutants is associated with health problems and mortality, especially due to respiratory problems (WHO, 2016). To draw the world's attention, the UN sustainable development goal programme emphasized that clean air for everybody is a priority.

Environmental air quality is a worldwide concern and has no boundaries. The reduction of air pollutants is of the utmost importance. These air pollutants can travel a great distance and affect air quality and public health locally and regionally, therefore control strategies are required at the local level. Some common air pollutants and the relative main sources are presented in Table 10.1 (ICS UNIDO, 2007). The preventive air pollution measures include elimination, reduction and capture of air pollutants to attain

---

* Corresponding author: singhravi004@gmail.com

Table 10.1. Common air pollutants and their sources. (From ICS UNIDO, 2007.)

Pollutant	Main source
Suspended particulate matter (SPM)	Power plants, automobiles, boilers, cement industries
Sulfur oxides (SO_2)	Power plants, boilers, sulfuric acid manufacture, petroleum refining
Nitrogen oxides (NO_2)	Automobiles, power plants, nitric acid manufacturing
Carbon monoxide (CO)	Automobiles
Hydrogen sulfide (H_2S)	Pulp and paper, petroleum refining
Hydrocarbon (HC)	Automobile, petroleum refining
Ammonia (NH_3)	Fertilizer plant
Lead	Automobile, battery manufacturing

an air-quality standard. One way to reduce air pollution is by making people sensitive towards the environment by using fewer motor vehicles, encouraging the use of public transport and cycling, and the use of clean fuel and a change in industrial combustion processes to minimize pollution. In addition to this approach, there is a need for air-pollution control techniques that can coexist with economic viability. The existing commercial air-pollution control techniques, such as the settling chamber, cyclone and electrostatic precipitator wet scrubber, are widely used in power plants and industries to reduce the air pollutant concentration in the environment (Boamah *et al.*, 2012). The electrostatic precipitators used in many industries trap the particulate matters by applying charges to them, and use scrubbers, during which process exhaust is forced through a liquid spray. However, although particulate control through this mechanism is efficient, the high capital cost is the concern. The increasing pressure of population and its infrastructural demand, alongside a completely unpolluted environment without any cost, is a dream. Policies to address air pollution generate a number of benefits to human health and several countries have established restrictive regulatory laws to control air pollutant emissions in the environment. Currently, the commercially used air-pollution control techniques are efficient in reducing pollutants generated by the

power plant industry. Some recent technologies such as biofiltration, phytoremediation and carbon capture technologies have shown potential as cost-effective technologies, but there is still the question of efficiency in comparison to established air-pollution control technologies.

10.2 Technologies Available to Curb Air Pollutants

The role of the scientist and engineering professional is to develop an air-pollutant control measure aimed towards the protection of public health and welfare. The major challenges in control measures are their efficiency and cost-effectiveness. Air-pollutant prevention depends upon the use of cleaner raw fuel material, emission control and improved process design, for the reduction of pollutants from sources. The approaches for the prevention of air pollutants are described as follows.

10.2.1 Elimination approach

Proper fuel-firing practices allowing combustion improvement, along with adequate air supply, can help in lowering incomplete combustion products. The complete burning of fuel reduces gaseous emissions like carbon monoxide, which is harmful to human health.

10.2.1.1 Choice of fuel

The major source of sulfur oxides (SO_2) and nitrogen oxides (NO_2) are power plants, which emit them through a high sulfur coal-based fired combustion process. The selection of cleaner fuel is a primary step towards air-pollutant reduction measures. Gas-based combustion emits negligible amounts of particulate matters in comparison to oil- and coal-based combustion processes. The use of low sulfur coal and oil in power plants and transport can be considered as a control option for the reduction of SO_2 emissions. The use of natural gas can also reduce the NO_x, SO_2 and volatile organic compound (VOC) emissions. However, the choice of fuel mainly depends upon economics as well as environmental regulation.

10.2.1.2   Fuel cleaning

Suspended particulate matter (SPM) is emitted from coal-based power plants and oil-based transport systems. The cleaning of fuel is an important step towards clean technology. The fuel-cleaning process reduces the generation of particulate emission. The physical cleaning of coal through washing minimizes its ash and sulfur content. The alternative to coal washing is coal co-firing, with higher and lower ash content. The cleaning of fuel not only reduces the particulate emission and lowers the ash, but also increases the life of the boiler.

10.2.1.3   Choice of technology and processes

The use of more efficient technologies and processes can reduce the production of incomplete combustion (PIC) emissions. Advanced coal combustion technologies such as coal gasification, fluidized-bed combustion and enclosed coal crushers and grinders are examples of cleaner processes that may lower PICs by approximately 10% (CEC, 2005).

10.2.2   Emission-control approaches

Strategies towards controlling emissions of air pollutants are important in protecting public health and welfare. The essential planning-control activities include setting up an emission limit, identifying all emission sources, the scope of process modification, defining the control problem and the selection of a control system. But the major challenge in specifying air-pollution control for air pollutants has been to set an acceptable emission level. The selection of a suitable air-pollution control system is based on environmental implication and economic viability. The modification of a process to reduce or eliminate usually offers the most economical way to reduce emissions because it requires little capital to implement it. The control technologies can be grouped into physical and chemical processes used to separate pollutants from the carrier gas and are arranged into two classes: (i) particle control systems; and (ii) gas and vapour control systems. However, technology may differ on the basis of pollutant nature, as well as whether it is organic or inorganic. The following are some commercially used technologies to curb air pollutants.

10.2.2.1   Particle-control system

The reduction of particulate matter is an important challenge for industrial air-pollutant abatement technologies. The generated particles are collected through several mechanisms. Some of the available technologies are described below.

GRAVITY SETTLING CHAMBERS.   This particulate collection device is based on the gravity principle to settle the particulate matter in a gas stream, which is passing through its long chamber. The settling chambers are highly effective for larger particulate matter greater than 50 micrometres in an aerodynamic diameter (Wark et al., 1981; USEPA, 1998). The collection efficiency of gravity settling chambers for PM_{10} is very low, typically less than 10%. Despite the low collection efficiency of PM_{10} and $PM_{2.5}$, the gravity settling chamber has been used in the past on vast levels in refining power and heating plant industries to collect larger particles (Mycock et al., 1995). In this process, the initial requirement is to reduce the carrier gas velocity to settle the particulate matter at the bottom of the chamber through the action of gravity. The velocity of gas less than 0.5 m/s produces good results (Perry, 1984). The main advantages of a settling chamber are its low initial costs, simple construction and minimum maintenance, but it requires a large space. The settling chamber has been replaced by cyclones due to their having a lower space requirement and higher collection efficiency.

CYCLONES.   The settling chamber has limitations with regard to smaller particles. The cyclone technique exerts greater force in comparison with gravity force on the particle, so finer particles can be removed from the stream. Cyclones are based on the principle of centrifugal force to remove fine particles from a dirty air stream. The cyclones are efficient for coarse particle size, and have an efficiency of 90% for particle size greater than 20 micrometres (Boamah et al., 2012). The efficiency of the cyclone depends on its diameter; the smaller the diameter, the greater the efficiency. The efficiency can be enhanced by the use of multiple cyclones in parallel or in a series pattern. The initial cost and maintenance requirement is low, but efficiency for particles below 5–10 µ in diameter is also low and there are operational problems of corrosion and erosion (Coury *et al.*, 2004). To achieve stringent air-quality standards, more efficient air cleaning equipment, such as electrostatic precipitators, is required.

ELECTROSTATIC PRECIPITATORS.   It is difficult to remove particles with diameters less than 5 microns through the cyclone. Electrostatic precipitators (ESP) are used for this purpose, as these devices use electrostatic force to remove particles from the air stream. The principle behind this is to provide an electrostatic charge to particles in a given gas stream and then allow gas to pass through an electrostatic field that drives them to a collecting electrode. ESP can also remove the finer particles, as small as one-tenth of a micrometre, with almost 100% efficiency, but the disadvantages are that this process has high capital costs and large space requirements.

10.2.2.2   Gas and vapour phase control systems

WET SCRUBBERS.   Wet scrubbers are control devices for removing particles and/or gases from industrial exhaust streams. A wet scrubber is used when soluble gases are present and the contaminant cannot be removed in dry form. A wet scrubber operates by introducing a scrubbing liquid into the dirty gas stream – typically water. Particulates or gases are collected in the scrubbing liquid. There are various types of scrubbers but the selection of scrubber is based on factors such as the gas temperature, type of pollutant to be removed and desired efficiency. Some of the scrubbers are mainly designed to remove gaseous pollutants, e.g. packed towers and tray towers. In this application, when the dirty gas stream comes into contact with the solvent or liquid it gets adsorbed. This occurs due to physical and chemical phenomena. The absorption process is used to control hydrocarbons, NO_2, SO_2 and H_2S. Ammonia is used as an absorbent for controlling SO_2 emissions.

ACTIVATED CHARCOAL.   This is the most commonly used method, mainly in the case of emissions from small sources. It can be either physical adsorption or chemisorption. Physical adsorption works on the application of the van der Waals force, giving the advantage of reversibility and regeneration due to the weaker bonding of the gas and adsorbent material. Activated charcoal is a commonly used adsorbent due to its high surface area and material hardness. It has between 800 and 1200 m^2/g of surface area. Activated charcoal techniques are relatively expensive.

10.3   Cost-Effective Techniques

Although the big industrial plants are a major contributor of air pollutants, some small and medium enterprises, such as agricultural chemical applicators, bakeries, distilleries, printing shops and furniture manufacturers contribute to air pollution and should be subjected to air-pollution control. Although there are a number of commercial technology options available to curb the air-pollutant problem, the capital and operation of commercial technology are expensive in resolving an environmental problem. The associated operating costs typically increase as the process matures and the scale of equipment gets larger (ICS UNIDO, 2007). There are commercially used technologies to control the generic classes of air-pollutant emissions, such as CO, volatile organic compounds (VOCs), NO_x or SO_x. The use of specific control technology depends upon the need to meet regulatory compliance in a cost-effective manner. The cost-effectiveness of any technology depends on a number of factors, such as the scale of operation, the nature of the pollutant and its concentration, the rate of pollutant exhaust and existing environmental regulation. There are many determining factors when choosing which emission-control option has the best available technology (BAT). From an environmental perspective, the best option is one that reduces the total emission levels of the pollutant for the lowest cost. Economic viability can vary depending on a number of factors: the pollutant(s) for which the area is in non-attainment; the precursor of pollutants; the relative size of pollutant inventories; the existing pollutant sources; and the level of control measures. It is also important to know that economic viability does not necessarily correspond with overall effectiveness because pollutant reaction properties change spatially and temporarily. The cost-effectiveness is referred to as the cost per ton of emissions reduced, which depends upon the source of pollutant, control measures and efficiency. The analysis of any cost-effective technology requires an assessment of technologies and quantification of their potential today and in the future. Stringent air-quality standard emission reductions are more difficult to achieve because the acceptable cost of achieving those reductions could increase. For instance, a technology may be very cost-effective at reducing VOCs in an ozone non-attainment area, but if the technology only applies to very few emissions sources, or if the air-quality chemistry in the ozone non-attainment area is NO_x-dependent, the overall effectiveness in reducing ozone may be quite limited (USEPA, 2007). Some of the cost-effective technologies used to curb air-pollutant emissions are described below.

10.3.1   Biofiltration

Biofiltration technology refers to the biological removal of air contaminants from off-gas streams in a solid phase reactor, and is a well-established Air Pollution Control (APC) technology in several countries, most widely in the Netherlands and Germany. In biofiltration technology, off-gases, which contain biodegradable VOCs, or inorganic air toxics are passed through a biologically active material. It has been successfully implemented in Germany and the Netherlands in a wide range of industrial and public-sector sources to control odours with 90% efficiency. The lowered operating cost of biofiltration can provide a significant economic edge over other air-pollution control technologies (Leson and Winer, 2012). Biofilters are currently competing with APC technologies, including thermal and catalytic oxidation (incineration) techniques, and adsorption by activated carbon and condensation by refrigeration in the field of VOC removal from off-gases. Although APC techniques are efficient in controlling organic emissions, they suffer from economic disadvantages.

10.3.1.1   Basic design of biofiltration techniques

A typical biofilter for the control of air pollutants consists of biologically active material, primarily mixtures based on compost, peat or soil. The height of the filter bed is normally 1 metre. A conceptual design of an open biofilter is shown in Fig. 10.1 in which contaminated off-gas is vented from the emitting source through the filter, and air contaminants will diffuse into a wet, biologically active material layer which surrounds the filter particles. Microorganisms (mainly bacteria) metabolize the target pollutant through aerobic biodegradation, and CO_2, water and microbial biomass are produced as end products. Further, the oxidation of reduced sulfur compounds and chlorinated organic

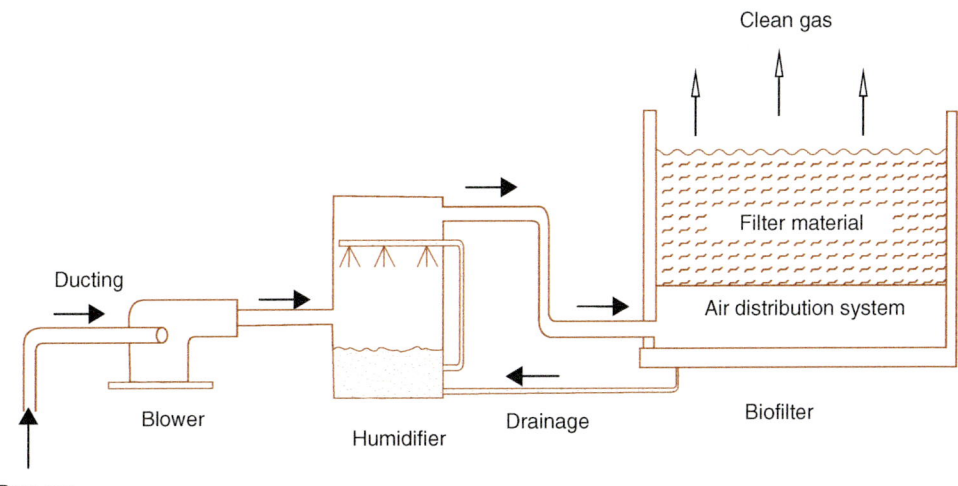

Fig. 10.1. Schematic of an open single-bed biofilter system. (From Leson and Winer, 2012.)

compounds also generates inorganic acids. The municipal waste, wood, bark, leaves and soil bed are used in biofilters. The degradation result depends upon filter material, temperature and presence of specific microorganisms. Although chlorinated organic compounds have lower rates of degradation, amines and sulfide compounds are also effectively controlled by biofilters.

The efficiency of biofilters for the abatement of air pollutants effectively depends primarily on loading rate, the concentration of these compounds in the off-gas and the biodegradation rate per volume. On the economic side, the filter cost increases with filter size. Although the biofilters have efficient capability at low gas concentration with biodegradable constituents, the biofilters also have some limitations with high organic loading and poorly biodegradable compounds of chlorinated organics.

10.3.1.2   Economic considerations

An economic comparison of available air-pollution control technology and biofiltration is carried out on a case-by-case basis and initial knowledge of physical and chemical properties of the off-gases is necessary, as biofilter size and associated capital cost are mainly determined by pollutant load. The initial capital cost of biofiltration techniques is much lower than APC techniques, making it a cost-effective abatement technique.

10.3.2   Phytoremediation

Phytoremediation is a process to mitigate pollution in air, water and soil environments, in relation to plants, and this concept has been widely and effectively applied to treat pollutants in soil and water (Cunningham and Ow, 1996; Schröder et al., 2002). But newer development and understanding of molecular and biochemical mechanisms has enhanced the metabolism of various chemicals in plants at a wider level. Phytoremediation has also been developed as a more cost-effective technology than a conventional commercial technique to curb air pollutants (Schnoor, 1997). Among all types of biological treatment of air contaminants, phytoremediation stands out for its benefits yielded from self-maintaining, soil stabilization and other advantages to

meet greater public approval (Doty et al., 2007). An understanding of the basic plant physiology and biochemistry is a prerequisite in improving the applicability of this plant-based method. The basic processes of phytoremediation involve several steps: concentration and accumulation of soil, heavy metals and organic contaminants in the root zone, which is known as phytostabilization, and the transformation and degradation of organics, including PAHs, TPHs and PCBs, and inorganics, including atmospheric nitrogen oxides and sulfur oxides, by the plant, called phytodegradation (Schnoor et al., 1995). The phytoremediation techniques to curb air pollutants are as follows.

10.3.2.1   Phytoremediation of particulate matters

Vegetation is used as a protective cover against dust through the process of phytofiltration and it is widely used in many countries. Normally, an 8-metre wide green belt may reduce dust fall by 2–3 times (Singh and Tripathi, 2007). The phytofiltration process depends upon the morphology of the plant, for example, the orientation of leaf, size, shape, wax deposition, etc., to capture dust particles from ambient air. The morphology of plant leaves affects the particulate matter capture capacity; some leaves are stickier, which is better for particle collection. The plant *Chlorophytum comosum* accumulates all fractions of particulates, but finer particulates are phytostabilized at a higher rate in comparison with coarser particulates, and through this process, it reduces the risk to human health from finer particulate matter (Gawrońska and Bakera, 2015).

10.3.2.2   Inorganic pollutants

NO_x. NO_x is an important air pollutant and a precursor of tropospheric ozone. NO_x can enter a plant system through wet/dry deposition to leaf or root systems. Penetration of NO_x through foliar systems depends upon the type of plant species, age of the plant, concentration of NO_x, and other environmental conditions. Once NO_x enters the plant system, most of the NO_2 is metabolized into organic compounds, e.g. an amino acid, through the nitrate assimilation pathway. Some plant enzymes, such as nitrate reductase, nitrite reductase or glutamine synthetase, play an important role in this process. The removal of the NO_x compound through biofiltering involves a biological process where nitrification/denitrification takes place: first NO (g) is turned to NO_2 (g) and then NO_2 (g) to NO_3 (g), and finally it is converted into nitrogen through the denitrification process. Takahashi (2001) studied the genetically engineered Arabidopsis plant, where the enzyme nitrite reductase (NiR) gene was overexpressed, which is responsible for the NO_2 assimilation. He found a positive correlation between overexpressed NiR genes and NO_2 assimilation. The results indicate that the increase in expression of the NiR level increases the NO_2-assimilation ability of plants. This study supported the proposal that further advanced research through genetic engineering can turn plants into a magic sink for NO_x, by causing overexpression of the enzymes.

SO_2. The major source of SO_2 in the atmosphere comes from fossil fuel combustion in power plants and industry (Lu *et al.*, 2010). SO_2 enters plant systems through the foliar system via leaf stomata, and further becomes involved in the reductive sulfur cycle. After entering the plant, it changes into SO_4^{2-} or SO_3^{2-} in the cell wall. In the reductive cycle, it passes through the various intermediate stages of adenyl-5-phosphosulfate and carrier protein and is converted into final product cysteine protein or other organic compounds. It was found that about 42.62 Mg of sulfur dioxide was removed through the planting of urban trees in Guangzhou, China, in 2000 (Zhang *et al.*, 2013).

10.3.2.3   Organic Pollutants

FORMALDEHYDE.   Formaldehyde is a ubiquitous air pollutant, and is considered a harmful carcinogen and mutagen. At low levels of concentration in the air, formaldehyde could be removed by plant leaves alone, while in the case of higher concentrations, the toxic chemical can be filtered through activated carbon first, and after that the plant root system and its associated microorganisms degrade and assimilate the remaining chemicals (Wolverton, 1988). When the formaldehyde concentration is as low as 8.5 mg/m^3, a spider plant (*Chlorophytum comosum* L.) shoot can effectively metabolize it into amino acids, organic acids, lipid free sugars and cell-wall components (Giese *et al.*, 1994).

BENZENE AND TOLUENE.   A volatile organic compound (VOC), benzene is genotoxic and carcinogen in nature, while toluene acts as a neurotoxic chemical (Pariselli *et al.*, 2009). It has been reported that benzene and toluene can be removed from the air and be assimilated by plants (Porter, 1994; Ugrekhelidze, 1997). An indoor ornamental plant has the capability of reducing the concentration of benzene and toluene in the air. The entire process includes several steps.

1. A plant-screening stage, where the aim is to acquire a plant which is able to remove the mixture of benzene and toluene, for example, *Chrysanthemum morifolium*, rhododendron hybrids or *Hosta plantaginea*. All these plants have the ability to remove benzene or toluene by 30%.

2. A fumigation experiment performed to investigate the concentration of benzene and toluene required for phytoremediation. The fumigation system consists of a cylindrical chamber, which is coated with Teflon to avoid loss of benzene and toluene. Simultaneously, benzene and toluene, along with compressed purified air enter into the chamber, and after around 2 hours, the concentration of benzene and toluene is measured at the outlet.

It was observed that the plant *Chrysanthemum morifolium* significantly reduced the benzene and toluene concentration (Yang *et al.*, 2010).

10.3.2.4   Economic considerations

Although phytoremediation as a technology in soil and water pollution is in wide application, with regard to air pollutant mitigation, it is still in its infant stage. Some studies of green belt plantations around thermal power plants (Govindaraju, 2012) and urban plantations have shown a significant reduction of air pollutants in the atmosphere, in a cost-effective manner (Jim and Chen, 2007). Phytoremediation involves the process of phytoextraction, phytofiltration, phytodegradation and phytovolatilization. For the physical and chemical air pollutants, phytoremediation of NO_x, SO_2 formaldehyde, benzene, toluene and particulate matter have been reported but there is still a need for research into the promotion of phytoremediation as a future technology.

10.3.3   Cost-effective technology for CO$_2$ capture

The Energy Information Administration reported that in 2005 about 83% of total US greenhouse gas (GHG) emissions in the atmosphere were from combustion and nonfuel uses of fossil fuels (EIA, 2005), and it is estimated that consumption of fossil fuels (coal, petroleum and natural gas) will increase by 27% over the next 20 years in the US (Figueroa *et al.*, 2008). It is also well known that GHG emissions are a global issue and no single nation can sufficiently reduce GHGs to stabilize their atmospheric concentrations. The effort for the reduction of GHGs must be cooperative and cost-effective to sustain both domestic and global economic growth, while reducing GHG emissions. One significant approach that holds great potential for reduction of GHG emissions is carbon capture and sequestration (CCS). Under this

mechanism, CO$_2$ is captured from large point sources, such as power plants, and injected into geological formations, such as depleted oil and gas fields and saline formations (Klara *et al.*, 2003). This method has the ability to sequester the CO$_2$ for thousands of years and this can be cost-effectively achievable and environmentally sound.

10.3.3.1   Carbon capture technologies

There are several CO$_2$ capture technologies available, some of which are commercially established and some of which are still in the infancy stage. For effective CO$_2$ capture from coal-derived power generation, there are three technological methods that can be pursued: (i) post-combustion carbon capture; (ii) pre-combustion carbon capture; and (iii) oxy-combustion, as shown in Fig. 10.2. In the post-combustion capture method, the CO$_2$ is separated from other chimney gas constituents originally present in the air, or it may be produced during combustion. In the pre-combustion capture method, carbon is removed from the fuel before the combustion process, and in

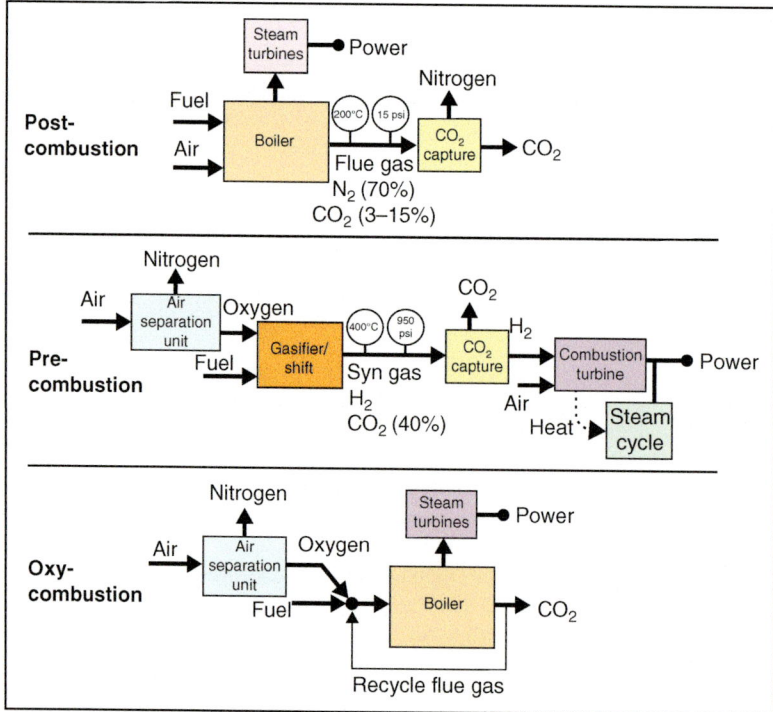

Fig. 10.2. Block diagrams illustrating post-combustion, pre-combustion and oxy-combustion systems. (From Figueroa *et al.*, 2008.)

oxy-combustion, the fuel is burned in an oxygen stream that contains little or no nitrogen.

Post-combustion CO_2 capture for power plants takes place after combustion of air and fuel to generate electricity. The most important advantage of post-combustion CO_2 capture technology is that it does not require any major modifications and can be retrofitted to existing plants through the installation of the necessary capturing equipment (Fig. 10.3).

The major technical challenge for the development of cost-effective advanced capture processes is the higher thermodynamic driving force which is required for CO_2 capture from chimney gas. In spite of this difficulty, post-combustion carbon capture technology still has the greatest near-term potential for the reduction of GHG emissions, by retrofitting the existing units that can generate two-thirds of the CO_2 emissions in the power sector.

In recent times, emerging technologies have involved a combination of products and processes that have demonstrated, either in the field or the laboratory, significant improvements reported in efficiency and cost-effectiveness over state-of-the-art technologies. Some novel approaches are discussed below.

1. Amine-based systems. In the amine-based system, amines react with CO_2 to form a water-soluble compound. At the formation of this compound, amines are able to capture CO_2 from streams with a low CO_2 partial pressure, but capacity is equilibrium limited. Through this, the amine-based systems are able to recover CO_2 from the chimney gas of conventional pulverized coal (PC)-fired power plants; however, efficiency is a disadvantage but significant cost reduction is an advantage.

Although amines have been used for many years, mainly in the removal of acid gases from natural gas, there is still the need for improvement of processes to achieve the highest efficiency.

2. Carbonate-based systems. Carbonate systems are a geological carbon sequestration process, based on the ability of a soluble carbonate to react with CO_2 to form a bicarbonate, which, when heated, releases CO_2 and reverts back to a carbonate. The lower energy requirement for regeneration in carbonate-based systems is a major advantage over amine-based systems. A K2CO3 based system has been developed by the University of Texas in Austin, USA, in which the solvent is promoted with catalytic amounts of piperazine (PZ). The K2CO3/PZ system (5 molar K; 2.5 molar PZ) shows a higher absorption rate of around 10–30% faster than a 30% solution of monoethanolamine (MEA) under favourable equilibrium conditions. Further analysis has indicated that around 5% lower energy is required with a higher loading capacity of 40% versus about 30% for MEA. The improvement in structured packing and multi-pressure stripping can enhance the additional 5–15% energy

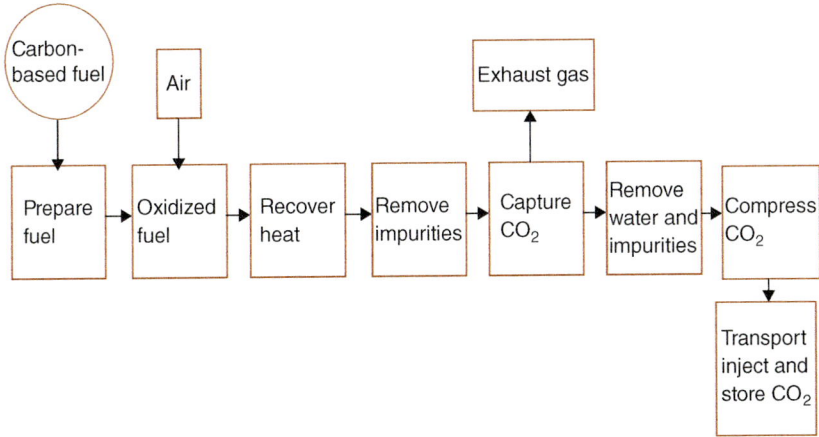

Fig. 10.3. Post-combustion CO_2 capture. (From Moazzem *et al.*, 2012.)

savings to make it an economically efficient technology (Rochelle *et al.*, 2006).

3. *Aqueous ammonia.*

The ammonia-based wet-scrubbing method is similar in operation to amine systems. The characteristics of ammonia and its derivatives are to react with CO_2 via various mechanisms, one of which is the reaction of ammonium carbonate (AC), CO_2 and water to form ammonium bicarbonate (ABC). The advantage of this system is the lower heat of reaction than amine-based systems, resulting in energy savings and greater cost-effectiveness. In addition to this, ammonia-based absorption has a number of other advantages over amine-based systems, such as the potential for high CO_2 capacity, lack of degradation during absorption/regeneration, tolerance to oxygen in the chimney gas, low cost, and potential for regeneration at high pressure. There is also the possibility of reaction with SO_x and NO_x – criteria pollutants found in flue gas – to form fertilizer (ammonium sulfate and ammonium nitrate) as a saleable by-product (Figueroa *et al.*, 2008).

There are a few concerns related to the higher volatility of ammonia compared to that of MEA. One is that the chimney gas must be cooled in the range of 60–80°F to enhance the CO_2 absorption of the ammonia compounds and also reduce the ammonia vapour emissions during the absorption process. The other major concerns are over ammonia losses during regeneration, which occur at higher temperatures. Further research and development of this process is required to increase CO_2 loading, and the application of other engineering techniques to minimize the ammonia vapour losses from the system during operation (Resnik *et al.*, 2004, 2006; Yeh *et al.*, 2005).

4. *Membrane technology.*

There are several methods for using membranes to recover CO_2 from chimney gas. In one method, the chimney gas would be passed through a bundle of membrane tubes, while an amine solution flowed through the shell side of the bundle. The CO_2 passes through the membrane absorbed in the amine, while other impurities would be blocked from the amine, thus leading to a reduction of the loss of amine as a result of stable salt formation. It should also be possible to achieve a higher loading differential between the rich amine and lean amine. The amine would be regenerated before being recycled after leaving the membrane bundle. There is a requirement for further research and development pathways in order to increase the membrane selectivity and permeability and decrease the operational cost (Falk-Pederson *et al.*, 2000). The CO_2 capture and its separation from large point sources, such as power plants, can be achieved through continued research and development.

PRE-COMBUSTION CO_2 CAPTURE. In this process, the CO_2 is removed from any industrial sources prior to combustion of the fuel such as coal, oil or gas, to produce energy. First, the fuel is converted into synthesis gas ($CO+H_2$). The produced CO reacts with water and produces CO_2, the CO_2 is separated from the hydrogen and compressed for transportation and storage, and the remaining hydrogen is combusted to produce energy (Fig. 10.4). Through the pre-combustion CO_2 pathway, about 90–95% of CO_2 emissions can be reduced. Currently, this technology is widely used in oil refineries but it still has limited application in power plants. Integrated Gasification Combined Cycle (IGCC) and Fluidized Bed Combustion (FBC) technology are involved in the pre-combustion CO_2 capture. IGCC technology is an application to produce electricity and minimize the CO_2 emissions from the power plant. In this process, fossil fuel is first converted into CO_2 and hydrogen gas (H_2), after which the H_2 and the CO_2 gas are separated from each other and electricity is produced by the combustion of hydrogen-rich gas. Around 90% of the CO_2 can be removed from a power plant through pre-combustion CO_2 capture technology, but this is not economically viable for existing power plants (MIT, 2007).

OXY-FUEL COMBUSTION TECHNOLOGY. In oxy-fuel combustion technology, the combustion of coal is carried out with oxygen in the place of air. The produced chimney gas mainly consists of highly concentrated CO_2 and water vapour. The CO_2 and water vapour are easily separated by a cooling process. The water is then condensed and a CO_2-rich gas stream is formed. Through this process, up to 100% of CO_2 can be removed from the chimney gas (Fig.10.5). But the major

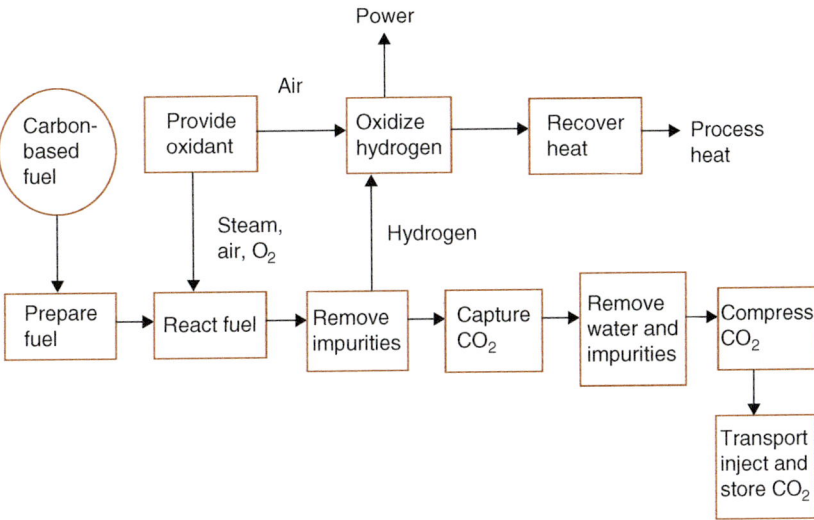

Fig. 10.4. Pre-combustion CO_2 capture. (From Moazzem *et al.*, 2012.)

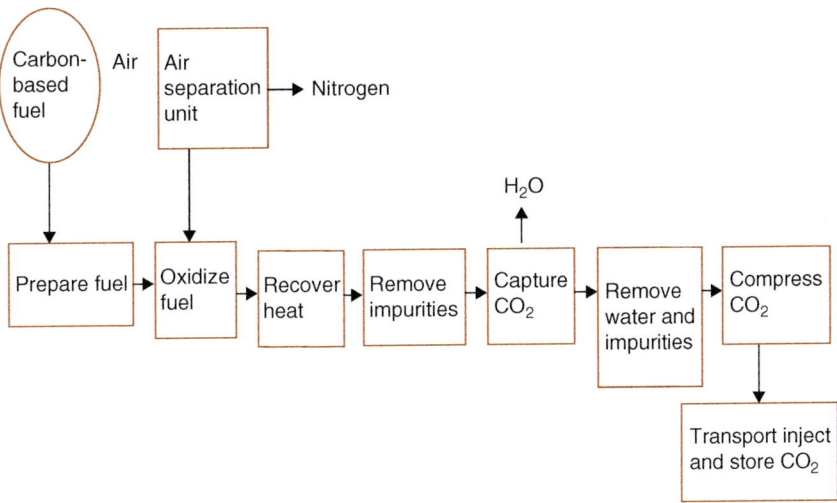

Fig. 10.5. Oxy fuel-combustion CO_2 capture. (From Moazzem *et al.*, 2012.)

challenge of this technology is the separation of oxygen from the air, because of its higher energy consumption, which ultimately affects the power generation cost. However, chemical looping combustion, which is under development, is able to potentially separate oxygen from the air, but the efficiency of the power plant will be affected and will be reduced by about 8% (MIT, 2007).

10.3.3.2 Economic considerations

The development of the technologies and processes for the efficiency of capture systems and a reduction in the overall cost is critical to creating a feasible GHG control implementation plan, which not only covers the facilities of power plants and industries, but also the required infrastructure to support that implementation. The sequestration of CO_2 through the above-discussed technology has proved to be more economically feasible and cost-effective than other established air-pollution control technologies.

10.4 Conclusions

The threat of global climate change and the damage caused by air pollution is a challenge in determining environmental policies. The cost of technology to curb air pollutants has always been a bone of contention between developing and developed countries. The adaptive technologies appear to be costly and require a cost-effective alternative. In this chapter, we have discussed biofiltration, phytoremediation and advances in CO_2 capture

technologies as major cost-effective and environmentally sound technologies. With some limitation of efficiency and specificity, these technologies have been widely accepted, although they are still in their infancy stage and require further research, development and promotion to make them viable. Assessing the cost for air pollution is not an easy task, since building the system or technology is again about financial investments, and developing and emerging countries struggle to locate the money for research and development, which seems to be a long-term goal. But unfortunately, these are the countries which are experiencing the air-pollution problem most acutely, therefore in the given situation, cost-effective technologies would be the best option to curb air pollution.

References

Boamah, P.O., Onumah, J., Takase, M., Osei Bonsu, P. and Salifu, T. (2012) Air pollution control techniques. *Global Journal of Bio-sciences and Biotechnology* 1(2), 124–131.

CEC (2005) *Best Available Technology for Air Pollution Control, Analysis Guidance and Case Studies for North America*. Commission for Environmental Cooperation, pp. 19–20.

Cunningham, S.D. and Ow, D.W. (1996) Promises and prospects of phytoremediation. *Plant Physiology* 110, 715–719.

Doty, S.L., Andrew, J.C., Moore, A.L., Vajzovic, A., Singleton, G.L. *et al.* (2007) Enhanced phytoremediation of volatile environmental pollutants with transgenic trees. *Proceedings of the National Academy of Sciences USA* 104(43), 16816–16821.

EIA (2005) *Emissions of Greenhouse Gases in the United States 2005*. DOE/EIA-0573. Energy Information Administration, Washington, DC.

Falk-Pederson, O., Dannstrom, H., Gronvold, M., Stuksrud, D. and Ronning, O. (2000) Gas treating using membrane gas/liquid contactors. In: Williams, D.J., Durie, R.A., McMullan, P., Paulson, C.A.J. and Smith, A.Y. (eds) *Fifth International Conference on Greenhouse Gas Control Technologies*. CSIRO Publishing, Cairns, Australia.

Figueroa, J.D., Fout, T., Plasynski, S., McIlvried, H. and Srivastava, R.D. (2008) Advances in CO_2 capture technology – The US Department of Energy's Carbon Sequestration Program. *International Journal of Greenhouse Gas Control* 12, 9–20.

Gawrońska, H. and Bakera, B. (2014) Phytoremediation of particulate matter from indoor air by *Chlorophytum comosum* L. plants. *Air Quality, Atmosphere & Health* 8, 265–272.

Giese, M., Bauer-Doranth, U., Langebartels, C. and Sandermann, H. (1994) Detoxification of formaldehyde by the spider plant (*Chlorophytum comosum* L.) and by soybean (*Glycine max* L.) cell-suspension cultures. *Plant Physiology* 104(4), 1301–1309.

Govindaraju, M., Ganeshkumar, R.S., Muthukumaran, V.R. and Visvanathan, P. (2012) Identification and evaluation of air-pollution-tolerant plants around lignite-based thermal power station for greenbelt development. *Environmental Science Pollution Research* 19(4), 1210–1223.

ICS UNIDO (2007) Pollutants and pollution sources. In: *Air Pollution Control Technologies Compendium*. International Centre for Science and High Technology, United Nations Industrial Development Organization (UNIDO), Trieste, Italy, pp. 9–12. Available at: https://institute.unido.org/wp-content/uploads/2014/11/25.-Air-Pollution-Control-Technologies-Compendium.pdf (accessed 1 August 2018).

Jim, C.Y. and Chen, W.Y. (2007) Assessing the ecosystem service of air pollutant removal by urban trees in Guangzhou. *Journal of Environmental Management* 88, 665–676.

Klara, S.M., Srivastava, R.D. and McIlvried, H.G. (2003) Integrated collaborative technology development program for CO_2 sequestration in geologic formations – United States Department of Energy R&D. *Energy Conversion and Management* 44(17), 2699–2712.

Leson, G. and Winer, A.M. (2012) Biofiltration: an innovative air pollution control technology for VOC emissions. *Journal of the Air & Waste Management Association* 41(8), 1–4.

Lu, Z., Streets, D.G., Zhang, Q., Wang, S., Carmichael, G.R.F. *et al.* (2010) Sulfur dioxide emissions in China and sulfur trends in East Asia since 2000. *Atmospheric Chemistry and Physics* 10, 6311–6315.

MIT (2007) *The Future of Coal*. Massachusetts Institute of Technology, Cambridge, MA.

Moazzem, S., Rasul, M.G. and Khan, M.M.K. (2012) A review on technologies for reducing CO_2 emission from coal fired power plants. In: Rasul, M. (ed.) *Thermal Power Plants*. InTech Europe University Campus STeP Ri Slavka Krautzeka, Croatia, pp. 232–233.

Mycock, J.C., McKenna, J. and Theodore, L. (1995) *Handbook of Air Pollution Control Engineering and Technology*. CRC Press, Boca Raton, FL.

Pariselli, F., Sacco, M.G., Ponti, J. and Rembges, D. (2009) Effects of toluene and benzene air mixtures on human lung cells. *Experimental and Toxicologic Pathology* 61(4), 381–386.

Perry, R.H. (1984) Gas-solid operations and equipment. In: Perry, R. and Green, D. (eds) *Perry's Chemical Engineers Handbook*. McGraw-Hill, New York.

Porter, J.R. (1994) Toluene removal from air by *Dieffenbachia* in a closed environment. *Advances in Space Research* 14, 99–103.

Resnik, K.P., Yeh, J.T. and Pennline, H.W. (2004) Aqua ammonia process for simultaneous removal of CO_2, SO_2, and NO_x. *International Journal of Environmental Technology and Management* 4, 89–104.

Resnik, K.P., Garber, W., Hreha, D.C., Yeh, J.T. and Pennline, H.W. (2006) A parametric scan for regenerative ammonia-based scrubbing for the capture of CO_2. In: *Proceedings of the 23rd Annual International Pittsburgh Coal Conference*, Pittsburgh, PA.

Rochelle, G., Chen, E., Dugas, R., Oyenakan, B. and Seibert, F. (2006) Solvent and process enhancements for CO_2 absorption stripping. In: *2005 Annual Conference on Capture and Sequestration*. Alexandria, VA.

Schnoor, J.R. (1997) *Phytoremediation*. Technical evaluation report. Ground-Water Remediation Technologies Analysis Center, Pittsburgh, PA.

Schnoor, J.L., Licht, L.A., McCutcheon, S.C., Wolfe, N.L. and Carreira, L.H. (1995) Phytoremediation of organic and nutrient contaminants. *Environmental Science & Technology* 29, 318A–323.

Schröder, P., Harvey, P.J. and Schwitzgébel, J.P. (2002) Prospects for the phytoremediation of organic pollutants in Europe. *Environmental Science and Pollution Research International* 9(1), 1–3.

Singh, S.N. and Tripathi, R.D. (2007) *Environmental Bioremediation Technologies*. Springer, New Delhi, India.

Takahashi, M. (2001) Nitrite reductase gene enrichment improves assimilation of NO_2 in Arabidopsis. *Plant Physiology* 126, 731–741.

Ugrekhelidze, D., Korte, F. and Kvesitadze, G. (1997) Uptake and transformation of benzene and toluene by plant leaves. *Ecotoxicology and Environmental Safety* 37(10), 24–29.

USEPA (1998) *Stationary Source Control Techniques Document for Fine Particulate Matter, Work Assignment No. 0-08*. US Environmental Protection Agency, Research Triangle Park, NC.

USEPA (2007) *The Cost-Effectiveness of Heavy-Duty Diesel Retrofits and Other Mobile Source Emission Reduction Projects and Programs*. US Environmental Protection Agency, Washington, DC.

Coury, R.C, Pisani Jr, R. and Hung, Y.T. (2004) Cyclones. In: Wang, L.K., Pereira, N.C. and Hung, Y.-T. (eds) Cyclones. In: *Air Pollution Control Engineering*, vol. 1, *Handbook of Environmental Engineering*, Humana Press, Totowa, NJ.

Wark, K., Warner, C.F. and Davis, W.T. (1981) *Air Pollution, its Origin and Control*. HarperCollins, New York.

WHO (2016) *Ambient Air Pollution: A Global Assessment of Exposure and Burden of Disease*. World Health Organization, Geneva, Switzerland. Available at: http://www.who.int/phe/publications/air-pollution-global-assessment/en/ (accessed 31 July 2018).

Wolverton, B.C. (1988) Foliage plants for improving indoor air quality. A paper presented at *National Foliage Foundation Interiorscape Seminar*, Hollywood, FL, pp. 29–45.

Yang, H., Liu, Z.Y., Ge, H., Yang, S.H., Ge, W.Y. and Liu, Y.J. (2010) Performance-testing in removing benzene-toluene binary gas using new lines of chrysanthemum. *Northern Horticulture* 226 (October), 5–8.

Yeh, J.T., Resnik, K.P., Rygle, K. and Pennline, H.W. (2005) Semi-batch absorption and regeneration studies for CO_2 capture by aqueous ammonia. *Fuel Processing Technology* 86, 1533–1546.

Zhang, X.Z., Zhou, P., Zhang, W.Q., Zhang, W.H. and Wang, Y.F. (2003) Selection of landscape tree species of tolerant to sulfur dioxide pollution in subtropical China. *Open Journal of Forestry* 3, 104–108.

11 Atmospheric Contaminants: Sources, Chemical Characterization and Hazards

Vineet Goswami[1]* and Naveen Chandra[2]
[1]*AIRIE Program, Colorado State University, Fort Collins, USA;* [2]*Research and Development Center for Global Change, JAMSTEC, Yokohama, Japan*

Abstract

As per the World Health Organization (WHO) estimates, around 7 million people died in the year 2012 due to exposure to air pollution. This estimate suggests that one in eight of total global deaths are caused by exposure to air pollution. Thus, air pollution, along with its direct and indirect impacts, can be regarded as the biggest environmental health hazard. While the natural emissions of atmospheric contaminants/pollutants have remained relatively stable over the past century, the emission fluxes of air pollutants from various anthropogenic activities have now surpassed their natural discharge. Over the last century, increased combustion of fossil fuels for meeting growing energy demands and several other anthropogenic activities (such as farming, deforestation, mining, biomass burning and waste incineration) have led to a substantial rise in emissions of various atmospheric pollutants, resulting in a rapid change in the basic chemistry of the atmosphere. Evidence from numerous epidemiological studies suggests that the quality of air inhaled is directly linked to health. Escape of hazardous air contaminants, their dispersal in the atmosphere, and exposure are linked to various cardiovascular and respiratory diseases and cancer. Both long-term as well as short-term exposure to air pollution is hazardous to the nourishment and sustenance of life on Earth. Exposure to air pollution is largely prevalent and unavoidable in the contemporary world. Thus, reduction and eventual elimination of fossil fuel usage and a drive towards cleaner energy resources are some of the steps that are required to reduce atmospheric pollution, and to mitigate the associated health and environmental hazards. In addition, educating the population regarding deleterious direct and indirect impacts of air pollution, efficient application of adequate policies to limit air pollution, and sustainable efforts in constraining various sources of air pollution are some of the steps that are needed to be taken. Further, accidental release of harmful short-lived radionuclides (from e.g. the Chernobyl and Fukushima Daiichi nuclear power plants) and their transport in the atmosphere pose serious health and environmental hazards. The increase in the atmospheric concentration of pollutants (for example ozone and black carbon) has also been linked to adverse effects on crop production and yield, eventually putting more stress on food resources. Here, in the present chapter, we discuss various sources of air contaminants and their chemistry in the atmosphere. Further, various health and environmental hazards of air pollution have also been highlighted and discussed.

11.1 Introduction

The study of the atmosphere and its chemistry initially started in the 18th century with the primary aim of understanding the abundance of major constituents of the Earth's atmosphere. Later, in the late 19th and early 20th centuries, the focus shifted to understanding the quality and quantity

---

* Corresponding author: vineet.goswami@colostate.edu

of trace constituents of the atmosphere (species with concentration <1 ppm of air by volume). The trace constituents occupy a relatively small volume in the atmosphere than major constituents; however, their abundance plays significant and diverse roles, ranging from urban smog to climate change (Hansen *et al.*, 2007). On a longer time scale, the chemistry of the atmosphere is susceptible to change due to natural processes such as wildfires (Gavin *et al.*, 2007) and volcanic eruptions (Robock, 2000). With the human population on Earth reaching new records in the Anthropocene, several significant rural and urban anthropogenic activities are rapidly influencing the abundance of trace constituents of the atmosphere. The ice core records from Antarctica reveal a marked increase in the atmospheric concentrations of trace greenhouse gases, namely carbon dioxide (CO_2), nitrous oxide (N_2O) and methane (CH_4), over the last decade (MacFarling Meure *et al.*, 2006). Several anthropogenic activities have also contributed largely to the significant increase in the concentrations of tropospheric ozone (O_3) and carbonaceous aerosols in the northern hemisphere over the last century (Hough and Derwent, 1990; Ramanathan *et al.*, 2001). Thus, there is enough evidence to suggest that human-induced activities are rapidly changing the basic chemistry of the Earth's atmosphere.

Over the last century, a large increase in the human population on the Earth has led to a significant increase in several anthropogenic activities (such as farming, deforestation, mining and transport, etc.) that have resulted in the release of numerous contaminants to the air, water and land. The contamination of air by human-induced activities has a significant and larger impact on the sustenance and nourishment of life on Earth due to rapid mixing dynamics of the atmosphere and potential intake of contaminants by living beings with the air inhaled. Thus, the presence of clean air in the atmosphere is vital for maintaining the quality of life on Earth. While a rough understanding of the correlation between poor air quality and human health has been known for a long time, the conscience to quantify and understand the direct as well as indirect effects of air contamination on health started in the 20th century.

Since the industrial revolution, the human population and anthropogenic emissions of atmospheric pollutants have greatly increased. Emissions of air pollutants from anthropogenic activities have now surpassed natural emissions. The upsurge in the abundance of air pollutants can be largely attributed to increased combustion of fossil fuels, to meet the growing demands for energy and transportation, for supporting the growing human population on the Earth. Hazardous pollutants/contaminants can escape to the atmosphere by sporadic accidental events, or can be released by numerous anthropogenic activities. These air contaminants or pollutants have the potential to be hazardous to the nourishment and health of life (animals and plants) on the Earth. Notable incidents such as toxic smog over the Meuse valley, Belgium (Nemery *et al.*, 2017) and London (Bell *et al.*, 2004), resulted in immediate acknowledgement of the lethality of air pollution on human health. It is unquestionable that the increased concentration of atmospheric contaminants can be linked to an increase in mortalities and/or serious ailments, and poses serious hazards to human health (Bell *et al.*, 2004; Sriramachari, 2004; Anderson *et al.*, 2012). With the human population on Earth setting new records in the Anthropocene, the demand for energy has also increased tremendously over the last century, resulting in increased combustion of fossil fuels (Fig. 11.1) and emission of various atmospheric pollutants (Crutzen, 2006). Thus, considerable efforts are needed to reduce emissions of atmospheric contaminants due to their adverse effects on the environment and human health. The aim of this chapter is to outline and discuss the sources and chemistry of atmospheric contaminants, and their health hazards and environmental impacts.

11.2 Sources and Chemical Characterization of Atmospheric Contaminants

With reference to their sources, atmospheric contaminants/pollutants can be characterized as (i) primary pollutants, or (ii) secondary pollutants. Various sources of primary and secondary air contaminants and their chemistry in the atmosphere are discussed in the following sections.

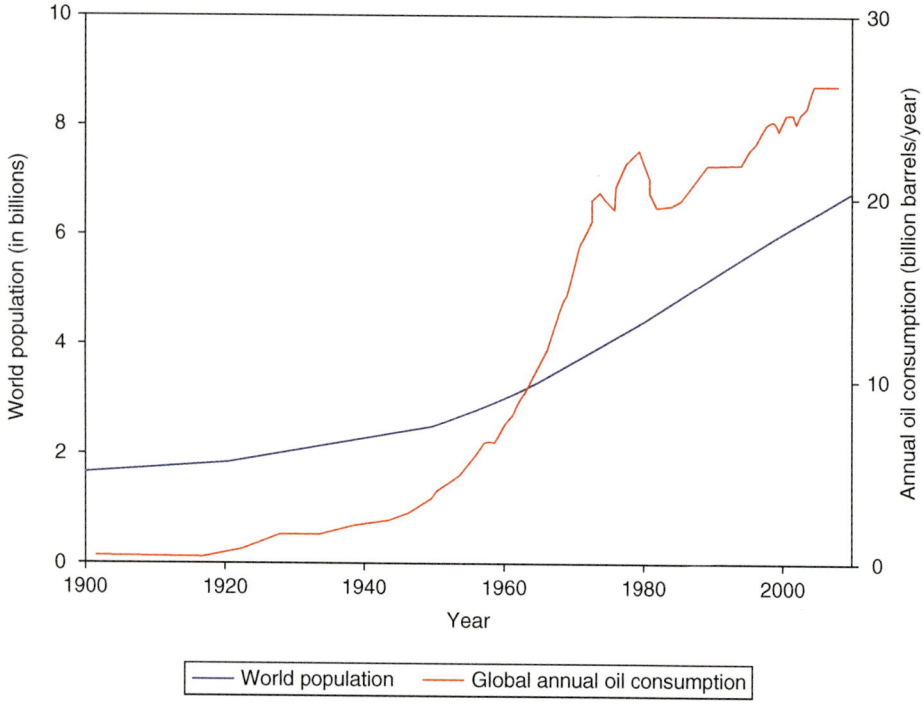

Fig. 11.1. Historical trends in global population and global oil consumption. The rise in global population has been concurrently matched by increased oil (fossil fuel) consumption to meet increasing energy demands. (From UN, 1999, Table 1; Owen *et al.*, 2010.)

11.2.1 Primary air pollutants

Primary air pollutants are the ones that are released directly into the air from their sources. Examples of primary air pollutants include carbon monoxide (CO), nitrogen oxides (NO_x), sulfur dioxide (SO_2), volatile organic compounds (VOC) and particulate matter (PM; dust, ash). Details on the sources and chemistry of primary pollutants are given below.

Carbon monoxide is a colourless and odourless gas that is generated as a by-product of partial combustion of carbon in carbonaceous fuels (gasoline, oil, natural gas, coal, wood, etc.). During this process, the lack of oxygen inhibits the formation of carbon dioxide and subsequently results in the formation of carbon monoxide instead. Figure 11.2 shows the annual global mean surface emissions of CO (Yin *et al.*, 2015). The CO is substantially released to the atmosphere by the incomplete combustion of fossil fuels. Thus, in developing countries, significant atmospheric CO emissions are sourced from vehicular emissions

(Chandra *et al.*, 2016a), combustion of fossil fuels, biomass burning, and combustion of biofuels (Edwards *et al.*, 2004). Evidently, higher atmospheric loading of CO in the northern hemisphere (in contrast to the southern hemisphere; Fig. 11.2) is the result of a larger human population in the northern hemisphere and a linkage of primary sources of CO release to anthropogenic activities. Fig. 11.2 shows high emissions of CO from urban and industrial zones over east China, eastern USA, India and Europe, in addition to significant CO sources over the equatorial landmasses of Africa and South America due to dry season burning of biomass.

The residence time of CO in the atmosphere ranges from 0.1 to 2.7 years varying both in space and time (Weinstock, 1969); however, the global average residence time has been estimated to be ~2 months. Carbon monoxide is a very toxic gas on inhalation by living beings. On inhalation, carbon monoxide competes with oxygen in the bloodstream and binds with haemoglobin in red blood cells, consequently forming

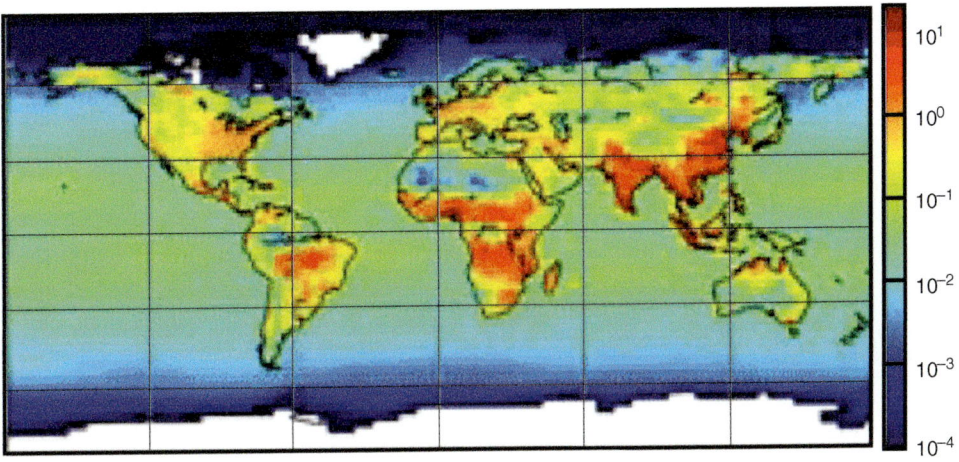

Fig. 11.2. Distribution of annual mean global surface CO emissions from 2002 to 2011 in units of Tg/yr. Significantly higher CO emissions are seen over the northern hemisphere due to a greater human population and the association of CO emissions with anthropogenic activities. Significant CO emissions can be seen over east China, India, eastern USA, Europe and equatorial areas of Africa and South America. (From Yin *et al.*, 2015.)

carboxyhaemoglobin. This eventually starves the key organs (brain, liver, central nervous system, heart) of oxygen and inhibits their ability to function properly.

Nitrogen oxides (NO$_x$) form during the process of high temperature combustion, on reaction of oxygen with nitrogen. This high temperature process results in the generation of nitrogen oxide (NO). During the process, to a lesser degree, some other by-products (nitrogen dioxide (NO$_2$) and other oxides of nitrogen (NO$_x$)) are also generated. Nitrogen oxide can further be converted to nitrogen dioxide by an oxidation reaction that involves oxygen (O$_2$), ozone (O$_3$) and organic compounds (Glasson and Tuesday, 1963).

The natural sources of nitrogen oxides (NO$_x$), namely lightning, wildfires, biogenic emissions by bacterially mediated nitrification in soil and marine environments, account for a portion of global NO$_x$ emissions; however, the relative contribution of these natural sources remains uncertain (Reis *et al.*, 2009). Since the initiation of the industrial revolution, anthropogenic activities have contributed largely to NO$_x$ emissions and they have now exceeded natural emissions (Galloway *et al.*, 2004). The prime source of nitrogen oxide and nitrogen dioxide to the ambient air is emissions from motor vehicles, but power plants (coal-, oil- and natural gas-based)

and the fuel-burning industry also contribute significantly (Felix *et al.*, 2012). The amount of NO$_x$ emissions from different sources is highly uncertain; however, NO$_x$ emissions contributed by fossil fuel combustion amount to more than 50% of the overall emissions (Beirle *et al.*, 2003).

Fig. 11.3 shows the annual global mean vertical distribution of NO$_2$ in the troposphere. The vertical atmospheric loading of NO$_2$ is more prominent in the northern hemisphere than the southern hemisphere: an effect of a greater human population and significant NO$_2$ emissions due to anthropogenic activities. Like CO emissions, another important aspect is the large plume of NO$_2$ emitting from the areas of larger human population and anthropogenic activities: east China, eastern USA and parts of western USA, India, Europe and densely populated areas of Africa and South America (Fig. 11.3). In the atmosphere, the NO$_x$ species are very reactive as shown by their short mean residence time of the order of ~0.5–2 days. The hazard and toxicity posed by NO$_x$ is primarily due to its oxidative potential. Exposure to low doses of NO$_2$ (~0.04 ppm) has been shown to cause a number of pulmonary and systematic effects in living beings, while short-term exposures to higher doses (0.3–0.5 ppm NO$_2$ or 1 ppm NO) result in bronchoconstriction and blood biochemical changes

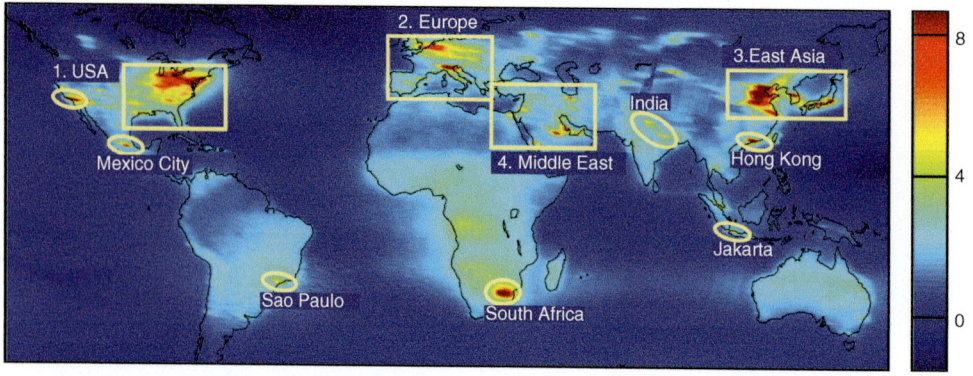

Fig. 11.3. Annual global mean NO$_2$ vertical column distribution in the troposphere (in units of 10^{15} molecules/cm^2). The plot shows some key 'hotspots' of high NO$_2$ emissions. Significant emissions of NO$_2$ from other densely populated areas can also be seen. Substantial tropospheric loading of NO$_2$ over the northern hemisphere is an effect of the greater human population in the northern hemisphere and linkage of NO$_2$ emissions to anthropogenic activities. (From Beirle *et al.*, 2003.)

(Kagawa, 1985). Thus, higher concentrations of NO$_x$ gases in the ambient air are liable to have adverse effects on health, and in part contribute to respiratory symptoms and illness along with reduced pulmonary functions.

Sulfur dioxide (SO$_2$) is discharged into the atmosphere from natural sources, and as emissions from various anthropogenic activities. Sulfur compounds are naturally emitted as SO$_4$ aerosols from sea spray and H$_2$S from decomposing organic matter in marshes, swamps and peat-bogs. Once emitted, H$_2$S is quickly oxidized to SO$_2$ in the troposphere by ozone via a heterogeneous reaction on surfaces of aerosol particles. The average residence time of H$_2$S species in the atmosphere ranges from ~2 hours in urban areas to ~2 days in remote uninhabited areas (Robinson and Robbins, 1970). The anthropogenic emission of SO$_2$ results from the combustion of sulfur-rich fossils fuels (for example coals and heavy oil) and smelting of ores that contain significant amounts of sulfur (smelting of copper, lead and zinc ores). The global annual emissions of SO$_2$ due to various anthropogenic activities have been estimated to be ~146 Tg/y (Robinson and Robbins, 1970). In contrast, the global annual SO$_2$ emissions to the atmosphere from volcanic discharge stand at ~15–20 Tg/y (Halmer *et al.*, 2002).

Combustion of coal is a major contributor of SO$_2$ emissions to the atmosphere. Sulfur is removed from euxinic waters of the peat-forming mires during the deposition of coal. This drawdown of sulfur from the water column of mires results in its enrichment in the coal. High-sulfur coal generally contains sulfur in the mineral form as iron pyrite (FeS), but may also contain sulfides and/or sulfates of other metals and a small amount of elemental sulfur (Calkins, 1994). In order to reduce the SO$_2$ emissions from coal, considerable efforts have been made in the past towards removal of the sulfur from it, but all have failed to yield any economical and practical methods (Calkins, 1994). Of the global annual emissions of SO$_2$ to the atmosphere due to anthropogenic activities, ~70% is estimated to be sourced from the combustion of coal, while ~16% can be estimated to result from the combustion of petroleum products and residual oil at refineries, the remaining being accounted for by the extraction and purification of metals (copper, lead, zinc) by smelting of their ores (Robinson and Robbins, 1970). The greater northern hemispheric population results in greater (~93%) emissions of anthropogenic SO$_2$ in contrast to a lower atmospheric loading of SO$_2$ from southern hemisphere (~7%).

Fig. 11.4 shows the global mean vertical column distribution of anthropogenic SO$_2$ release into the atmosphere (Fioletov *et al.*, 2016). The figure clearly marks the significant areas for the atmospheric loading of SO$_2$ released from anthropogenic activities. Most SO$_2$ enrichments exist over areas of significant industrial activity or downwind of these areas. The highest

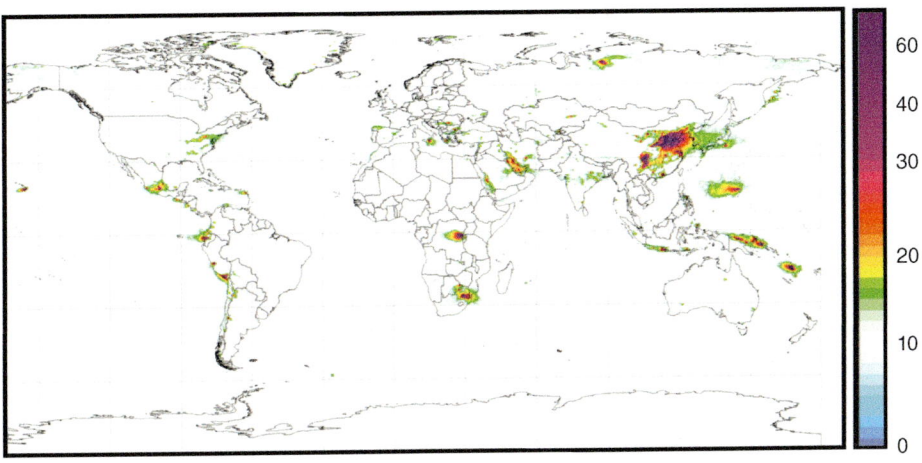

Fig. 11.4. Annual global mean (2005–2007) spatial vertical column distribution of SO_2 emissions in units of 10^{15} molecules/cm². The highest atmospheric loading of SO_2 can be seen over and downwind of active industrial zones in the eastern United States, eastern China and northern-central India. The highest emissions (>30 x 10^{15} molecules/cm²) can be seen over eastern China, a consequence of coal combustion for meeting energy demands. (From Fioletov *et al.*, 2016.)

emissions of SO_2 can be seen over eastern China, while areas in eastern USA, northern-central India and Europe also contribute significantly (Fig. 11.4). It has been estimated that anthropogenic activities contribute to >60% of total SO_2 atmospheric loading in air columns over large regions of all continents (Lee *et al.*, 2011). The major contributors to SO_2 emissions in the atmosphere are from Asia, North America and Europe, with areas over the Sahara and the Amazon having a negligible contribution of SO_2 loading to the atmosphere (Fig. 11.4).

The health hazard posed by SO_2 is due to cardiovascular and pulmonary effects of this pollutant. In a study from Europe, it was found that an increase of 50 µg/cm³ in the concentration of SO_2 was associated with a ~3% elevation in daily mortality (Katsouyanni *et al.*, 1997). The long-term effects of SO_2 pollution are still ambiguous; however, the short-term effects of relatively low abundance of SO_2 in the present-day atmosphere are unquestionable (Katsouyanni *et al.*, 1997).

Volatile organic compounds (VOC) are a major class of compounds that have a range of chemical and physical properties. VOCs are organic compounds that possess high vapour pressure at normal temperature and pressure conditions, resulting in a low boiling point and emissions from solid or liquid organic compounds. While the pure hydrocarbons constitute C and H, VOCs contain O, Cl and other elements in addition to C and H. In addition to a broad spectrum of chemical compounds, sources of VOCs are also quite diverse (Sahu and Saxena, 2015; Sahu *et al.*, 2017). These include: (i) incomplete combustion of fossil fuels; (ii) production, handling, storage and transport of fossil fuels; (iii) usage of solvents containing VOCs; (iv) industrial processes; (v) emissions from vegetation; and (vi) emissions from wildfires.

Methane (CH_4) is the simplest and most long-lived VOC in the atmosphere (average residence time in the atmosphere ~10 years; Cicerone and Oremland, 1988), and is significantly emitted from both anthropogenic as well as natural sources (Bousquet *et al.*, 2006). Excluding methane from the suite of VOCs, the remaining pool can be clubbed together as non-methane volatile organic compounds (NMVOCs). VOC emissions have substantially decreased over the last 10–15 years, due to tighter constraints on their sources (Simon *et al.*, 2015). A high concentration of VOCs poses a significant threat in the formation of tropospheric ozone near the Earth's surface, as a consequence of photochemical interaction of CO, VOC and NO_x (NO + NO_2) species near their emission sources (Finlayson-Pitts and Pitts, 1993). Tropospheric ozone can cause a range of respiratory problems, and is harmful to the growth of plants.

Particulate matter (PM) is an intricate mixture of miniscule particles and liquid droplets composed of acids, organic compounds, metals and dust particles that are suspended in the air (Anderson *et al.*, 2012). PM can be sourced from both natural as well as anthropogenic processes. Some of the natural sources of PM are wildfires, volcanoes, dust/sand storms, sea-salt spray and bio-aerosols (fungi and pollens). Anthropogenic sources of PM comprise construction activities, vehicle emissions, tobacco burning, and combustion of fuels in domestic and commercial sectors. PM is also transformed in the atmosphere as a result of complex interactions involving other pollutants (SO_2, NO_x) that are emitted from industries, power plants and vehicles.

PM has a large size distribution; some particles (for example dust and soot) can be seen with the naked eye, whereas others are too small and can only be seen under powerful microscopes. Generally, in terms of their size, PM is characterized by the *aerodynamic equivalent diameter (AED)*, a measure of their size. Particles of different size are largely subdivided into AED fractions. Particles with the same AED are poised to have the same settling velocities in the air. Accordingly, in terms of the AED, particles are subdivided based on their genesis and deposition in the human respiratory system, as (i) PM_{10}: diameter <10 μm; (ii) $PM_{2.5}$: diameter <2.5 μm; and (iii) $PM_{0.1}$: diameter <0.1 μm. The size fraction of particles having a typical size between 10 μm and 2.5 μm ($PM_{10-2.5}$) is defined as the *coarse* fraction, less than 2.5 μm ($PM_{2.5}$) as the *fine* fraction, and less than 0.1 μm ($PM_{0.1}$) as the *ultrafine* fraction. In a mixed environment consisting of a distribution of PM particles, the aggregated surface area of the particles decreases exponentially with the particle size, while the total particulate mass of the particles increases exponentially with size (Anderson *et al.*, 2012).

An increase in the concentration of PM_{10} particles in the ambient air poses a great threat to health in humans and animals as they can be inhaled deep into the lungs, causing serious respiratory problems. Further, smaller fraction particles may even get into the bloodstream by their diffusion from the lungs to the blood and cause acute cardiovascular symptoms. Acute exposure to PM_{10} particles results in changes in coagulation and platelet stimulation, causing coronary artery ailments. Major factors that influence the toxicity of the PM are: (i) their bulk chemical composition; (ii) trace element abundance and composition of the PM; (iii) acid content of the PM; (iv) sulfate content of the PM; and (v) their particle size distribution (Harrison and Yin, 2000).

Heavy trace metal pollution: heavy metals (such as Cd, Pb, Hg, As, V, Cr) are naturally found in the Earth's crust. At higher concentrations (than required by the bodily metabolic processes), the accumulation of heavy metals in the body becomes toxic, leading to serious damage to various systems (cardiovascular, nervous, renal), and causes cancer. A large increase in the human population and intensity of various anthropogenic activities has led to the contamination of air with heavy metals. Emissions of heavy metal contaminants to the air occur by a host of processes that involves combustion of fuels, industrial activities, and extraction and processing of metals. Trace metals are abundant in various raw materials (for example fossil fuels, metal ores) that are used in a range of anthropogenic activities, including transport, industries and power generation. Some heavy trace metals are partially or completely volatilized from the raw materials during high-temperature combustion of fuels and production of industrial goods. Additionally, incineration of waste materials (metropolitan and industrial) leads to the emission of heavy metals to the ambient air in exhaust gases. Once in the ambient air, heavy metals can be transported in the atmosphere as a function of their particle size distribution, leading to their short- to long-range transport, causing health hazards and perturbing their geochemical distribution on Earth on a regional to global scale. While most heavy metal pollutants are enriched in the vicinity of their emission sources, some of them may also be enriched on a regional to global scale, depending on their particle size. Sporadic long-range transport of heavy metal contaminants results in the enrichment of these metals in soils and sediments in regions far from their source regions (e.g. Renberg *et al.*, 2001). Removal of heavy metal contaminants from the atmosphere and their deposition to the land takes place by either dry deposition (based on their settling velocities, a function of AED) or wet scavenging. Discharge of heavy metals to the atmosphere poses a great health hazard because of the quantity of the contamination involved and widespread dispersion and ensuing exposure. This section discusses the sources and atmospheric chemistry of three heavy metals (Cd, Pb and Hg) as atmospheric contaminants.

Cadmium (Cd) is released into the atmosphere by various natural and anthropogenic processes. Natural sources of Cd emissions are windblown dust, wildfires and volcanogenic particles (Nriagu, 1979). The anthropogenic sources of Cd emissions to the atmosphere include non-ferrous metal extraction and purification, fossil-fuel combustion, and production of iron, steel and cement (Nriagu and Pacyna, 1988; Järup, 2003). In a study from Poland, it was also observed that brake lining and tyre wear also contribute to significant Cd (and some other heavy metals) air contamination in urban areas (Adamiec et al., 2016). Upon its release from various anthropogenic sources, Cd is discharged into the atmosphere as oxide, sulfide, chloride or in the elemental state. Anthropogenic activities are largely responsible for the emission of CdO, while elemental Cd is released by high temperature combustion of fossil fuels. However, the last few decades have seen a drop in the atmospheric loading of Cd from anthropogenic sources (WHO, 2007). Cd released into the atmosphere is bound to fine fraction particles ($\sim$1–1.25 μm in size; Kuloglu and Tuncel, 2005); however, in a study from Hungary, it was found that Cd is also susceptible to be bound onto particles of smaller ($\sim$0.1 μm) size (Molnár et al., 1995).

Lead (Pb) is emitted into the atmosphere from a spectrum of natural and anthropogenic activities. Significant natural sources for the atmospheric loading of Pb are wind-induced particle suspension and sea-salt sprays (Nriagu, 1989). On a global scale, major anthropogenic processes that are responsible for significant atmospheric loading of Pb are vehicular emissions, non-ferrous metal extraction and purification, and steel and cement production (Pacyna and Pacyna, 2001). Emitted atmospheric Pb is in a bound state to the particulate matter, with $\sim$90% of the mass residing in particles of the size range $\sim$0.2–1.0 μm (AED), with the contribution of Pb bound to finer particles being largely insignificant (Dillner et al., 2005). Upon emission, Pb is primarily found in the atmosphere in chloride, oxide and sulfate species. The process of oil combustion releases Pb as lead oxide (PbO), whereas coal combustion and waste burning releases Pb in the chloride form ($PbCl_2$). Non-ferrous metal extraction and purification releases Pb in the sulfate form ($PbSO_4$).

Analysis of laminated lake sediments and peat deposits from northern Europe and ice cores from Greenland suggests concurrent temporal changes in the abundance of Pb and $^{206}Pb/^{207}Pb$ ratio, linked to anthropogenic activities (Brännvall et al., 1999; Renberg et al., 2001). Significant peaks in abundance of Pb and associated synchronous dips in $^{206}Pb/^{207}Pb$ ratio were observed during the periods of significant anthropogenic activities: (i) during the height of the Roman empire ($\sim$0 AD); (ii) the medieval period ($\sim$900–1600 AD), due to metallurgical activities; (iii) the 18th century, during the industrial revolution; and (iv) after the Second World War, due to a rapid increase in the use of automobiles and leaded gasoline in addition to amplified industrial emissions. However, in the last few decades, mitigation of the usage of leaded gasoline and tighter constraints on the utilization of Pb has resulted in a reduction in atmospheric Pb pollution (Brännvall et al., 1999). The sediment archives also reveal that atmospheric Pb pollution during medieval times (well before the industrial revolution) was comparable to present-day levels (Brännvall et al., 1999).

Mercury (Hg) is a highly toxic environmental contaminant that is discharged to the atmosphere from both natural and anthropogenic sources, and is among the most readily bioaccumulated trace metals in the human food chain, especially in aquatic environments. The natural sources of Hg emission to the atmosphere are volatilization from water bodies, volcanoes, diffusion from the Earth's crust and re-emission from top soil and vegetation (Pirrone et al., 2001). Human-induced activities have significantly perturbed the biogeochemical cycling of Hg. The largest anthropogenic source of global atmospheric Hg is the combustion of coal, fossil fuels, and municipal and industrial wastes (Pacyna and Pacyna, 2001; Wong et al., 2006). Additionally, a significant contribution of Hg emissions are sourced from the process of gold production (Pacyna and Pacyna, 2001). Minor sources of Hg emissions are metal and cement production. The estimated emissions of Hg from anthropogenic sources to the atmosphere represent around two-thirds of the total emissions, while natural sources characterize the remaining one-third of the total global discharge (Lamborg *et al.*, 2002).

The most significant characteristics of mercury that sets it apart from other trace metals that are normally found in the atmosphere are: (i) the tendency to be readily re-emitted to the atmosphere after being deposited onto surfaces; and (ii) the fact that elemental mercury can exist in the ambient air predominantly in the vapour phase, whereas other metals are almost exclusively bound to the solid phases (Bergan *et al.*, 1999). Being a volatile element, atmospheric transport is an important phenomenon in global biogeochemical cycling of mercury (Nriagu, 1979; Bergan *et al.*, 1999). Upon its release to the atmosphere, Hg is dispersed in the air largely in its elemental form (Hg^0), which is the most abundant compound of mercury in the atmosphere (Bergan and Rodhe, 2001), contributing to >95% of its atmospheric budget. In addition, trace amounts of mercury in the atmosphere are found in another speciation: Hg^{2+} or Hg(II) commonly known as *reactive gaseous mercury (RGM)*, which constitutes ~1– 5% of the atmospheric Hg (Lindberg and Stratton, 1998). Atmospheric RGM compounds are associated with particles or gases in the form $Hg(OH)_2$, $HgCl_2$ or compounds of other halides (Lindberg and Stratton, 1998). The largest contribution to the natural emissions of Hg are from the land and sea in the elemental form of mercury (Hg^0), while anthropogenic emissions contain both Hg^0 and Hg^{2+}, with their relative fractions variable in both time and space (Bergan *et al.*, 1999). A schematic representation of sources and atmospheric cycling of mercury is shown in Fig. 11.5.

Elemental mercury (Hg^0), in its vapour state, is relatively inert to chemical interactions with other species present in the atmosphere and is frugally soluble in water. In return, atmospheric Hg^0 is characterized by a long residence time (~1 year) enabling its transport on a global scale (Schroeder and Munthe, 1998). In contrast, the RGM compounds are highly water soluble (of the order of 10^5 times more than elemental mercury), strongly affecting their removal and deposition from the atmosphere (Lindberg *et al.*, 1992). Recent studies have shown that the presence of hydroxyl (OH) radicals in the atmosphere acts as a significant gas-phase oxidant for

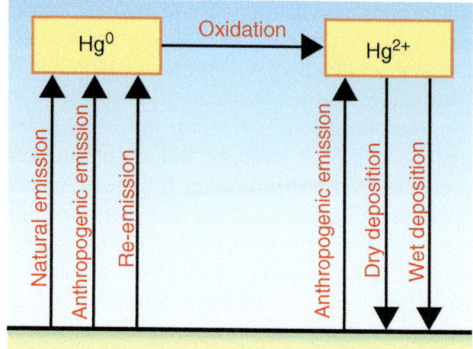

Fig. 11.5. Schematic diagram representing the sources, atmospheric cycling and pathways of deposition of mercury. (Adapted from Bergan *et al.*, 1999.)

the oxidation of Hg^0 (Bergan and Rodhe, 2001; Sommar *et al.*, 2001). In the presence of OH radicals in the atmosphere, elemental mercury gets oxidized to RGM, facilitating another pathway for the removal of Hg^0 via its transformation to Hg^{2+} (Bergan and Rodhe, 2001).

11.2.2   Secondary air pollutants

Intricate chemical reactions in the atmosphere convert primary pollutants to secondary pollutants. Thus, the secondary air pollutants are not directly emitted from their polluting sources, but form in the atmosphere on chemical interactions between primary air pollutants, or on the reaction of primary air pollutants with pre-existing atmospheric constituents. Examples of secondary air pollutants are tropospheric ozone (O_3), sulfur trioxide (SO_3) and acid mist/rain.

Tropospheric ozone (O_3): the atmospheric budget of ozone is defined by ~90% of the ozone lying in the stratosphere, with the remaining 10% in the troposphere (Lelieveld and Dentener, 2000). The increase in the concentration of tropospheric ozone (also known as ground-level ozone or surface ozone) is hazardous to the growth of plants and human health (Lal *et al.*, 2017). Continued inhalation of ozone results in the breakdown of the respiratory system and damages extrapulmonary organs. In addition, tropospheric ozone regulates the oxidation of various gaseous species in the Earth's atmosphere (Levy, 1971), dictating their atmospheric residence time. Figure 11.6 shows a schematic diagram for the formation, chemistry and cycling of ozone in the atmosphere.

Important sources of tropospheric ozone are: (i) stratosphere–troposphere exchange across the extratropical tropopause and the associated downward transport from the stratosphere (Junge, 1962); and (ii) *in situ* photochemical production in the troposphere. While the leakage of stratospheric ozone has remained constant over time (Lelieveld and Dentener, 2000), *in situ* production of ozone in the troposphere is caused by photochemical interactions between atmospheric contaminants, and can be highly variable on spatial and temporal scales. In the presence of sunlight, primary air pollutants, namely CO, VOCs and NO$_x$ react to produce tropospheric ozone (Finlayson-Pitts and

Pitts, 1993). A significant increase in anthropogenic activities (such as combustion of fossil fuels, biomass burning) over the last century has led to increased atmospheric loading and abundance of air pollutants, consequently resulting in an increase in the concentration of tropospheric ozone (Logan, 1985; Lelieveld and Dentener, 2000). While urban areas are significant hotspots for producing the pollutants that act as seeds for tropospheric ozone (Chandra *et al.*, 2016a), rural areas that lie downwind of such hotspots can also be affected. On production of ozone from such urban hotspots, air masses that are rich in ozone and its precursors can rise and advect to remote sites, influencing the distribution of ozone in the troposphere (Lal *et al.*, 2014; Chandra *et al.*, 2016b). An elevated concentration of tropospheric ozone has the potential to affect future trends in

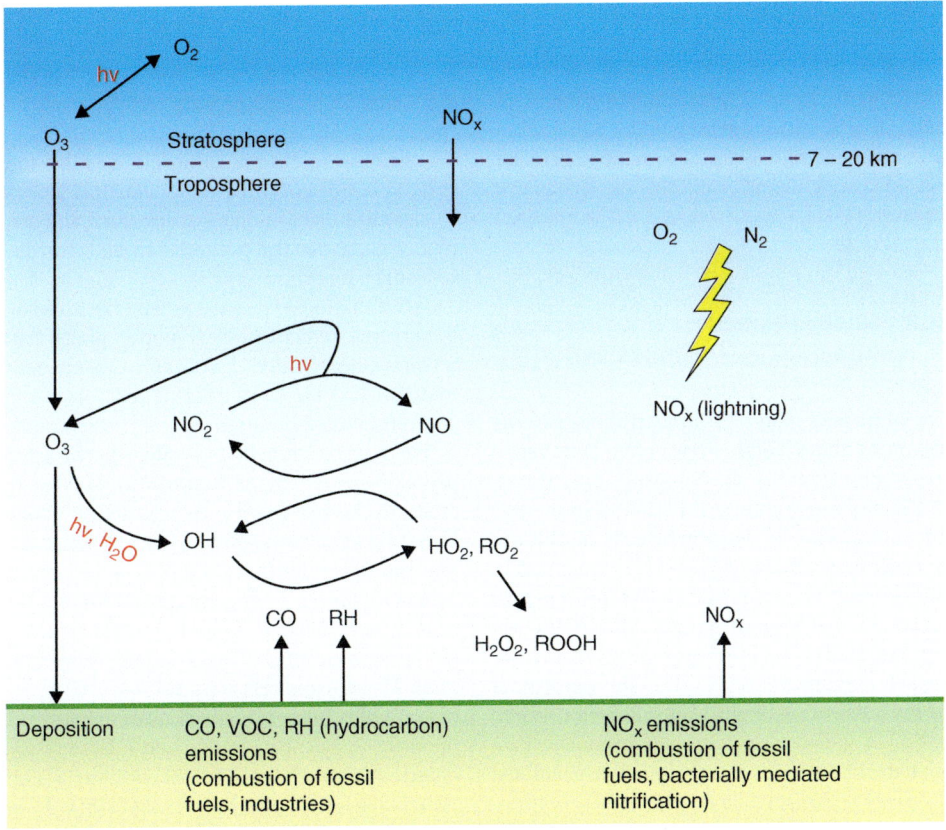

Fig. 11.6. Schematic diagram representing the formation, chemistry and cycling of ozone in the atmosphere. (Adapted from Jacob, 2000.)

climate, human health, plant growth, agriculture and crop yield.

Acid mist/rain: combustion of coal (primarily in coal-fired power plants) and fossil fuels results in the emission of nitrogen and sulfur-based pollutants (NO_x and SO_2). The formation of these pollutants affects the environment due to their linkage to the formation of acid rain and haze. Furthermore, the presence of these pollutants in the atmosphere is hazardous to human, animal and plant health. Once emitted from their polluting sources to the ambient air, the oxidation (SO_2 to SO_3 and particulate sulfate; NO_x to particulate nitrates) is regulated by urban air photochemistry, with rates of oxidation being a significant factor in governing the concentration of the secondary pollutants. In the presence of OH radicals, SO_2 reacts to form SO_3, and leads to the formation of sulfuric acid (Stockwell and Calvert, 1983). Nitric acid is formed in the atmosphere by the homogeneous hydrolysis of N_2O_5 (which is formed by the reaction between NO_2 and NO_3 radical), the rate of formation being limited by the presence of tropospheric ozone (Russell *et al.*, 1986). The formation and precipitation of acid rain is detrimental for stone sculptures and buildings, as it results in their corrosion and enhanced chemical weathering (Gauri and Holdren, 1981).

11.3 Nuclear Pollution: Environmental Impacts and Health Hazards

The increased usage of conventional energy resources (coal, fossil fuels, etc.) has had a major contribution in polluting the atmosphere, and endangering the environment and ecosystem. The various drawbacks of the use of fossil fuels as energy resources have led to other energy resources (such as nuclear energy) being explored to meet the ever-increasing energy demands. Currently, a total of 449 active nuclear reactors in 30 countries are operational for the generation of electricity, contributing to ~16% of global electricity generation. France holds the highest proportion of nuclear-generated electricity (~72%), while nuclear energy contributes ~20% of US electricity generation. Although cleaner than the conventional energy resources

with no atmospheric emissions, the usage of nuclear energy to generate power is marred by the possibility of ionizing radiation leakage from nuclear reactors. Short-lived radionuclides such as I-131 ($t_{1/2}$ = 8.02 days), Sr-90 ($t_{1/2}$ = 28.8 years) and Cs-137 ($t_{1/2}$ = 30.17 years), if leaked from nuclear reactors, are deleterious to human health, largely due to air contamination (Anspaugh *et al.*, 1988; Chino *et al.*, 2011). Furthermore, the radioactive contamination of air results in soil and water contamination with radioisotopes, due to air-fallout of radionuclides. Once released due to accidental leakage from a nuclear reactor, the airborne radionuclides can be advected away from their source to distant sites through their long-range transport in the atmosphere, resulting in radioactive fallout posing additional hazards. Acute short-term exposure to the ionizing radiation from radioactive isotopes results in damage to the cells and can impair functioning of the tissues and organs, producing severe effects, such as radiation burns, hair loss and acute radiation syndrome (also known as radiation toxicity). Chronic long-term exposure to radiation from radioisotopes can result in cancer, benign tumours, cataracts and harmful genetic changes. Thus, the accidental leakage of harmful radionuclides from nuclear power plants possesses the potential to be hazardous to human health.

Notable incidents such as the Chernobyl and Fukushima Daiichi nuclear power plant disasters have highlighted the hazards of radioactive material leakage from nuclear reactors on the environment (Anspaugh *et al.*, 1988; Koizumi *et al.*, 2012). Immediately after the disasters, people were evacuated from the vicinity of the reactors. An exclusion zone was set up around Chernobyl encompassing an area of around 30 km radius (~2800 km^2), and approximately 116,000 people were evacuated (Steinhauser *et al.*, 2014). Later, in order to mitigate the accidental exposure of radionuclides, the zone of prohibition was expanded to ~4300 km^2. In the case of the Fukushima Daiichi disaster, immediately upon the accidental leakage and release of hazardous radionuclides from the reactor, a 'complete exclusion no-go zone' was set up in a 20-km radius around the Fukushima reactor (Yoshida and Takahashi, 2012). Long-range atmospheric transport of airborne radionuclide contaminants also resulted in radioactive fallout in remote and distal sites in Russia, the USA and the Pacific Ocean (Bolsunovsky and Dementyev, 2011; Leon *et al.*, 2011; Povinec *et al.*, 2012). In the case of the Chernobyl incident, the radioactive fallout was detected in most of Europe (Steinhauser *et al.*, 2014).

11.4 Health Hazards of Atmospheric Pollution

The contamination of ambient (outdoor) air has a considerable impact on human health. Contact with air pollution occurs primarily by inhalation and ingestion. Contamination of ambient air and subsequent fallout of pollutants further contributes to the contamination of water and food, posing additional hazards regarding the ingestion of pollutants. Dermal contact with pollutants may represent a supplementary method of exposure. Short-term as well as long-term exposure to air pollution has the potential to damage the respiratory and cardiovascular systems. Contact with air contaminants also severely affects other bodily organs and systems (Genc *et al.*, 2012).

There is no doubt over the association of air pollution with the health hazards in living beings. Some of the notable sporadic *high pollution intensity* incidents, such as toxic smog over the Meuse valley, Belgium (Nemery *et al.*, 2017) and London (Bell *et al.*, 2004) and leakage of highly toxic methyl isocyanate (MIC) during the Bhopal gas tragedy (Sriramachari, 2004), have already highlighted the lethal effects of air pollution on health. It has been well established that fatalities and hospital admissions are linked proportionately to the quality and quantity of pollutants in the ambient air (Brunekreef and Holgate, 2002; Fischer *et al.*, 2004; Liu *et al.*, 2013). The hazard and lethality of air pollution are primarily governed by these key factors: (i) composition (quality) of air pollution; (ii) concentration (quantity) of pollutants in the air; (iii) duration of exposure to air pollutants. The health effects of air pollution can vary from minor (such as breathlessness, nausea, skin and respiratory tract irritation) to major (such as asthma, bronchitis and cancer). Air pollution can also drive some indirect health effects, such as birth defects and acute delays in development of children (Ritz *et al.*, 2002). Exposure to contaminated air may also suppress the immune system, causing other ailments (Calderón-Garcidueñas *et al.*, 2008). Both long-term as well as short-term exposure to air pollution is hazardous to health, leading to various diseases and mortality in the most acute cases.

The hazardous effects of air pollution on various bodily systems are discussed below.

11.4.1 Respiratory system

Good air quality is of utmost importance for maintaining healthy functioning of the lungs and respiratory system in human beings, as the inhalation of polluted air is the primary bodily route of exposure to toxic airborne gaseous and particulate pollutants. There have been various studies that describe the harmful impacts of air pollution on the air intake pathways in the human body (e.g. Kelly and Fussell, 2011; Wright and Brunst, 2013). These studies describe how the low to high abundance of air contaminants in the ambient air, and short- to long-term exposures to them have the potential to affect the respiratory system and are connected to the onset of asthma, lung dysfunction, inflammation and exacerbation in the respiratory system, severely affecting human health. Short- to long-term exposure to gaseous air pollutants such as sulfur dioxide (SO_2), nitrogen oxides (NO_x) and certain heavy metals (such as As, Ni, V) may lead to signs and symptoms such as irritation to the nose, throat, trachea and major bronchi, which may be followed by dyspnea and bronchoconstriction (Liu *et al.*, 2013). Chronic exposure to heavy metal pollution has also been linked to the generation of free radicals, resulting in lung tumorigenesis and cancer (Kuo *et al.*, 2006; Nawrot *et al.*, 2006).

Exposure to PM causes pulmonary oxidative stress and inflammation in the respiratory system. When exposed to high concentrations of PM, human epithelial cells in the respiratory system release inflammatory cytokines and chemokines (Quay *et al.*, 1998). The size of the particles is a factor that decides their lethality on inhalation. Large particles (>10 µm) are generally held and filtered by nasal epithelium. However, smaller PM_{10} particles have the capacity to reach the trachea, bronchi and bronchioles in the lungs. Thus, the smaller the particles, the higher is their capacity to reach deeper into the lungs, ultimately reaching the alveoli. Upon exposure to PM, inflammatory cells, especially alveolar macrophages and polymorphonuclear leukocytes, release reactive oxygen compounds and proteolytic enzymes, damaging the lung tissue and cells (Driscoll *et al.*, 1990).

Exposure to and inhalation of powerful oxidants, such as ozone and NO_x present in the ambient air, results in oxidation of cell membrane lipids and proteins and causes inflammation of the lungs (Menzel, 1994; Uysal and Schapira, 2003). Exposure to these oxidative pollutants results in the accumulation of peroxides (instead of aldehydes or carboxylic acids) due to oxidative degradation of fatty acids (lipid peroxidation), damaging the tissue and cells in the respiratory system. Chronic exposure to ozone is linked to a reduced functioning efficiency of the lungs (Tager *et al.*, 2005), with the younger population (children and adolescents) at higher risk of lung and airway damage due to ozone exposure and inhalation. Further, inhalation of NO_x air pollution has the potential to cause viral respiratory infection, resulting in asthma exacerbations (Chauhan *et al.*, 2003).

11.4.2 Cardiovascular system

There is enough epidemiological evidence to suggest that the quality of the air we breathe is linked to the health of the heart and cardiovascular (circulatory) system (see, for example, Brook *et al.*, 2004; Pope *et al.*, 2004; Anderson *et al.*, 2012). The cardiovascular system is primarily affected by the inhalation of CO, PM and heavy metal air pollution.

CO is a tasteless, odourless, non-irritating but highly toxic gas. Because of the combined effect of these factors, CO poisoning is often difficult to detect, yet intoxication by CO remains one of the most common types of poisoning. Thus, true instances of CO poisoning sometimes may go unnoticed, and rightly so, CO can be dubbed a 'silent killer'. As the route of CO exposure is inhalation, it is quite easy to get intoxicated and poisoned by CO in enclosed spaces

with high CO concentration (e.g. garages, factories, homes), eventually leading to death. The acuteness of the CO intoxication depends on various natural and anthropogenic factors: concentration of CO in the ambient air, rate of removal of CO from the ambient air, atmospheric pressure, the medical condition of the person subjected to CO inhalation, duration of exposure and rate of metabolism (Sykes and Walker, 2016). Once inhaled, CO enters the lungs and diffuses to the blood in the pulmonary capillaries across the alveolo-capillary membrane (Prockop and Chichkova, 2007). Once in the bloodstream, CO is capable of rapidly binding to haemoglobin, which leads to the formation of carboxyhaemoglobin (Prockop and Chichkova, 2007). Formation of carboxyhaemoglobin in the bloodstream suppresses the oxygen carrying capacity of the blood, resulting in depletion of oxygen in the vital organs (heart and brain) and causes tissue hypoxia. Haemoglobin has a much higher affinity (by a factor of ~250) to bind to CO than oxygen (Rodkey *et al.*, 1974). Thus, once in the bloodstream, CO is taken up by haemoglobin at a very high rate, resulting in the partial pressure of CO in capillaries to be very low.

Along with affecting the respiratory system, inhalation of PM also results in degradation of the cardiovascular system. Long-term chronic exposure to both $PM_{2.5}$ and PM_{10} particulates has been linked to elevation in C-reactive protein (CRP), a marker for increased inflammation and infection in the body (Hoffmann *et al.*, 2009; Schicker *et al.*, 2009). Chronic exposure to *ultrafine* PM ($PM_{0.1}$) results in changes to blood coagulation and platelet activation causing deposition of foreign particles in arteries (atherosclerosis), eventually leading to coronary artery disease (CAD). Exposure to high concentration of PM_{10} particles has been associated with an increase in fibrinogen glycoprotein levels and other blood coagulating factors, leading to various cardiovascular diseases such as cardiac arrhythmias, myocardial infarction (MI), vascular dysfunction, ischaemic stroke, hypertension, atherosclerosis and cardiac arrest (Du *et al.*, 2016). Chronic exposure and inhalation of PM_{10} particles sourced from the handling of hay, has been linked to increased CRP and platelet aggregation in the bloodstream (Schicker *et al.*, 2009). Finally, exposure to heavy metal pollution (such as Cd, Hg, As) has also been linked to cardiovascular symptoms such as elevated resting heart rate and blood pressure (Huang and Ghio, 2006). Chronic exposure to heavy metal pollutants may also result in anaemic conditions due to suppressed production of red blood cells and platelets.

11.4.3   Central nervous and renal systems

Exposure to increased concentrations of pollutants (especially CO and heavy metals) in the air has been linked to impairment of the central nervous system (e.g. Duruibe *et al.*, 2007; Singh *et al.*, 2011; Sykes and Walker, 2016). Several studies have shown an impairment of the cognitive functions on blood carboxyhaemoglobin levels as low as ~5% (Sykes and Walker, 2016). People with mild to moderate exposure of CO showed neuropsychiatric symptoms (such as anxiety, apathy, mood disorder, hallucination, delusion, behavioural change) even 4 weeks after the exposure (Sykes and Walker, 2016). Chronic CO intoxication has also been shown to lead to the development of neurological symptoms such as memory loss, dizzy spells, tremor, sleep pattern alterations, headaches and emotional instability (Myers *et al.*, 1998). Inhalation of CO and the formation of carboxyhaemoglobin in the bloodstream results in depletion of the oxygen supply to the brain, inhibiting its proper functioning. This results in brain hypoxia and leads to elevated levels of excitatory amino acids (glutamate) and nitrite in the brain, causing lipid peroxidation and damage to the central nervous system (Thom, 1992).

Chronic exposure to air contaminated with toxic heavy metals (such as Cd, Pb, Hg, As) has the potential to be hazardous to the central nervous system. Increased uptake of toxic heavy metals has been linked to neurological symptoms such as peripheral neuropathy, decreased efficiency of brain function with cognitive deficits, and diseases such as Alzheimer's, Parkinson's, Down's syndrome, autism, lupus, and amyotrophic lateral sclerosis (Bressler and Goldstein, 1991; Zahir *et al.*, 2005). Inhalation and ingestion of heavy metals causes the generation of highly reactive oxygen compounds in the body (such as hydrogen peroxide (H_2O_2), and hyperoxide (O_2^-) and hydroxyl (OH) radicals) and causes peroxidation of lipids, resulting in damage to several neural components such as proteins, myelin sheath and nucleic acid (Hsu and Guo, 2002). High Pb exposure causes substitution of Ca in calmodulin (calcium-modulated protein), an important messenger protein in the body responsible for maintaining the short- and long-term memory and cognitive functions, resulting in delayed learning (Bressler *et al.*, 1999). Inhalation of air contaminated with Hg results in cell cytotoxicity and severe oxidative stress in the body and increased secretion of β-amyloid 1–40 and 1–42 peptides, leading to Alzheimer's and Parkinson's diseases (Olivieri *et al.*, 2002).

In addition to potential hazards to the central nervous system, uptake of toxic heavy metals is also deleterious to the renal/urinary system. For example, chronic uptake of Cd (even in small amounts) results in renal dysfunction by tubular proteinuria, linked to increased elimination of low molecular weight proteins and low glomerular filtration rate (GFR), eventually leading to the failure of the kidneys (Johri *et al.*, 2010). Additionally, uptake of Pb results in damage to the urinary tract resulting in bloody urine (Duruibe *et al.*, 2007).

11.5   Impact of Air Pollution on Plant Growth and Crop Production

The world faces a great challenge today in the adequate production of food resources to feed the global population of nearly 7.5 billion people. The stress and challenges in adequate and sustainable production of food resources are further aggravated considering the fact that around 76 million people are added to the global population every year. Working with those numbers, the current estimates suggest that adequate food resources must be in place to support a future global population of over 8 billion by 2025 and 9.5 billion by 2050. However, increases in emissions and concentrations of atmospheric pollutants (ozone and black carbon (BC)) have been shown to have adverse direct as well as indirect (through climatic feedbacks) impacts on agriculture and crop production (Burney and Ramanathan, 2014; Lal *et al.*, 2017). The presence of BC (an absorbing aerosol) in the atmosphere has a direct impact on crop growth due to a reduction in both the quality and quantity of direct and diffused light, potentially resulting in lower crop growth and yield (Burney and Ramanathan, 2014). The increase in the concentration of tropospheric ozone has had an adverse direct impact on crop growth and yield as ozone is toxic to plants (Ainsworth *et al.*, 2012). Ozone enters the plant system from the leaves through their stomatal openings. Once inside the plant cells, ozone generates other reactive oxygen species and results in oxidative stress leading to reduced photosynthesis and a decline in overall plant growth. When subjected to a severe dose of ozone, the plants suffer an initial decrease in photosynthesis followed by cell death and subsequent damage that is visible as chlorotic or necrotic lesions (Sandermann, 1996). When exposed to present ambient air surface ozone levels, northern temperate trees have shown a reduction in their net photosynthesis by ~11% and tree biomass by ~7% (Ainsworth *et al.*, 2012).

A study based on the impact of ozone on the growth of wheat averaged over India showed a reduction in wheat production by as much as ~36% (in 2010) due to the trending increase in pollutants (ozone and BC) in the atmosphere (Burney and Ramanathan, 2014). The study also highlighted that relative wheat production yields were much lower (by around 50%) over densely populated zones, an effect of greater atmospheric loading and abundance of atmospheric pollutants over such regions. In another study to evaluate the crop yield loss over India due to tropospheric ozone, it was estimated that wheat loss amounts to 4.0–14.2 million tons (4.2–15.0%; Lal *et al.*, 2017). The amount of loss in rice production due to surface ozone was estimated to be 0.3–6.7 million tons (0.3–6.3%). Lower crop yield losses for rice (than wheat) can be attributed to the fact that the rice-growing season in India starts after the initiation of the Indian summer monsoon, a season with relatively lower surface ozone levels. In contrast, wheat is cultivated in India during the winter season and the harvest is reaped in March and April, a season with a relatively higher concentration of surface ozone. Incidentally, the largest estimated losses of crop yield are from the areas of highest crop cultivation: ~9 million tons for

wheat in northern India; ~2.6 million tons for rice in eastern India. The impact of increased surface ozone concentrations in the ambient air on other dicotyledonous and C_3 annual plants (such as soybean) has also been assessed (Morgan *et al.*, 2003). Chronic exposure of soybean plants to ozone doses of <60 ppb resulted in a significant decrease in biomass and seed production due to a reduction in photosynthesis. Exposure to higher levels of surface ozone resulted in larger losses in production, associated with a reduction in photosynthesis and leaf area. Ambient (50–62 ppb) and elevated (63–75 ppb) ozone exposure to soybean plants resulted in ~15–25% decrease in grain/seed yield, ~8–15% decrease in grain/seed weight and ~17% decrease in the number of pods per plant (Morgan *et al.*, 2003).

With the rapid increase in the global population and emission of air pollutants, tropospheric ozone levels are also expected to grow in the near future (Wild *et al.*, 2012; Stevenson *et al.*, 2013). In view of this, adequate measures must be taken to safeguard crop production from the adverse impacts of ozone exposure. Some of these measures include: a shift in cultivation to seasons with lower levels of tropospheric ozone; development of rice, wheat and other crops and plants that are more resistant to surface ozone exposure; adequate policies for the reduction in emission of air pollutants; and technological advances in the energy and transportation sectors for a decreased reliance on fossil fuels.

11.6 Concluding Remarks

This chapter presents a discussion on various atmospheric contaminants/pollutants, their sources and chemical characterization, and discusses various health hazards of atmospheric pollution on human beings. The environmental impacts and health hazards of nuclear pollution and the adverse impacts of air pollution on plant growth and crop yield have also been detailed. While exposure to air pollutants is ubiquitous and inevitable in the present world, it is difficult to define a *safe limit* of exposure to air pollution. Thus, educating the population on the harmful impacts of air contamination and methods to curb air pollution emissions seems to be a suitable strategy. The reliance on fossil fuels to meet ever-increasing global energy demands has contributed largely to contamination of the atmosphere with various pollutants, presenting a significant hazard to the health and sustenance of humans, animals and plants on the Earth. In view of this, adequate policies should be made and efficiently applied to alleviate and constrain air-pollution emissions in order to mitigate their direct and indirect effects. Reduction and eventual elimination of the usage of fossil fuels to meet various energy requirements and a drive towards more efficient and cleaner energy resources are the primary steps that are needed to be taken.

References

Adamiec, E., Jarosz-Krzemińska, E. and Wieszała, R. (2016) Heavy metals from non-exhaust vehicle emissions in urban and motorway road dusts. *Environmental Monitoring and Assessment* 188, 369.

Ainsworth, E.A., Yendrek, C.R., Sitch, S., Collins, W.J. and Emberson, L.D. (2012) The effects of tropospheric ozone on net primary productivity and implications for climate change. *Annual Review of Plant Biology* 63, 637–661.

Anderson, J.O., Thundiyil, J.G. and Stolbach, A. (2012) Clearing the air: a review of the effects of particulate matter air pollution on human health. *Journal of Medical Toxicology* 8, 166–175.

Anspaugh, L.R., Catlin, R.J. and Goldman, M. (1988) The global impact of the Chernobyl reactor accident. *Science* 242, 1513–1519.

Beirle, S., Platt, U., Wenig, M. and Wagner, T. (2003) Weekly cycle of NO_2 by GOME measurements: a signature of anthropogenic sources. *Atmospheric Chemistry and Physics* 3, 2225–2232.

Bell, M.L., Davis, D.L. and Fletcher, T. (2004) A retrospective assessment of mortality from the London smog episode of 1952: the role of influenza and pollution. *Environmental Health Perspectives* 112, 6–8.

Bergan, T. and Rodhe, H. (2001) Oxidation of elemental mercury in the atmosphere; constraints imposed by global scale modelling. *Journal of Atmospheric Chemistry* 40, 191–212.

Bergan, T., Gallardo, L. and Rodhe, H. (1999) Mercury in the global troposphere: a three-dimensional model study. *Atmospheric Environment* 33, 1575–1585.

Bolsunovsky, A. and Dementyev, D. (2011) Evidence of the radioactive fallout in the center of Asia (Russia) following the Fukushima nuclear accident. *Journal of Environmental Radioactivity* 102, 1062–1064.

Bousquet, P., Ciais, P., Miller, J.B., Dlugokencky, E.J., Hauglustaine, D.A. *et al.* (2006) Contribution of anthropogenic and natural sources to atmospheric methane variability. *Nature* 443, 439.

Brännvall, M.-L., Bindler, R., Renberg, I., Emteryd, O., Bartnicki, J. and Billström, K. (1999) The medieval metal industry was the cradle of modern large-scale atmospheric lead pollution in Northern Europe. *Environmental Science & Technology* 33, 4391–4395.

Bressler, J.P. and Goldstein, G.W. (1991) Mechanisms of lead neurotoxicity. *Biochemical Pharmacology* 41, 479–484.

Bressler, J., Kim, K., Chakraborti, T. and Goldstein, G. (1999) Molecular mechanisms of lead neurotoxicity. *Neurochemical Research* 24, 595–600.

Brook, R.D., Franklin, B., Cascio, W., Hong, Y., Howard, G. *et al.* (2004) Air pollution and cardiovascular disease. *Circulation* 109, 2655–2671.

Brunekreef, B. and Holgate, S.T. (2002) Air pollution and health. *The Lancet* 360, 1233–1242.

Burney, J. and Ramanathan, V. (2014) Recent climate and air pollution impacts on Indian agriculture. *Proceedings of the National Academy of Sciences USA* 111, 16319–16324.

Calderón-Garcidueñas, L., Solt, A.C., Henríquez-Roldán, C., Torres-Jardón, R., Nuse, B. *et al.* (2008) Long-term air pollution exposure is associated with neuroinflammation, an altered innate immune response, disruption of the blood–brain barrier, ultrafine particulate deposition, and accumulation of amyloid β-42 and α-synuclein in children and young adults. *Toxicologic Pathology* 36, 289–310.

Calkins, W.H. (1994) The chemical forms of sulfur in coal: a review. *Fuel* 73, 475–484.

Chandra, N., Lal, S., Venkataramani, S., Patra, P.K. and Sheel, V. (2016a) Temporal variations of atmospheric CO_2 and CO at Ahmedabad in western India. *Atmospheric Chemistry and Physics* 16, 6153–6173. Available at: https://www.atmos-chem-phys.net/16/6153/2016 (accessed 18 September 2017).

Chandra, N., Venkataramani, S., Lal, S., Sheel, V. and Pozzer, A. (2016b) Effects of convection and long-range transport on the distribution of carbon monoxide in the troposphere over India. *Atmospheric Pollution Research* 7, 775–785. Available at: http://www.sciencedirect.com/science/article/pii/S1309104215300167 (accessed 18 September 2017).

Chauhan, A.J., Inskip, H.M., Linaker, C.H., Smith, S., Schreiber, J. et al. (2003) Personal exposure to nitrogen dioxide (NO_2) and the severity of virus-induced asthma in children. The Lancet 361, 1939–1944.

Chino, M., Nakayama, H., Nagai, H., Terada, H., Katata, G. and Yamazawa, H. (2011) Preliminary estimation of release amounts of 131I and 137Cs accidentally discharged from the Fukushima Daiichi nuclear power plant into the atmosphere. Journal of Nuclear Science and Technology 48, 1129–1134.

Cicerone, R.J. and Oremland, R.S. (1988) Biogeochemical aspects of atmospheric methane. Global Biogeochemical Cycles 2, 299–327. DOI: 10.1029/GB002i004p00299.

Crutzen, P.J. (2006) The 'Anthropocene'. In: Ehlers, E. and Krafft, T. (eds) Earth System Science in the Anthropocene. Springer, Berlin, Heidelberg, pp. 13–18. DOI: 10.1007/3-540-26590-2_3.

Dillner, A.M., Schauer, J.J., Christensen, W.F. and Cass, G.R. (2005) A quantitative method for clustering size distributions of elements. Atmospheric Environment 39, 1525–1537.

Driscoll, K.E., Lindenschmidt, R.C., Maurer, J.K., Higgins, J.M. and Ridder, G. (1990) Pulmonary response to silica or titanium dioxide: inflammatory cells, alveolar macrophage-derived cytokines, and histopathology. American Journal of Respiratory Cell and Molecular Biology 2, 381–390.

Du, Y., Xu, X., Chu, M., Guo, Y. and Wang, J. (2016) Air particulate matter and cardiovascular disease: the epidemiological, biomedical and clinical evidence. Journal of Thoracic Disease 8, E8–E19.

Duruibe, J.O., Ogwuegbu, M.O.C. and Egwurugwu, J.N. (2007) Heavy metal pollution and human biotoxic effects. International Journal of Physical Sciences 2, 112–118.

Edwards, D.P., Emmons, L.K., Hauglustaine, D.A., Chu, D.A., Gille, J.C. et al. (2004) Observations of carbon monoxide and aerosols from the Terra satellite: Northern Hemisphere variability. Journal of Geophysical Research: Atmospheres 109, D24202. DOI: 10.1029/2004JD004727.

Felix, J.D., Elliott, E.M. and Shaw, S.L. (2012) Nitrogen isotopic composition of coal-fired power plant NO_x: influence of emission controls and implications for global emission inventories. Environmental Science & Technology 46, 3528–3535.

Finlayson-Pitts, B.J. and Pitts Jr, J.N.P. (1993) Atmospheric chemistry of tropospheric ozone formation: scientific and regulatory implications. Air Waste 43, 1091–1100. DOI: 10.1080/1073161X.1993.10467187.

Fioletov, V.E., McLinden, C.A., Krotkov, N., Li, C., Joiner, J. et al. (2016) A global catalogue of large SO_2 sources and emissions derived from the Ozone Monitoring Instrument. Atmospheric Chemistry and Physics 16, 11497–11519. Available at: https://www.atmos-chem-phys.net/16/11497/2016 (accessed 18 September 2017).

Fischer, P.H., Brunekreef, B. and Lebret, E. (2004) Air pollution related deaths during the 2003 heat wave in the Netherlands. Atmospheric Environment 38, 1083–1085.

Galloway, J.N., Dentener, F.J., Capone, D.G., Boyer, E.W., Howarth, R.W. et al. (2004) Nitrogen cycles: past, present, and future. Biogeochemistry 70, 153–226. DOI: 10.1007/s10533-004-0370-0.

Gauri, K.L. and Holdren, G.C. (1981) Pollutant effects on stone monuments. Environmental Science & Technology 15, 386–390.

Gavin, D.G., Hallett, D.J., Hu, F.S., Lertzman, K.P., Prichard, S.J. et al. (2007) Forest fire and climate change in western North America: insights from sediment charcoal records. Frontiers in Ecology and the Environment 5, 499–506.

Genc, S., Zadeoglulari, Z., Fuss, S.H. and Genc, K. (2012) The adverse effects of air pollution on the nervous system. Journal of Toxicology 2012. Article ID 782462. DOI: 10.1155/2012/782462.

Glasson, W.A. and Tuesday, C.S. (1963) The atmospheric thermal oxidation of nitric oxide. Journal of the American Chemical Society 85, 2901–2904.

Halmer, M.M., Schmincke, H.-U. and Graf, H.-F. (2002) The annual volcanic gas input into the atmosphere, in particular into the stratosphere: a global data set for the past 100 years. Journal of Volcanology and Geothermal Research 115, 511–528.

Hansen, J., Sato, M., Kharecha, P., Russell, G., Lea, D.W. and Siddall, M. (2007) Climate change and trace gases. Philosophical Transactions of the Royal Society A 365, 1925 LP-1954.

Harrison, R.M. and Yin, J. (2000) Particulate matter in the atmosphere: which particle properties are important for its effects on health? Science of the Total Environment 249, 85–101.

Hoffmann, B., Moebus, S., Dragano, N., Stang, A., Möhlenkamp, S. et al. (2009) Chronic residential exposure to particulate matter air pollution and systemic inflammatory markers. Environmental Health Perspectives 117, 1302–1308.

Hough, A.M. and Derwent, R.G. (1990) Changes in the global concentration of tropospheric ozone due to human activities. Nature 344, 645–648.

Hsu, P.-C. and Guo, Y.L. (2002) Antioxidant nutrients and lead toxicity. Toxicology 180, 33–44.

Huang, Y.-C.T. and Ghio, A.J. (2006) Vascular effects of ambient pollutant particles and metals. *Current Vascular Pharmacology* 4, 199–203.

Jacob, D.J. (2000) Heterogeneous chemistry and tropospheric ozone. *Atmospheric Environment* 34, 2131–2159. Available at: http://www.sciencedirect.com/science/article/pii/S1352231099004628 (accessed 18 September 2017).

Järup, L. (2003) Hazards of heavy metal contamination. *British Medical Bulletin* 68, 167–182.

Johri, N., Jacquillet, G. and Unwin, R. (2010) Heavy metal poisoning: the effects of cadmium on the kidney. *BioMetals* 23, 783–792. DOI: 10.1007/s10534-010-9328-y.

Junge, C.E. (1962) Global ozone budget and exchange between stratosphere and troposphere. *Tellus* 14, 363–377.

Kagawa, J. (1985) Evaluation of biological significance of nitrogen oxides exposure. *Tokai Journal of Experimental and Clinical Medicine* 10, 348–353.

Katsouyanni, K., Touloumi, G., Spix, C., Schwartz, J., Balducci, F. *et al.* (1997) Short term effects of ambient sulphur dioxide and particulate matter on mortality in 12 European cities: results from time series data from the APHEA project. *British Medical Journal* 314, 1658.

Kelly, F.J. and Fussell, J.C. (2011) Air pollution and airway disease. *Clinical & Experimental Allergy* 41, 1059–1071.

Koizumi, A., Harada, K.H., Niisoe, T., Adachi, A., Fujii, Y. *et al.* (2012) Preliminary assessment of ecological exposure of adult residents in Fukushima Prefecture to radioactive cesium through ingestion and inhalation. *Environmental Health and Preventive Medicine* 17, 292–298.

Kuloglu, E. and Tuncel, G. (2005) Size distribution of trace elements and major ions in the Eastern Mediterranean atmosphere. *Water, Air, & Soil Pollution* 167, 221–241.

Kuo, C.-Y., Wong, R.-H., Lin, J.-Y., Lai, J.-C. and Lee, H. (2006) Accumulation of chromium and nickel metals in lung tumors from lung cancer patients in Taiwan. *Journal of Toxicology and Environmental Health, Part A* 69, 1337–1344.

Lal, S., Venkataramani, S., Chandra, N., Cooper, O.R., Brioude, J. and Naja, M. (2014) Transport effects on the vertical distribution of tropospheric ozone over western India. *Journal of Geophysical Research: Atmospheres* 119, 10012–10026.

Lal, S., Venkataramani, S., Naja, M., Kuniyal, J.C., Mandal, T.K. *et al.* (2017) Loss of crop yields in India due to surface ozone: an estimation based on a network of observations. *Environmental Science and Pollution Research* 24, 20972–20981.

Lamborg, C.H., Fitzgerald, W.F., O'Donnell, J. and Torgersen, T. (2002) A non-steady-state compartmental model of global-scale mercury biogeochemistry with interhemispheric atmospheric gradients. *Geochimica et Cosmochimica Acta* 66, 1105–1118.

Lee, C., Martin, R.V., van Donkelaar, A., Lee, H., Dickerson, R.R. *et al.* (2011) SO_2 emissions and lifetimes: estimates from inverse modeling using in situ and global, space-based (SCIAMACHY and OMI) observations. *Journal of Geophysical Research: Atmospheres* 116, D06304. DOI:10.1029/2010JD014758.

Lelieveld, J. and Dentener, F.J. (2000) What controls tropospheric ozone? *Journal of Geophysical Research: Atmospheres* 105, 3531–3551.

Leon, J.D., Jaffe, D.A., Kaspar, J., Knecht, A., Miller, M.L. *et al.* (2011) Arrival time and magnitude of airborne fission products from the Fukushima, Japan, reactor incident as measured in Seattle, WA, USA. *Journal of Environmental Radioactivity* 102, 1032–1038.

Levy, H. (1971) Normal atmosphere: large radical and formaldehyde concentrations predicted. *Science* 173, 141–143.

Lindberg, S.E. and Stratton, W.J. (1998) Atmospheric mercury speciation: concentrations and behavior of reactive gaseous mercury in ambient air. *Environmental Science & Technology* 32, 49–57.

Lindberg, S.E., Meyers, T.P., Taylor, G.E., Turner, R.R. and Schroeder, W.H. (1992) Atmosphere-surface exchange of mercury in a forest: results of modeling and gradient approaches. *Journal of Geophysical Research: Atmospheres* 97, 2519–2528.

Liu, H.-Y., Bartonova, A., Schindler, M., Sharma, M., Behera, S.N. *et al.* (2013) Respiratory disease in relation to outdoor air pollution in Kanpur, India. *Archives of Environmental & Occupational Health* 68, 204–217.

Logan, J.A. (1985) Tropospheric ozone: seasonal behavior, trends, and anthropogenic influence. *Journal of Geophysical Research: Atmospheres* 90, 10463–10482.

MacFarling Meure, C., Etheridge, D., Trudinger, C., Steele, P., Langenfelds, R. *et al.* (2006) Law dome CO_2, CH_4 and N_2O ice core records extended to 2000 years BP. *Geophysical Research Letters* 33, L14810. DOI: 10.1029/2006GL026152.

Menzel, D.B. (1994) The toxicity of air pollution in experimental animals and humans: the role of oxidative stress. *Toxicology Letters* 72, 269–277.

Molnár, A., Mészáros, E., Polyák, K., Borbély-Kiss, I., Koltay, E. *et al.* (1995) Atmospheric budget of different elements in aerosol particles over Hungary. *Atmospheric Environment* 29, 1821–1828.

Morgan, P.B., Ainsworth, E.A. and Long, S.P. (2003) How does elevated ozone impact soybean? A meta-analysis of photosynthesis, growth and yield. *Plant, Cell & Environment* 26, 1317–1328. DOI: 10.1046/j.0016-8025.2003.01056.x.

Myers, R.A.M., DeFazio, A. and Kelly, M.P. (1998) Chronic carbon monoxide exposure: a clinical syndrome detected by neuropsychological tests. *Journal of Clinical Psychology* 54, 555–567.

Nawrot, T., Plusquin, M., Hogervorst, J., Roels, H.A., Celis, H. *et al.* (2006) Environmental exposure to cadmium and risk of cancer: a prospective population-based study. *The Lancet Oncology* 7, 119–126.

Nemery, B., Hoet, P.H.M. and Nemmar, A. (2017) The Meuse Valley fog of 1930: an air pollution disaster. *The Lancet* 357, 704–708.

Nriagu, J.O. (1979) Global inventory of natural and anthropogenic emissions of trace metals to the atmosphere. *Nature* 279, 409–411.

Nriagu, J.O. (1989) A global assessment of natural sources of atmospheric trace metals. *Nature* 338, 47–49.

Nriagu, J.O. and Pacyna, J.M. (1988) Quantitative assessment of worldwide contamination of air, water and soils by trace metals. *Nature* 333, 134–139.

Olivieri, G., Novakovic, M., Savaskan, E., Meier, F., Baysang, G. *et al.* (2002) The effects of β-estradiol on SHSY5Y neuroblastoma cells during heavy metal induced oxidative stress, neurotoxicity and β-amyloid secretion. *Neuroscience* 113, 849–855.

Owen, N.A., Inderwildi, O.R. and King, D.A. (2010) The status of conventional world oil reserves – Hype or cause for concern? *Energy Policy* 38, 4743–4749. Available at: http://www.sciencedirect.com/science/article/pii/S0301421510001072 (accessed 9 September 2017).

Pacyna, J.M. and Pacyna, E.G. (2001) An assessment of global and regional emissions of trace metals to the atmosphere from anthropogenic sources worldwide. *Environmental Reviews* 9, 269–298.

Pirrone, N., Costa, P., Pacyna, J.M. and Ferrara, R. (2001) Mercury emissions to the atmosphere from natural and anthropogenic sources in the Mediterranean region. *Atmospheric Environment* 35, 2997–3006.

Pope, C.A., Burnett, R.T., Thurston, G.D., Thun, M.J., Calle, E.E. *et al.* (2004) Cardiovascular mortality and long-term exposure to particulate air pollution. *Circulation* 109, 71–77.

Povinec, P.P., Hirose, K. and Aoyama, M. (2012) Radiostrontium in the Western North Pacific: characteristics, behavior, and the Fukushima impact. *Environmental Science & Technology* 46, 10356–10363.

Prockop, L.D. and Chichkova, R.I. (2007) Carbon monoxide intoxication: an updated review. *Journal of the Neurological Sciences* 262, 122–130.

Quay, J.L., Reed, W., Samet, J. and Devlin, R.B. (1998) Air pollution particles induce IL-6 gene expression in human airway epithelial cells via NF-kB activation. *American Journal of Respiratory Cell and Molecular Biology* 19, 98–106.

Ramanathan, V., Crutzen, P.J., Kiehl, J.T. and Rosenfeld, D. (2001) Aerosols, climate, and the hydrological cycle. *Science* 294, 2119 LP-2124.

Reis, S., Pinder, R.W., Zhang, M., Lijie, G. and Sutton, M.A. (2009) Reactive nitrogen in atmospheric emission inventories. *Atmospheric Chemistry and Physics* 9, 7657–7677.

Renberg, I., Bindler, R. and Brännvall, M.-L. (2001) Using the historical atmospheric lead-deposition record as a chronological marker in sediment deposits in Europe. *The Holocene* 11, 511–516.

Ritz, B., Yu, F., Fruin, S., Chapa, G., Shaw, G.M. and Harris, J.A. (2002) Ambient air pollution and risk of birth defects in Southern California. *American Journal of Epidemiology* 155, 17–25.

Robinson, E. and Robbins, R.C. (1970) Gaseous sulfur pollutants from urban and natural sources. *Journal of the Air Pollution Control Association* 20, 233–235.

Robock, A. (2000) Volcanic eruptions and climate. *Reviews of Geophysics* 38, 191–219.

Rodkey, F.L., O'Neal, J.D., Collison, H.A. and Uddin, D.E. (1974) Relative affinity of hemoglobin S and hemoglobin A for carbon monoxide and oxygen. *Clinical Chemistry* 20, 83–84.

Russell, A.G., Cass, G.R. and Seinfeld, J.H. (1986) On some aspects of nighttime atmospheric chemistry. *Environmental Science & Technology* 20, 1167–1172.

Sahu, L.K. and Saxena, P. (2015) High time and mass resolved PTR-TOF-MS measurements of VOCs at an urban site of India during winter: Role of anthropogenic, biomass burning, biogenic and photochemical sources. *Atmospheric Research* 164, 84–94. Available at: http://www.sciencedirect.com/science/article/pii/S0169809515001386 (accessed 18 September 2017).

Sahu, L.K., Tripathi, N. and Yadav, R. (2017) Contribution of biogenic and photochemical sources to ambient VOCs during winter to summer transition at a semi-arid urban site in India. *Environmental Pollution* 229, 595–606. Available at: http://www.sciencedirect.com/science/article/pii/S0269749117309946 (accessed 18 September 2017).

Sandermann, H. (1996) Ozone and plant health. *Annual Review of Phytopathology* 34, 347–366. DOI: 10.1146/annurev.phyto.34.1.347.

Schicker, B., Kuhn, M., Fehr, R., Asmis, L.M., Karagiannidis, C. and Reinhart, W.H. (2009) Particulate matter inhalation during hay storing activity induces systemic inflammation and platelet aggregation. *European Journal of Applied Physiology* 105, 771–778.

Schroeder, W.H. and Munthe, J. (1998) Atmospheric mercury – an overview. *Atmospheric Environment* 32, 809–822.

Simon, H., Reff, A., Wells, B., Xing, J. and Frank, N. (2015) Ozone trends across the United States over a period of decreasing NO_x and VOC emissions. *Environmental Science & Technology* 49, 186–195.

Singh, R., Gautam, N., Mishra, A. and Gupta, R. (2011) Heavy metals and living systems: an overview. *Indian Journal of Pharmacology* 43, 246–253.

Sommar, J., Gårdfeldt, K., Strömberg, D. and Feng, X. (2001) A kinetic study of the gas-phase reaction between the hydroxyl radical and atomic mercury. *Atmospheric Environment* 35, 3049–3054.

Sriramachari, S. (2004) The Bhopal gas tragedy: an environmental disaster. *Current Science* 86, 905–920.

Steinhauser, G., Brandl, A. and Johnson, T.E. (2014) Comparison of the Chernobyl and Fukushima nuclear accidents: a review of the environmental impacts. *Science of the Total Environment* 470, 800–817.

Stevenson, D.S., Young, P.J., Naik, V., Lamarque, J.-F., Shindell, D.T. *et al.* (2013) Tropospheric ozone changes, radiative forcing and attribution to emissions in the Atmospheric Chemistry and Climate Model Intercomparison Project (ACCMIP). *Atmospheric Chemistry and Physics* 13, 3063–3085. Available at: https://www.atmos-chem-phys.net/13/3063/2013 (accessed 9 September 2017).

Stockwell, W.R. and Calvert, J.G. (1983) The mechanism of the $HO\text{-}SO_2$ reaction. *Atmospheric Environment* 17, 2231–2235.

Sykes, O.T. and Walker, E. (2016) The neurotoxicology of carbon monoxide – historical perspective and review. *Cortex* 74, 440–448.

Tager, I.B., Balmes, J., Lurmann, F., Ngo, L., Alcorn, S. and Künzli, N. (2005) Chronic exposure to ambient ozone and lung function in young adults. *Epidemiology* 16, 751–759.

Thom, S.R. (1992) Dehydrogenase conversion to oxidase and lipid peroxidation in brain after carbon monoxide poisoning. *Journal of Applied Physiology* 73, 1584–1589.

UN (1999) *The World at Six Billion*. Population Division, Department of Economic and Social Affairs, United Nations Secretariat, New York.

Uysal, N. and Schapira, R.M. (2003) Effects of ozone on lung function and lung diseases. *Current Opinion in Pulmonary Medicine* 9, 144–150.

WHO (2007) *Health Risks of Heavy Metals from Long-Range Transboundary Air Pollution*. WHO Regional Office for Europe, Copenhagen.

Weinstock, B. (1969) Carbon monoxide: residence time in the atmosphere. *Science* 166, 224–225.

Wild, O., Fiore, A.M., Shindell, D.T., Doherty, R.M., Collins, W.J. *et al.* (2012) Modelling future changes in surface ozone: a parameterized approach. *Atmospheric Chemistry and Physics* 12, 2037–2054. Available at: https://www.atmos-chem-phys.net/12/2037/2012 (accessed 9 September 2017).

Wong, C.S.C., Li, X. and Thornton, I. (2006) Urban environmental geochemistry of trace metals. *Environmental Pollution* 142, 1–16. Available at: http://www.sciencedirect.com/science/article/pii/S0269749105004811 (accessed 16 November 2017).

Wright, R.J. and Brunst, K.J. (2013) Programming of respiratory health in childhood: influence of outdoor air pollution. *Current Opinion in Pediatrics* 25, 232–239.

Yin, Y., Chevallier, F., Ciais, P., Broquet, G., Fortems-Cheiney, A. *et al.* (2015) Decadal trends in global CO emissions as seen by MOPITT. *Atmospheric Chemistry and Physics* 15, 13433–13451. Available at: https://www.atmos-chem-phys.net/15/13433/2015 (accessed 18 September 2017).

Yoshida, N. and Takahashi, Y. (2012) Land-surface contamination by radionuclides from the Fukushima Daiichi nuclear power plant accident. *Elements* 8, 201–206.

Zahir, F., Rizwi, S.J., Haq, S.K. and Khan, R.H. (2005) Low dose mercury toxicity and human health. *Environmental Toxicology and Pharmacology* 20, 351–360.

12 Air Pollution Control: Policies and Legislations

Ruchi Singh[1]* and Amit Kumar[2]

[1]University of Delhi, India; [2]Georg-August University of Göttingen, Germany

Abstract

Emerging issues such as climate change, ozone layer depletion and global warming have driven policy makers to put a cap on the emission of pollutants through various legislations. Policies and legislations are the most important instrument for the proper functioning of any sector of society. Through legislative framework and regulatory policies, the fissure between the 'natural environment' and 'human society' can be bridged, which is the need of the hour in order that sustainable development and protection of the environment can progress in parallel. Legislation serves as a valuable tool for educating institutions about their responsibility in maintaining a healthy environment. Legislation has already been put forth at national and international levels but its proper implementation and enforcement has gone unseen. Witnessing the present scenario, more stringent laws need to be framed, especially by global leaders. This chapter highlights the importance of laws and policies and their role in reducing air pollution. It also critically analyses various regulatory measures and policies in the most polluted countries of the world, and what has been achieved through treaties and conventions organized by the United Nations, such as the Kyoto Protocol and the Montreal Protocol.

Introduction

The 20th century has articulated the history of air pollution throughout the world. This era of industrial revolution and a boom in globalization in every sector of the economy has resulted in deteriorating air quality. The complaints and concerns about air pollution are deeply rooted in ancient Athens and Rome, mainly at local and regional levels; scientific studies have revealed that people suffered from anthracosis because of indoor pollution. But historical and disastrous events such as 'the Great Smog of London' in the mid-20th century cried out for more stringent regulatory measures for air pollution (Brimblecombe,

2008). The decade of the 1970s was the most important in formulating environment-related laws and the establishment of government organizations such as USEPA (Andreen, 2012). In 1972, the United Nations (Stockholm declaration) for the first time recognized the 'human environment' as a global agenda over environmental concerns and brought the first-world and third-world countries onto a common platform (Brundtland *et al.*, 1987). The story of invisible threats such as acid rain, ozone layer depletion and global warming as not only transnational, but global, issues has been continuously narrated by scientific experts (Brimblecombe, 2008). Events such as the Bhopal gas

---

* Corresponding author: ruchisingh1907@gmail.com

tragedy in India (1984) and the Chernobyl disaster in the USSR (1986) on the one hand, and the rapidly growing desire for energy in every sector for the developed and developing nations on the other hand, has raised a question mark over the future existence of this planet (Brundtland report, 1987).

Since 1972 (Stockholm declaration), a number of conferences, meetings and treaties have taken place, but what has been concealed is the worsening of old environmental problems and the emergence of new challenges (Galizzi, 2005). The emissions of greenhouse gases (GHGs) have increased rapidly since the 1970s, despite existing laws and regulations (Fig. 12.1). Figure 12.1 shows CO_2 from fossil fuel combustion and industrial processes; CO_2 from Forestry and Other Land Use (FOLU); methane (CH_4); nitrous oxide (N_2O); and fluorinated gases covered under the Kyoto Protocol (F-gases). On the right side of the figure, GHG emissions in 2010 are shown again broken down into these components with the associated uncertainties (90% confidence interval) indicated by the error bars. Global CO_2

emissions from fossil fuel combustion are known within 8% uncertainty (90% confidence interval). CO_2 emissions from FOLU have very large uncertainties attached in the order of ± 50%. Uncertainty for global emissions of CH_4, N_2O and the F-gases has been estimated as 20%, 60% and 20%, respectively. 2010 was the most recent year for which emission statistics on all gases as well as assessment of uncertainties were essentially complete at the time of the data cut-off for this report. Emissions are converted into CO_2-equivalents based on GWP1006 from the IPCC Second Assessment Report (Watson et al., 1996). The emission data from FOLU represents land-based CO_2 emissions from forest fires, peat fires and peat decay that approximate to net CO_2 flux from FOLU. Average annual growth rate over different periods is highlighted with brackets.

Global leaders such as the USA, UK and Japan are becoming global carbon emitters, and fast-growing countries such as China and India are also contributing the same amount because of their population size. The demand for energy in these countries is increasing rapidly and their

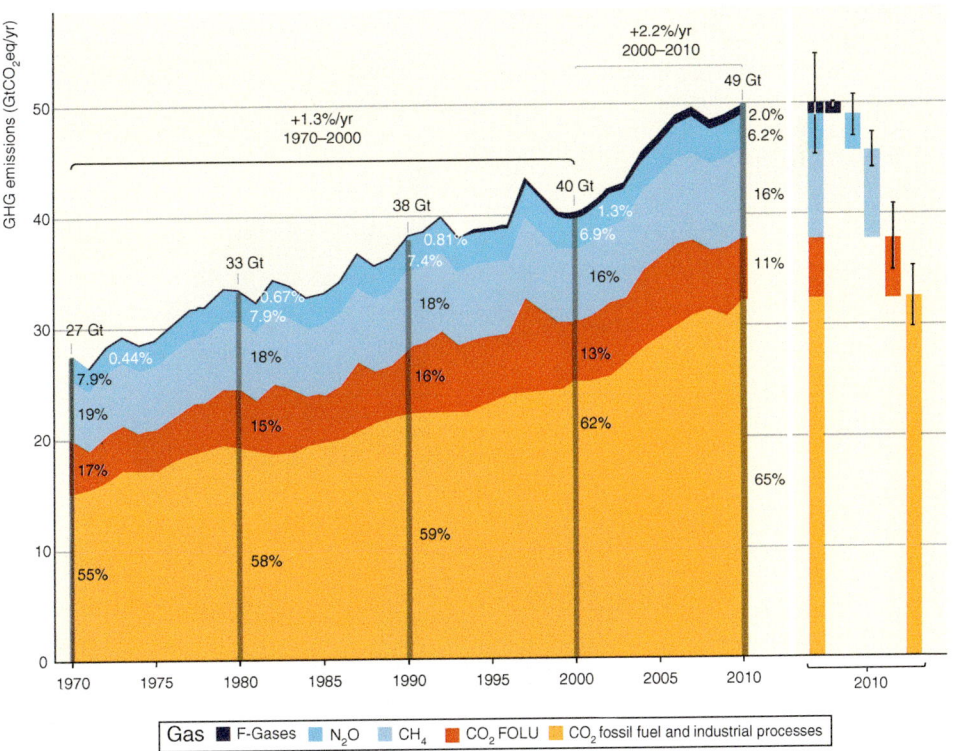

Fig. 12.1. Total annual anthropogenic greenhouse gas (GHG) emissions (GtCO$_2$eq/yr) by groups of gases, 1970–2010. (From Boden et al., 1995; IPCC, 2014.)

primary source of energy consumption is non-renewable resources such as fossil fuels, thus contributing to GHG emissions and therefore degrading air quality. In 2014, the top carbon dioxide (CO_2) emitters were China, the United States, the European Union, India, the Russian Federation and Japan. These data include CO_2 emissions from fossil fuel combustion, as well as cement manufacturing and gas flaring. Together, these sources represent a large proportion of total global CO_2 emissions (IPCC, 2014) (Fig. 12.2).

According to the reports of 'Our Common Future' (Brundtland *et al.*, 1987), by 2025 there will be an increase of 40% in energy consumption per capita compared to that of 1980. Therefore, small and developing countries look to global policy makers as role models in formulating legislation and regulatory measures to control air pollution (Brundtland *et al.*, 1987). This requires common but differentiated responsibilities as stated in the Earth Summit held in Rio de Janeiro, 1992, targeting the 21st century. Climate change is a long-term, global problem, which has significant international and intergenerational implications, and it should be treated as a principal concern (IPCC, 2001). Hence a collective and responsive role has to be played by both developed and developing nations.

According to the IPCC (2001), small steps taken by the international community have contributed to a lowering of greenhouse gases through conventions and treaties such as UNFCCC and the Kyoto Protocol, thus trying to avoid 'dangerous anthropogenic interference with climate change'. Similarly, protocols like Montreal and Kyoto called for equal participation in phasing out ozone-depleting resources as well as greenhouse gases, and coordinated interdependencies of developed and developing nations, for example the sharing of technical and environmental information, affordable financing and access to technology (UNEP, 1992).

Policy principles such as the 'precautionary principle' and 'polluter pays principle' were adopted in the Rio Declaration to formulate regulatory measures for the risks where scientific knowledge is incomplete. The precautionary principle stated in Principle 15 of the Earth Summit 1992 that

> In order to protect the environment, the precautionary approach shall be widely applied by States according to their capabilities. Where there are threats of serious or irreversible damage, lack of full scientific certainty shall not be used as a reason for postponing cost-effective measures to prevent environmental degradation. (UNEP, 1992)

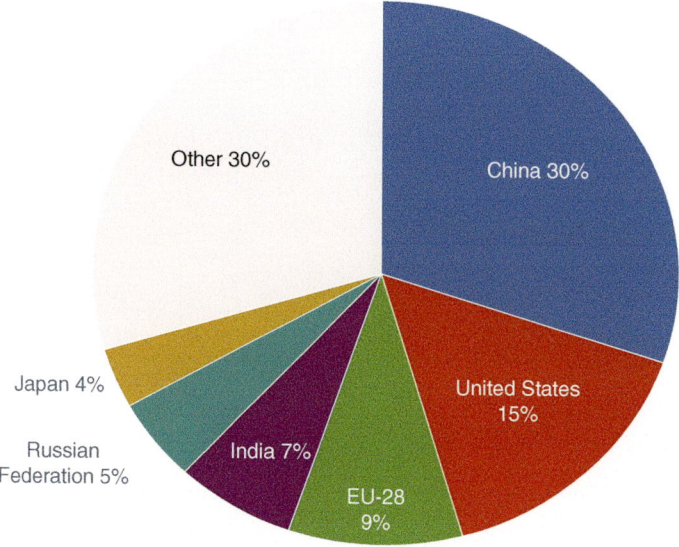

Fig. 12.2. 2014 global CO_2 emissions from fossil fuel combustion and some industrial processes. (From Boden *et al.*, 2017.)

The precautionary principle can be analysed as 'the effect on traditional decision-making', of 'the interaction of irreversibility' and 'uncertainty' (Viriyo, 2012). Europe (Alemanno, 2007) and India (Ghosh, 2016) have approved Principle 15 of the Rio Declaration as a statutory requirement in law related to the environment, showing remarkable progress. The polluter pays principle stated in Principle 16 of the Earth Summit 1992 states

> National authorities should endeavour to promote the internalization of environmental costs and the use of economic instruments, taking into account the approach that the polluter should, in principle, bear the cost of pollution, with due regard to the public interest and without distorting international trade and investment. (UNEP, 1992)

The polluter pays principle is an environmental policy which places the onus of the cost of environmental damage on its producer (Luppi and Rajagopalan, 2009). The Organization for Economic Cooperation and Development (OECD) had recommended the adoption of this policy in the 1970s and made PPP an established legal principle (OECD, 1992).

To date, numerous environmental policies have been framed and adopted by member countries, but other efforts that have been undertaken by global polluters in the direction of protecting the environment and improving its air quality are discussed below.

India

India is one of the fastest-growing economies in the world. Its industrial activities and ever-increasing number of vehicles and population has made India the biggest contributor of air pollution among South Asian countries (World Bank, 2010). Almost 60% of its energy needs are fulfilled by coal, a main source of air pollution in the energy sector. Traditional energy resources such as fuel, wood and animal dung, which is still a common practice for cooking and heating, etc., contribute to air pollution (MoEFCC, 2010). According to the WHO, four out of ten cities in India are among the most polluted cities in the world. The air quality of major Indian cities doesn't meet the global guideline value of $PM_{2.5}$, i.e. 10 µg/m³ (OECD/IEA, 2016). The ambitious

goals of India for economic development will further put its environment at stake (MoF, 2016).

After independence in 1972, India enacted its first environment statute as an immediate effect of the Stockholm declaration on the 'human environment', and since then numerous environment-related laws have been enacted (Ghosh, 2016). A remarkable effort made by India in 1976 was in amending its constitution (the 44th amendment) to provide constitutional provisions to protect the environment under section 48A and 51A (g). Article 48A, the directive principle of state policy, mandates the state 'to protect and improve the environment and safeguard the forests and wildlife of the country'. Article 51A (g) imposes a fundamental duty and a social obligation on every citizen 'to protect and improve the natural environment including forests, lakes, rivers and wildlife and to have a compassion for living creatures' (Gill, 2013).

A proper legislative framework was enacted in 1981 as the Air (Prevention and Control of Pollution) Act 1981, also known as the Air Act to regulate pollutant emission. The law was further amended in 1986 and more comprehensive and stringent laws were enacted after India witnessed the Bhopal gas tragedy in 1984, with the death of hundreds of people in one night (Ghosh, 2016). The Air Act 1981 demarcates various functions of the Central Pollution Control Board (CPCB) under section 15 and the State Pollution Control Boards (SPCB) under section 16 of the act (The Air Act, 1981). The CPCB and SPCB set out the permissible limit of pollutants through their annex monitoring agency NAAQS (National Ambient Air Quality Standards). To regulate emissions in the transport sector, the Motor Vehicles Act, 1939 was amended in 1988; prior to the amendment, emissions were regulated under the Air Act only (CSE, 2017).

At present, each sector has regulatory measures set by the policy makers of India (GoI, 2011; MoEFCC, 2015; UNEP, 2015; MNRE, 2016; Amann et al., 2017). For example, in the transport sector:

1. Ministry of Road Transport and Highways (MoRTH) aimed to jump from BS-IV to BS-VI emission standards by 1 April 2020.
2. Vehicular emissions limit to be revised from Euro III (Euro IV in 11 major cities) to Euro VI from 2020 onwards.

3. Promote and improve public transport, mass transit and non-motorized transport.
4. Government of India is also taking incentives for the manufacturing and promotion of electric vehicles.

In the industrial sector:

1. Policy makers are promoting investment in renewable energy, for example, the National Solar Mission.
2. Encouragement for purchasing energy-efficient and pollution-free technologies.
3. Finance Act 2015 provides inducements for developing facilities to be set up in rural areas to reduce industry emissions.

In the household sector:

1. The National Green tribunal has announced a ban on open burning of rubbish and directed all the concerned authorities for the regulation of the same.
2. Government of India is also promoting off-grid/grid electrification, and cleaner cooking fuels.

Other than this national, state and local level legislation, India has been actively participating and contributing to international conferences and conventions and has been a signatory member in treaties (such as the Montreal Protocol) for abating air pollutants. India became the third country to establish a green court, i.e. the National Green Tribunal in 2010.

In spite of established laws and legislations, regulatory policies, ambitious targets and governmental organizations, such as the MoEFCC (Ministry of Environment, Forest and Climate Change) and MNRE (Ministry of New and Renewable Energy), the air quality of India is degrading at a very fast rate. The present scenario calls for the introduction of reforms and proper enforcement of existing regulations (Ghosh, 2016). A separate department for air quality planning and development of standards could be created by CPCB under the direction of MoEFCC. Delegation of powers to local bodies could have been done by the governmental organizations to enforce plans for improving air quality at its root level (CSE, 2017). A more competent regulatory authority having access to human, technical and financial resources would be far better to formulate regulatory policies

than the judiciary (Ghosh, 2016). Punitive powers should be given to the existing regulatory authorities to levy fines and tax environmental damage, followed by actual restrictive action. A lot has been done and a lot still has to be done, as the air quality of some major Indian cities has reached an alarming stage.

United States

By the end of the 1700s, some cities like Allegheny City, Pittsburgh started to experience air pollution, as the use of bituminous coal turned these cities into a 'smoke city'. During wartime, the consumption of coal was rapidly increasing, hence the deteriorating air quality. In 1881 Chicago came up with an ordinance against coal smoke, followed in 1930 by further cities. In the 1940s, Los Angeles witnessed photochemical smog, or what it is known today as 'ozone pollution'. It was so bad that people experienced immediate health problems such as eye irritation, sneezing and coughing. This scenario led to the enactment of an ordinance in 1944, putting a cap on smoke emissions. During the 1950s and 1960s American industry started to participate voluntarily in smoke emissions controls through 'restrictive legislations'. Afterwards, the Air Pollution Control District (APCD) was created by Los Angeles County in 1947 which demanded technology-based requirements to curb the industrial emitters. In the 1950s there was a ban on the open burning of rubbish, and regulatory measures were taken to reduce sulfur emissions from oil refineries, but there was no focus on the control of automobile emissions, which later led to the formation of the California Motor Vehicle Pollution Control Act in 1960. In 1964 the Act called for modified engines with more exhaust devices, catalytic convertors and flame afterburners.

The congress government finally came up with the Clean Air Act in 1963. The first legislation to regulate air pollution through the Motor Vehicle Air Pollution Control Act was amended in 1965 (Andreen, 2012). Before 1970 it was the responsibility of states, countries and municipalities to maintain environmental regulations. In 1970 the Environment Protection Act (EPA) was enacted and the Clean Air Act was amended with the goal of attainment of NAAQS by all the countries, giving the responsibility to set national air-quality standards, and acceptable levels for pollutants, including particulate matter, which are associated with health risks and degrade the environment (Huang and Brook, 2011). The Clean Air Act was further amended in 1977, but it failed to reduce air pollution, which led to the amendment of the Act in 1990. The new Act included notable provisions like 'enhanced' motor vehicle inspection and maintenance programmes and environmental regulations for coal-fired power plants; and the Acid Rain Programme was included under title IV. In 1992 the EPA revised its Act and made it more stringent with several mandatory inspections (Toshiyuki and Goto, 2010), but in 2001 it entirely abandoned emissions testing (USEPA, 1994, 1995, 1996, 2000, 2001). There were some noticeable reductions in the emission of pollutants such as hydrocarbons, oxides of nitrogen and carbon dioxide after the implementation of the state laws. These were less stringent than the EPA (Holmes and Cicerone, 2002) with the help of remote sensing devices (RSD) and computer-based on-board diagnostics (OBD) (Eisinger and Wathern, 2008).

The transport and power sectors in the US are among the highest-emitting zones of the economy, but there is still no legal regulation of carbon dioxide emissions by the Clean Air Act (CAA), as carbon dioxide didn't come under the NAAQS, and was considered as a non-polluter product. It was in 2007 that the US Supreme Court decided that GHGs should be included under the legal scope of the CAA (Sueyoshi and Goto, 2010). The CAA proved to be very useful and effective in controlling NO_2 and SO_2 emissions under the Acid Rain programme. Therefore, CO_2 emissions should also be legally considered under the CAA and as stated by the US EPA, government should expand the legal scope of the CAA for GHGs (Sueyoshi and Goto, 2010). GHGs from other sectors are still unregulated and existing legislation follows traditional regulatory measures. The economic argument for carbon taxing would create a major effect (Richardson, 2016).

The United States is counted among the largest emitters in the world and has an effect worldwide. The Clean Air Act of the US has played a very important role in its history in abating air pollution, and is still fighting for clean air. Section 115 of the Clean Air Act, entitled 'International

Air Pollution', is very interesting to note. This section invites the United States to play a central role with flexibility in reducing the GHG emissions with other nations. It also allows the states to implement a 'national carbon cap-and-trade or tax system'. This section could prove useful for legal and policy scholars to deal with international pollution. The national carbon tax or the cap-and-trade programme method could prove simpler and more cost-effective than the existing regulatory pathways of each emitter reducing carbon emissions rather than paying the tax (Richardson, 2016).

In 2011, the USEPA enforced the regulations affecting two major sectors of the economy: transport and industry. These regulations were new cooperate average fuel efficiency (CAFE) standards for transportation sectors and 'construction permitting for new and modified sources', for power and industries (Burtraw, 2011).

In the Copenhagen climate change meeting (2009), former president Barack Obama pledged that the United States would reduce emissions 'in the range of' 17% by 2020 from the levels of 2005; to achieve this target the Waxman-Markey legislation that was passed in June 2009 to reduce the emissions by 17% by 2020 was overloaded with an extra 3% to anticipate the international forest projects. The bill that was passed was expected to lower emissions by 33% from different sectors of the economy, for example, out of 33%, 5% was to be cut down from the domestic sector as shown in Fig. 12.3.

These targets can be achieved by focusing on these factors. First, the regulation of fuel efficiency standards for non-stationary sources, and making rules for preconstruction under the Clean Air Act. Second, secular trends in fuel prices and energy efficiency will lead to significant inclination towards the use of renewable energy resources. Third, subnational efforts, for example 'cap and trade' to reduce emissions.

Current trends in the United States are on the right track to achieve the target which was promised in the 2009 Copenhagen conference (Burtraw and Woerman, 2012). For the improvement of air quality, public health, longevity and to provide a quality of life, the EPA, in recent years, has enacted and developed national programmes, which when fully implemented, will have a significant impact on reducing air pollutants. According to reports of the EPA,

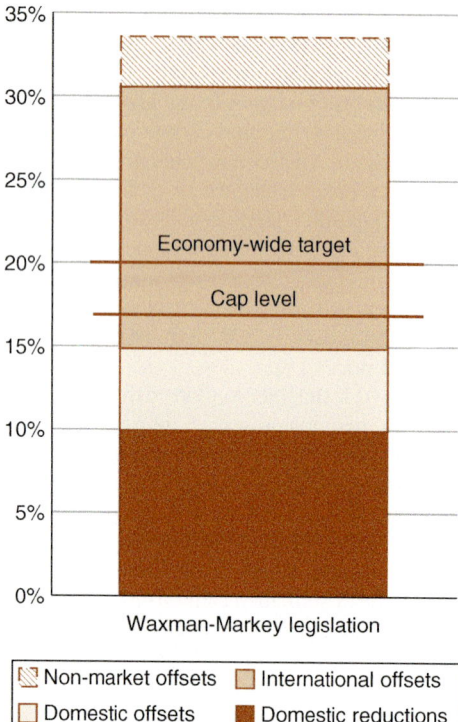

Waxman-Markey legislation

| Non-market offsets | International offsets |
| Domestic offsets | Domestic reductions |

Fig. 12.3. Emissions reductions in the USA between 2005 and 2020. (From EIA, 2009.)

2016 has seen a drop of 63% in the emissions of six major air pollutants (Table 12.1).

China

The air quality of China is worsening day by day. China is witnessing a blanket of smog almost every year, affecting health and the environment (Qu *et al.*, 2009). The energy sector (burning of coal) is the primary source of air pollution in China with industries also playing a vital role in its deteriorating air quality. The history of China's environmental legislation is not lengthy. Important environmental legislations have been framed and strictly implemented in recent years. The atmosphere of China has been severely affected by sulfur dioxide, total suspended particles (TSP, including particulate matters) and ground level ozone (Qu *et al.*, 2009; Lu *et al.*, 2010). According to the Global Burden of Disease study (GBD), air pollution in China has led to

Table 12.1. Change in air pollutant emissions since 1980 (%) (EPA, 2016).

	1980 vs 2014	1990 vs 2014	2000 vs 2014
Carbon monoxide (CO)	−69	−62	−46
Lead (Pb)	−99	−80	−50
Nitrogen oxides (NO$_x$)	−55	−51	−45
Volatile organic compounds (VOC)	−53	−38	−16
Direct PM$_{10}$	−58	−19	−33
Direct PM$_{2.5}$	---	−25	−33
Sulfur dioxide (SO$_2$)	−81	−79	−70

1. --- Trend data not available
2. Direct PM$_{10}$ emissions for 1980 are based on data since 1985
3. Negative numbers indicate reductions in emissions
4. Change in emissions based on thousand tons unit (%)

1.2 million premature deaths in 1 year (Lim *et al.*, 2012) and multiple studies have suggested that this has led to annual economic loss equivalent to between 1% and 7% of China's GDP (Zhang and Huang, 2011). In 1978, the environmental commission was included in the third Chinese constitution, followed by the enactment of environmental protection law in 1979, which was formalized in 1989. Since then China has established many environment-related laws, for example the Law on the Prevention and Control of Atmospheric Pollution. But even after establishing so many laws, China faced actual implementation difficulties (Wang, 2009a).

There was some attention during the formulation of Five-Year Plans (FYPs) but today environmental planning has become an important part of FYPs (Young *et al.*, 2015). Ten traditional regulatory measures were combined during the past few decades, known as 'old three, new five, target response and total emission control' ('old three' refers to 1–3 of the list, consolidated between 1972 and 1979; 'new five' refers to numbers 4–8, consolidated between 1980 and 1989). China has been able to quickly respond in certain situations that called for special actions; although temporary in nature, they are effective for immediate relief, i.e. local, short-term administrative measures at any cost (Zhang, 2004). For example: 'the environmental law Enforcement Inspection', 'Beijing Blue Sky Project' for the 2008 Olympics and 'the Action Plan' for PM$_{2.5}$ crisis (Zhou, 2014). China has adopted a number of 'state ideologies', which worked as soft policies in a broad sense, for example 'protect the environment', 'Green GDP', 'Beautiful China', 'Low Carbon Economy' and 'Resource conserving and environmental civilization'. These ideologies provided a win–win atmosphere for environmental protection and the economy (Young *et al.*, 2015). Between 2006 and 2012, there was a significant implementation of policies on air pollution such as the total emission control on SO_2 and the energy saving policy, which led to a 10% reduction of SO_2 emissions and a 20% reduction of per unit energy consumption from that of the 2005 level. A cap on coal consumption was the core strategy; to achieve this, the coal to gas policy was implemented in Beijing. In 2014 the environmental protection law was amended to include a penalty on illegal discharge of emissions beyond the cap (Yue, 2015). Pollution fees were changed to pollution taxes under the Environment Protection Tax Law (Wang, 2009b).

China has shown better results in 2015 in improved air quality, with the help of sustained regulatory measures towards air pollution prevention and control, and the targets that were promised in the energy sector were also achieved ahead of schedule. Of the six pollutants, the annual mean concentration of $PM_{2.5}$, PM_{10}, SO_2 and NO_2 was generally on the decline in 74 cities when compared with 2014, decreasing by 14.1%, 11.4%, 21.9% and 7.1% on average, respectively. But the concentrations of $PM_{2.5}$ and O_3 are showing an increasing trend in some major cities of China (Fu *et al.*, 2018).

Air pollution has regional impacts, therefore regional collaboration involving all the cities, local government and strategies should be synchronized and motivated by incorporating social, economic and environmental concerns. A shift from supervising enterprises to supervising governments is much needed. A time-bound roadmap and step-by-step actions are required for mid- to long-term improvement of air quality at the national as well as local level. For the pre-evaluation, tracking and post-implementation evaluation of policies, a scientific-based approach is required. It can help the government to select the most effective measures and can enhance the precision of control measures to improve air quality (Fu *et al.*, 2018). China has pledged to reach peak emissions by 2030. These targets can only be achieved by adopting policy measures and strict steps to impose a cap on coal consumption to pursue the target of achieving 20% of green electricity by 2030. Environmental governance is taking a shift towards a law-centred process, which is needed to strengthen the implementation of regulatory policies (Young *et al.*, 2015).

European Union

Environmental policies were beginning to develop in the 1990s in the EU. Momentum was gained by EU environmental policies when the single European Act of 1987 recognized environmental protection as a part of legal competence of the Act (Hildebrand, 1992).

The European Union Emission Trading System (EU ETS) is the world's largest carbon pricing organization (a cap and trade system of CO_2 allowances). It was adopted by the EU in 2003 and work started in 2005, covering 31 countries (Ellerman and Buchner, 2007). Emission sources such as transport, commercial and residential sectors were not covered in the programme. The EU ETS was framed to be enforced in different phases: Phase I: 2005 to 2007 – pilot phase with a penalty of €40 per tonne of CO_2; Phase II: 2008 to 2012 – Kyoto Protocol with increased penalty of €100 per tonne of CO_2 and so on with the gradual inclusion of some GHGs (Schmalensee and Stavins, 2015) (Fig. 12.4).

European leaders have pledged to 'increase energy efficiency in the EU so as to achieve the objective of saving 20% of the EU's energy consumption compared to projections for 2020'. The European Commission has also adopted an action plan in 2006, for achieving energy efficiency through strategies such as labelling, improving fuel efficiency in new cars and power-generating facilities, improving the ability of the taxation system to encourage and reward energy efficiency, increasing public awareness, etc. A directive was also passed in 2006, according to which, each member country was required to achieve an energy savings target of 9% within 9 years as part of the Energy Efficiency Action Plan (Dernbach and Tyrrell, 2012).

'Living well, within the limits of our planet' (EU, 2013) – the seventh environment action programme, which is still going on and runs until

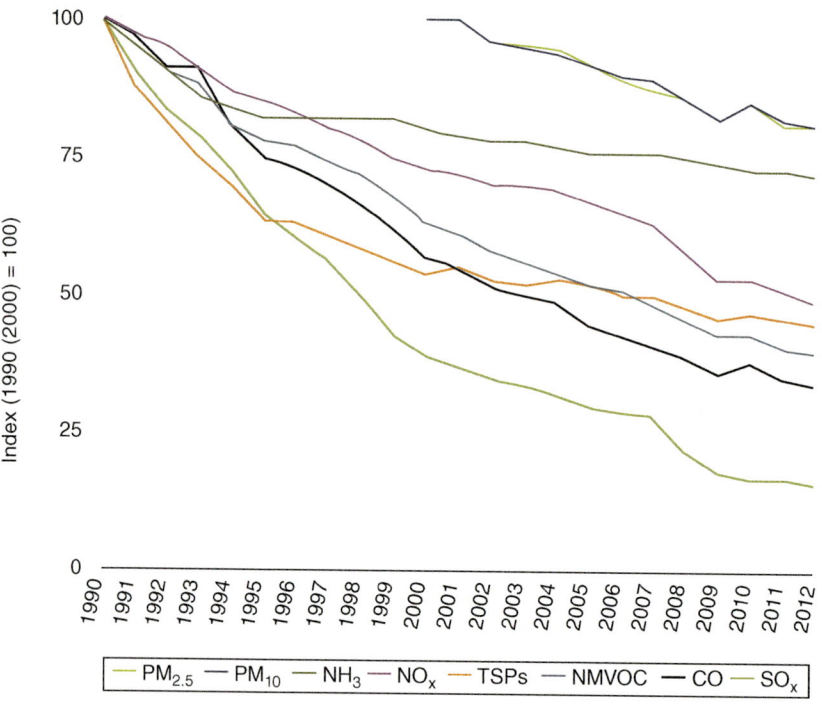

Fig. 12.4. Trends of key pollutants in Europe (1990–2012). (From EEA, 2015.)

2020, distinguishes the long-standing goal within the EU to accomplish 'levels of air quality that do not give rise to significant negative impacts on, and risks to, human health and the environment'. Short-term plans include improving outdoor air quality according to the recommendations of the WHO and reducing air pollution and its effect on the ecosystem and biodiversity. The Ambient Air Quality Directives (EU, 2005) and the National Emission Ceilings (NEC) Directive (EU, 2001), are included in the main policy instruments on air pollution of the EU. International conventions such as the United Nations Economic Commission for Europe (UNECE), Convention on Long-Range Transboundary Air Pollution (CLRTAP) and its various protocols also address the emissions beyond the EU (EEA, 2016). In 2013, the European Commission published 'The Clean Air Policy Package', which aims to certify the full submission with present rules and regulations by 2020 and aims to reduce premature deaths (by more than half of the number in 2005) by reducing air pollution (European Commission, 2013) (Table 12.2).

International Agreements

In 1972, the first United Nations conference on the human environment was held in Stockholm, Sweden. It was the UN's first conference on international environment issues and marked a turning point in the development of the international environment and politics. Also, it led to the foundation of the United Nations Environment Program (UNEP), which sets the global environmental agenda, promotes the comprehensible execution of the environmental aspect of sustainable development within the United Nations system, and serves as an authoritative advocate for the global environment. In 1992 the United Nations conference on Environment and Development (UNCED) was hosted in Rio de Janeiro, also known as the Earth Summit. It resulted in the formation of an international environmental treaty, UNFCCC (United Nations Framework Convention on Climate Change), an important legally binding agreement. The UNFCCC was adopted on 9 May 1992. It entered into force on 21 March 1994. To date, UNFCCC has 165 countries as its signatories and 197

Table 12.2. Air quality policies of major economies. (From CSE, 2016.)

Air quality policy target	Europe	USA	China	India
Deadline for meeting NAAQS	2015	2012 The areas that do not meet the US NAAQS or the area violating the US Clean Air Act requirements are implementing a State Implementation Plan (SIP)	2030 Interim target for key cities 2017	No target time frame set to achieve the NAAQS
Coverage of measures	Clean air for Europe action plan available	National air quality targets/plans approved at federal level and executed at state level	Action plan based with 5-year measurable targets	No measurable targets set
Online monitoring of air quality	1000 stations in 400 cities/towns	770 stations in 540 cities/towns	1500 stations in 900 towns	450 stations in 70 towns
Flue gas desulfurization (FGD) system in thermal power plants (TPP)	75 per cent of TPP have FGD	60 per cent of TPP have FGD	95 per cent of TPP have FGD	10 per cent of TPP have FGD
Consequences for missing targets	Legal action against cities/ country	States must adopt emission reduction measures into law that are demonstrated to enable meeting targets	Promotion of province governors depends on meeting targets	None

have ratified the treaty. The objective of UNFCCC is to 'stabilize greenhouse gas concentrations in the atmosphere at a level that would prevent dangerous anthropogenic interference with the climate system'. Article 3(1) of the convention states that parties should adopt 'common but differentiated responsibilities' for the protection of the climate system (Fig. 12.5).

UNFCCC

Party conferences are held yearly by the UNFCCC to address the issues and assess the progress in dealing with climate change. The first UN climate change conference (COP1) was held in Berlin in 1995 and to date, 23 have been held. Some of the conferences are marked with important achievements or events. For example,

during COP1 (1995) in Berlin, Germany, it was decided by the parties that the aim of Annex I parties to stabilize their emissions at 1990 levels by the year 2000 was 'not adequate'. COP3 (1997) in Kyoto, Japan, led to the formation of the Kyoto Protocol. COP8 (2002) in New Delhi, India, called for efforts by developed countries to transfer technology and minimize the impact of climate change on developing countries. COP18 (2012) in Doha, Qatar, led to the amendment of the Kyoto Protocol that introduced the second commitment period for Annex I parties to the protocol from 1 January 2013 to 31 December 2020. During COP 21 (2015) in Paris, France, parties pledged to the keep the global temperature rise well below 2°C and to strive further for 1.5°C above pre-industrial levels.

'To prevent "dangerous" anthropogenic interference of the climate system' is the ultimate objective of the UNFCCC as stated in Article 2 of the convention; to achieve the concentration of GHGs needed to be stabilized in the atmosphere to a level where ecosystems can adapt naturally to climate change, when food production is not threatened and economic development can progress in a sustainable manner (Fig. 12.6).

Although there is considerable uncertainty over future changes regarding anthropogenic GHG emissions, atmospheric GHG concentrations and related climate change, without mitigation policies and with the continuous increase in demand of energy and wide use of fossil fuels, there is a projected temperature increase of 3.7–4.8°C (in 2100) relative to pre-industrial levels (2.5°C–7.8°C) including all climate uncertainties. To reduce or limit global warming (in 2100) to below 2°C, GHG emissions would need to be capped to around 450 ppm CO_2-eq (Figs. 12.7 and 12.8).

Montreal Protocol

In 1985 and 1987, new research showed that 40% of the ozone layer had been depleted over Antarctica between 1957 and 1984, and a hole had grown to the size of the United States. This report led to the formation of international negotiations between the political leaders of the world, known as the Montreal Protocol, signed on 8 September 1987 at the Vienna convention. Its target was to cut 50% of chlorofluorocarbons (CFCs) by 1998, following a freeze on three major halons, and the Protocol was signed by 197 countries. Differentiated proposals were set for different countries, for example, a 95% reduction for the USA and a complete freeze on CFCs for the European Union. Article 5 of the Protocol allowed developing nations to increase emissions up to specified

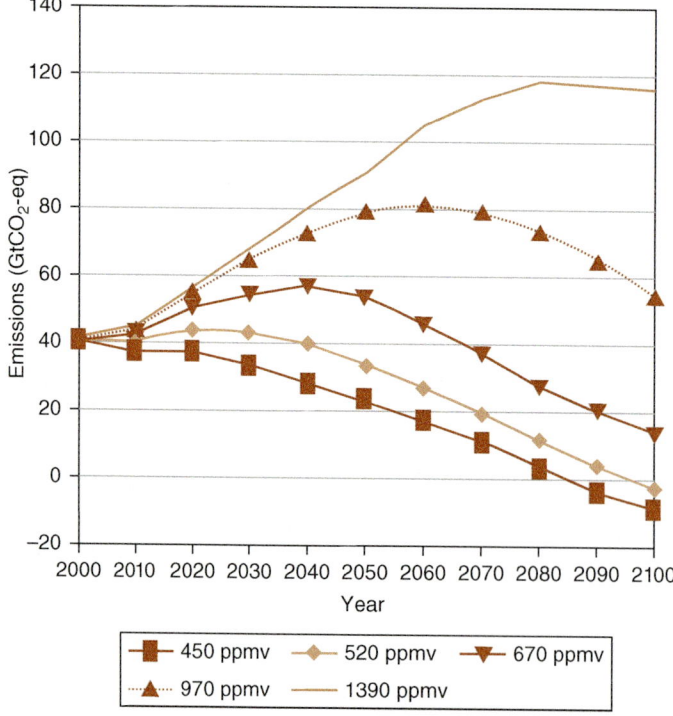

Fig. 12.5. This graph shows projected changes in annual human-caused greenhouse gas (GHG) emissions between the years 2000 and 2100 for a range of climate change mitigation scenarios. Emissions are measured in billion tonnes of carbon dioxide-equivalent (GtCO$_2$-eq). Projections for five scenarios are shown. Each scenario is designed to stabilize atmospheric GHG concentrations at a different level: 450, 520, 670, 970 and 1390 ppmv. In the scenarios, global emissions peak, then decline. Emissions peak sooner in the century for the lower stabilization scenarios. (From IIASA, 2018.)

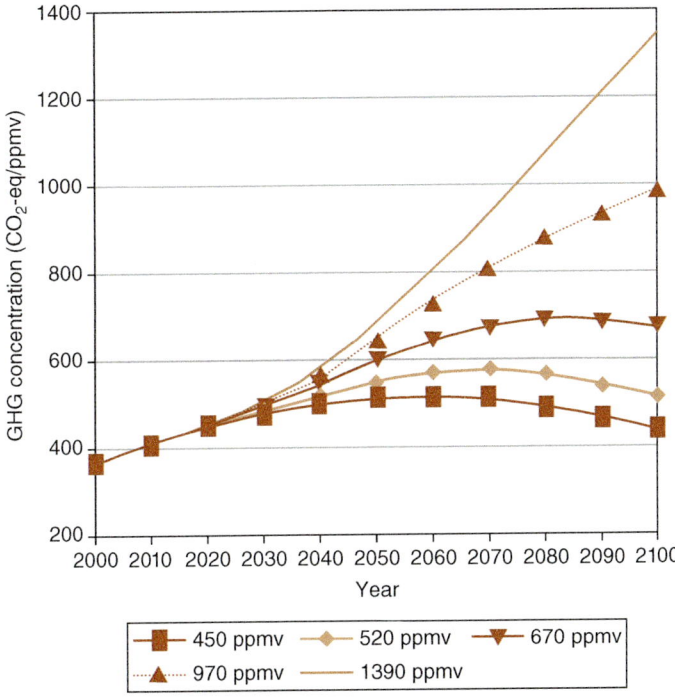

Fig. 12.6. This graph shows the projected changes in atmospheric greenhouse gas (GHG) concentrations between the years 2000 and 2100 for a range of climate change mitigation scenarios. GHG concentrations are measured in carbon dioxide-equivalents (GtCO$_2$-eq) in units of parts-per-million by volume (ppmv). Concentrations include the full basket of GHGs as well as other radiatively active gases, such as aerosol and aerosol precursors. Projections for five scenarios are shown. Each scenario is designed to stabilize atmospheric GHG concentrations at a different level: 450, 520, 670, 970 and 1390 ppmv. (From IIASA, 2018.)

levels to meet 'basic domestic needs' but for a limited period of 10 years, subject to a reduction of 50% for the next 10 years. The Montreal Protocol made a great impact on political leaders by highlighting the risk of skin cancer associated with ozone layer depletion, affecting primarily 'light-skinned people'. In 1990 London agreed to phase out 15 different CFCs worldwide, the European Community announced the phasing out of CFCs by 1997 and a total elimination of 'use and production' of CFCs by 2000. Meanwhile the USA announced a plan to phase out CFCs by 1996. The cost–benefit analysis provided was that 'costs of phasing out of ozone depleting resources would be much lower in comparison to benefits'. The rapid strengthening of the Montreal Protocol gave better results, which led to a cessation of new damage to the ozone layer and, as a result, the ozone hole is shrinking and expected to retain its natural levels by 2050

(Sunstein, 2006). The Kigali, Rwanda amendment of 15 October 2016 is aiming to phase out production and consumption of hydrofluorocarbons (HFCs) by more than 80% over the next 30 years (EPA, 2016). This protocol is marked by very successful international cooperation.

Kyoto Protocol

The Kyoto Protocol was adopted in 1997 in Japan as a result of the UNFCCC of 1992 held in Rio de Janeiro; the famous Earth Summit. The United States and major developed countries agreed to the Kyoto Protocol. According to the Protocol, the Annex I nations would reduce emissions by 5% from 1990 levels and these targets were to be achieved between 2008 and 2012. Different targets were set for developed nations, such as the USA at 7% reduction emissions, the

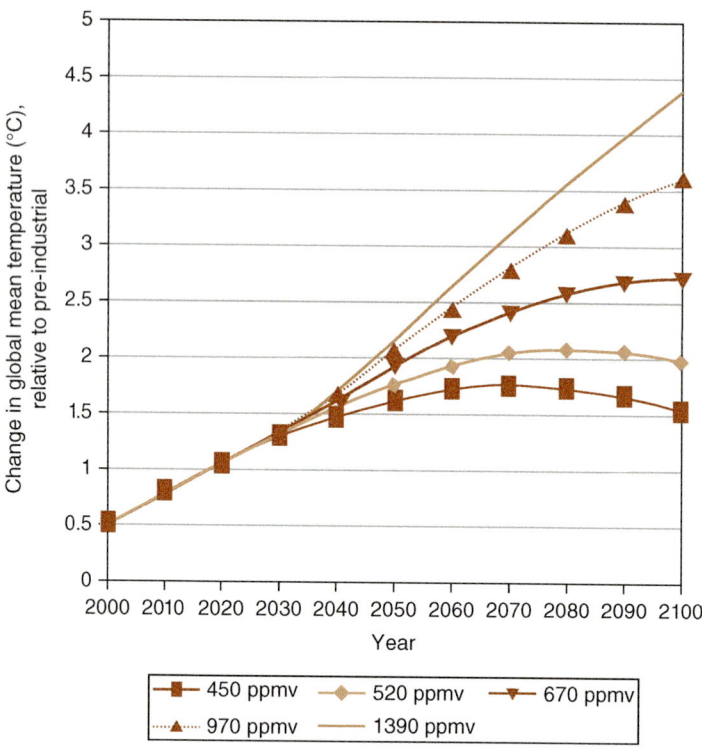

Fig. 12.7. This graph shows projected changes in global mean temperature between the years 2000 and 2100 for a range of climate change mitigation scenarios. Temperatures are measured in degrees Celsius, and are relative to pre-industrial levels. Projections for five scenarios are shown. Each scenario is designed to stabilize atmospheric GHG concentrations at a different level: 450, 520, 670, 970 and 1390 ppmv. (From IIASA, 2018.)

European Union by 8% and Japan at 6%. Although developing nations were not obliged to make any commitments, they were allowed to be involved in 'emissions trading with Annex I nations'. Countries such as China and India were among the largest emitters of GHGs, but they were not controlled by the Kyoto Protocol, whereas developed nations were facing difficulties in achieving the targets given to them. Some countries were already on the path of reducing GHGs by introducing subsidies on natural gas, for example the United Kingdom. On the other hand, the US argued for the inclusion of developing countries and suggested the need for 'meaningful participation' from developing countries, stating that they supported regulatory limits, and suggested that the base year should be set as 1995 instead of 1990 for a less stringent quantitative limit. The US was opposed to mandatory domestic measures such as energy

taxes, but was still less stringent than Japan and the European Union. Although the US signed the Protocol on 12 September 1998, in 2001 President Bush opposed the Protocol stating that 'it excludes 80% of the world, including major economies such as China and India', and that would 'cause serious harm to the US economy' (Sunstein, 2006). The Kyoto Protocol is based on three main mechanisms: the international emissions trading system; the Clean Development Mechanism; and the Joint Implementation.

In 2005, the Kyoto Protocol came into effect with the United States and Australia as non-ratifiers; in addition to this, Canada denounced the treaty in 2012. The current scenario states that there are numerous nations who are very far behind their targets. The Doha, Qatar amendment (2012) introduced a second commitment period for Annex I parties to the Kyoto Protocol from 1 January 2013 to 31 December 2020. Japan,

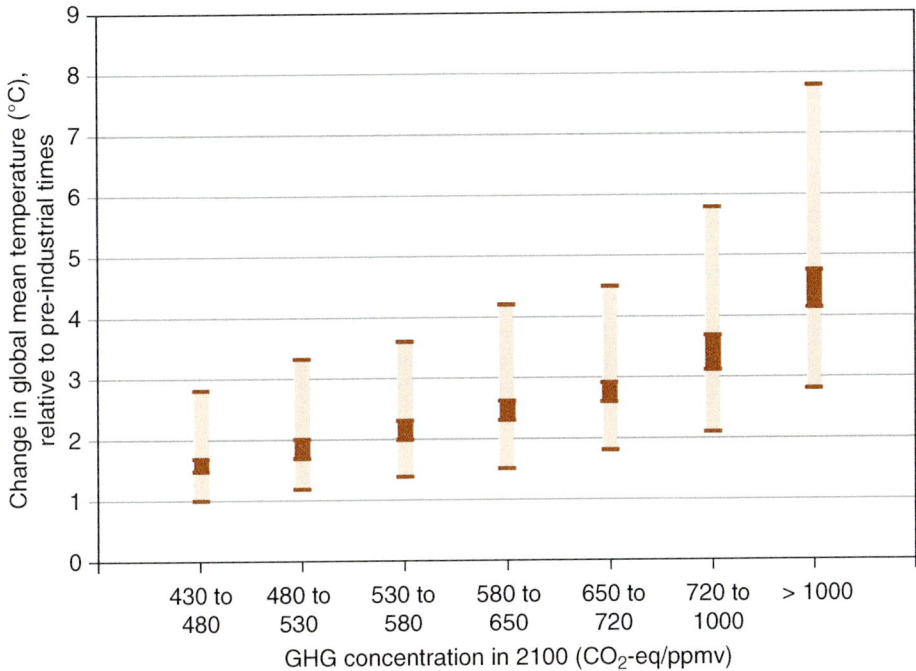

Fig. 12.8. This graph shows projected increases in global mean temperature (relative to pre-industrial times) in the year 2100, for a range of scenarios. Pre-industrial times are approximated as the years 1850 to 1900. Seven sets of temperature projections are given for different atmospheric concentrations of greenhouse gases (measured in parts per million of carbon dioxide equivalents) in the year 2100. On the graph, the dark brown area of the bar shows 10th to 90th percentile range of median estimates, while the light brown area of the bar shows the 16th to 84th percentile range of the full distribution of results. (From Enescot, 2014; IPCC, 2014.)

New Zealand and Russia participated in Kyoto's first round but refused to take on new targets in the second commitment period.

Conclusion

Environmental policies aim to maintain a balance between the environment and sustainable development. A number of environmental regulatory frameworks already exist but what is lacking is their proper implementation and a serious awareness among the general public (Eurostat, 2013). The main problem faced during the implementation of international environmental agreements is that they are based on the scientific evidence of a problem, which is somewhat undefined in manner, and because of the time-consuming and expensive nature of research, participating states face a dilemma in how to act in the face of uncertainty or not to act at all (Kelly, 2014). The climate change negotiations face challenges in offering policy recommendations that can avoid or minimize climate change impacts, while aligning the interests of the international community sufficiently to gain unanimous approval and minimize incentives for non-compliance. Such types of policy makers should focus on small-probability, high-magnitude climate events as well as the secondary consequences of climate change (Cole, 2007). A different but common objective approach is much needed in every sector of the economy. The environment is affected by the production and consumption behaviour of the market, therefore behavioural changes are needed: environmental tax, carbon tax, pigovian tax, etc. Such taxes can discourage behaviour that is potentially harmful for the environment and can provide incentives to decrease the burden on the environment and to preserve it in a very cost-effective manner, which is different from regulatory or administrative approaches. The EU ETS can be set as an example for implementation of environmental tax (Eurostat, 2013).

The last three decades have witnessed major environmental changes, including climate change, global warming, ozone layer depletion, etc., and the recognition of these problems has become the prime agenda. Thirty years of experience has given many lessons on the failure or success of any policy, treaty or agreement (Schmalensee and Stavins, 2015). The ambitious target of the Paris agreement to keep the rise in global temperature to 1.5°C requires cooperative participation of developed as well as developing economies. This would provide a regulatory framework for strengthening international action to mitigate climate change. Environmental sustainability has become a core pillar and a prerequisite for lasting socioeconomic development. Therefore, it is crucial to ensure that the development agenda for the future reflects the links between socioeconomic and environmental sustainability, and protects and reinforces the environmental pillar (UN, 2015).

References

Alemanno, A. (2007) The Shaping of the Precautionary Principle by European Courts: From Scientific Uncertainty to Legal Certainty. Bocconi Legal Studies Research Paper No. 1007404. Available at: https://ssrn.com/abstract=1007404 (accessed 23 August 2018).

Amann, M., Purohit, P., Bhanarkar, A.D., Bertok, I., Borken-Kleefeld, J. et al. (2017) Managing future air quality in megacities: a case study for Delhi. Atmospheric Environment 161, 99–111.

Andreen, W.L. (2012) Of fables and federalism: a re-examination of the historical rationale for federal environmental regulation. Environmental Law 42(3). University of Alabama Public Law Research Paper No. 2142077.

Boden, T.A., Marland, G. and Andres, R.J. (1995) Estimates of global, regional, and national CO_2 emissions from fossil-fuel burning, cement production, and gas flaring: 1950–1992. Carbon Dioxide Information Analysis Center, Oak Ridge National Laboratory, US Department of Energy, Oak Ridge, TN. DOI: 10.2172/207068. Available at: https://www.osti.gov/servlets/purl/207068 (accessed 23 August 2018).

Boden, T.A., Marland, G. and Andres, R.J. (2017) Global, regional, and national fossil-fuel CO_2 emissions. Carbon Dioxide Information Analysis Center, Oak Ridge National Laboratory, US Department of Energy, Oak Ridge, TN. DOI: 10.3334/CDIAC/00001_V2017. Available at: http://cdiac.ess-dive.lbl. gov/trends/emis/overview_2014.html (accessed 23 August 2018).

Brimblecombe, P. (2008) Air pollution history. In: Sohki, R.S. (ed.) World Atlas of Atmospheric Pollution. Anthem Press, London.

Brundtland, G., Khalid, M., Agnelli, Al-Athel, S., Chidzero, B. et al. (1987) Report of the World Commission on Environment and Development: Our Common Future. Oxford University Press, Oxford.

Burtraw, D. (2011) US Climate Change Policy Efforts. Policy Brief. Centre for European Policy Studies, Brussels, Belgium.

Burtraw, D. and Woerman, M. (2012) US Status on Climate Change Mitigation (16 October 2012). Resources for the Future Discussion Paper No. 12-48. Available at: http://www.rff.org/research/publications/us-status-climate-change-mitigation (accessed 23 August 2018).

Cole, D.H. (2007) Climate change and collective action. SSRN. DOI: 10.2139/ssrn.1069906. Available at: https://papers.ssrn.com/sol3/papers.cfm?abstract_id=1069906 (accessed 23 August 2018).

CSE (2016) Annual Report 2015–16. Centre for Science and Environment, New Delhi. Available at: https://cdn.cseindia.org/userfiles/annual-report2015-16.pdf (accessed 7 December 2017).

CSE (2017) 2016–2017 Annual Report: Knowledge Based Activism. Centre for Science and Environment, New Delhi. Available at: https://cdn.cseindia.org/userfiles/cse-annual-report-16-17.pdf (accessed 22 August 2018).

Dernbach, J.C. and Tyrrell, M. (2012) Federal energy efficiency and conservation laws. In: Gerrard, M.B. (ed.) The Law of Clean Energy: Efficiency and Renewables. American Bar Association, Chicago, IL.

EEA (2015) Annual Report 2015. European Environment Agency. Available at: https://www.eea.europa. eu//publications/annual-activity-report-2015 (accessed 23 August 2018).

EEA (2016) *Air Quality in Europe – 2016 Report*. EEA Report No. 28/2016. European Environment Agency, Luxembourg.

EIA (2009) *Energy Market and Economic Impacts of H.R. 2454, the American Clean Energy and Security Act of 2009*. US Energy Information Administration, Washington, DC.

Eisinger, B. and Wathern, P. (2008) Policy evolution and clean air: the case of us motor vehicle inspection and maintenance. *Transportation Research Part D: Transport and Environment* 13(6), 359–368.

Ellerman, A.D. and Buchner, B.K. (2007) The European Union emissions trading scheme: origins allocation and early results. *Review of Environmental Economics and Policy* 1, 66–87. DOI: 10.1093/reep/rem003.

Enescot (2014) Climate change mitigation scenarios (IIASA). Available at: https://commons.wikimedia.org/w/index.php?curid=36469315 (accessed 23 August 2018).

EPA (2016) *Global Greenhouse Gas Emissions Data*. Environmental Protection Agency, Cincinnati, OH.

EU (2001) Directive 2001/81/EC of the European Parliament and of the Council (2001) on national emission ceilings for certain atmospheric pollutants, Brussels. *Official Journal* L 309 , 27/11/2001 P. 0022–0030.

EU (2005) *Proposal for a Directive of the European Parliament and of the Council on Ambient Air Quality and Cleaner Air for Europe*. Commission of the European Communities, Brussels, Belgium. Available at: http://ec.europa.eu/environment/archives/cafe/pdf/cafe_dir_en.pdf (accessed 26 November 2017).

EU (2013) Living well, within the limits of our planet. *Environment for Europeans* 49 (March). Available at: https://ec.europa.eu/environment/efe/sites/efe/files/mag-efe-49-20130301_en.pdf (accessed 20 November 2017).

European Commission (2013) The clean air package. Available at: http://ec.europa.eu/environment/air/index_en.htm (accessed 25 December 2017).

Eurostat (2013) *Environmental Taxes: A Statistical Guide*. Eurostat Manuals and Guidelines. European Union, Luxembourg. Available at: https://ec.europa.eu/eurostat/en/web/products-manuals-and-guidelines/-/KS-GQ-13-005 (accessed 5 January 2018).

Fu, L., Wan, W., Zhang, W. and Cheng, H. (2018) *China Air 2016: Air Pollution Prevention and Control Progress in Chinese Cities*. Clean Air Asia, Beijing. Available at: http://cleanairasia.org/wp-content/uploads/2016/08/China-Air-2016-Report-Full.pdf (accessed 22 August 2018).

Galizzi, P. (2005) From Stockholm to New York via Rio and Johannesburg: has the environment lost its way on the global agenda? *Fordham International Law Journal* 29(5). Available at: https://ir.lawnet.fordham.edu/cgi/viewcontent.cgi?referer=https://www.google.co.in/&httpsredir=1&article=2024&context=ilj (accessed 15 December 2017).

Ghosh, S. (2016) Reforming the liability regime for air pollution in India. *Environmental Law & Practice Review* 4, 125–146.

Gill, G.N. (2013) Access to environmental justice in India with special reference to National Green Tribunal: a step in the right direction. *OIDA International Journal of Sustainable Development* 6(4). Available at: https://papers.ssrn.com/sol3/papers.cfm?abstract_id=2372921 (accessed 23 August 2018).

GoI (2011) *Census of India 2011*. Office of Registrar General and Census Commissioner, Government of India, New Delhi.

Hildebrand, P.M. (1992) The European Community's environmental policy, 1957 to 1992: from incidental measures to an international regime? *Environmental Politics* 1, 13–44.

Holmes, K.J. and Cicerone, R.J. (2002) The Road Ahead for Vehicle Emissions Inspection and Maintenance Programs. *EM: Air and Waste Management Association's Magazine for Environmental Managers*, 15–22 July. Available at: https://escholarship.org/uc/item/3998r0nw (accessed 23 August 2018).

Huang, Y.-C.T. and Brook, R.D. (2011) The clean air act: science, policy, and politics. *Chest* 140(1), 1–2.

IIASA (2018) *Annual Report 2017*. International Institute for Applied Systems Analysis. Available at: http://www.iiasa.ac.at/web/home/resources/publications/annual-report/ar17.pdf (accessed 23 August 2018).

IPCC (2001) Climate change 2001: impacts, adaptation & vulnerability: contribution of Working Group II to the third assessment report of the IPCC. In: McCarthy, J.J., Canziani, O.F., Leary, N.A., Dokken, D.J. and White, K.S. (eds). *Third Assessment Report: Climate Change 2001*. Cambridge University Press, Cambridge.

IPCC (2014) *Fifth Assessment Report: Climate Change 2014*. Cambridge University Press, Cambridge and New York.

Kelly, M. (2014) *Overcoming Obstacles to the Effective Implementation of International Environmental Agreements*. IPCC, Geneva, Switzerland.

Lim, S.S., Vos, T., Flaxman, A.D., Danaei, G., Shibuya, K. *et al.* (2012) *A Comparative Risk Assessment of Burden of Disease and Injury Attributable to 67 Risk Factors and Risk Factor Clusters in 21 Regions, 1990–2010: A Systematic Analysis for the Global Burden of Disease Study 2010.* US National Library of Medicine, Bethesda, MD.

Lu, Z., Streets, D.G., Zhang, Q., Wang, S., Carmichael, G.R. *et al.* (2010) Sulfur dioxide emissions in China and Sulfur trends in East Asia since 2000. *Atmospheric Chemistry and Physics* 10(6311–6331). Available at: http://www.atmos-chem-phys.net/10/6311/2010 (accessed 27 December 2017).

Luppi, B. and Parisi, F. and Rajagopalan, S. (2009) *Environmental Protection for Developing Countries: The Polluter-Does-Not-Pay Principle.* Berkeley Program in Law and Economics, University of California, Berkeley, CA.

MNRE (2016) *Annual Report 2015–16.* Ministry of New and Renewable Energy, Government of India, New Delhi.

MoEFCC (2010) *The Report to the People on Environment and People.* Ministry of Environment, Forest and Climate Change, Government of India, New Delhi.

MoEFCC (2015) *Environment (Protection) Amendment Rules, 2015.* Ministry of Environment, Forest and Climate Change, Government of India, New Delhi.

MoF (2016) *Economic Survey 2015–16.* Ministry of Finance, Government of India, New Delhi.

OECD (1992) *OECD Annual Report 1992.* OECD, Paris.

OECD/IEA (2016) *Energy and Air Pollution: World Energy Outlook Special Report.* International Energy Agency, Paris.

Qu, W.J., Arimoto, R., Zhang, X.Y. and Wang, Y.Q. (2009) Latitudinal gradient and interannual variation of PM_{10} concentration over eighty-six Chinese cities. *Atmospheric Chemistry and Physics Discussions* 9(6). DOI: 10.5194/acpd-9-23141-2009.

Richardson, N. (2016) The elephant in the room or the elephant in the mousehole? The legal risks (and promise) of climate policy under §115 of the Clean Air Act (31 October 2016). Resources for the Future Discussion Paper 16-41-REV. Available at: https://ssrn.com/abstract=2852000 (accessed 23 August 2018).

Schmalensee, R. and Stavins, R. (2015) Lessons learned from three decades of experience with cap and trade. *Review of Environmental Economic and Policy* 11(1), 59–79.

Sueyoshi, T. and Goto, M. (2010) Should the US clean air act include CO_2 emission control?: examination by data envelopment analysis. *Energy Policy* 38(10), 5902–5911.

Sunstein, C.R. (2006) Montreal versus Kyoto: a tale of two protocols. John M. Olin Program in Law and Economics Working Paper No. 302, 2006. University of Chicago Law School, Chicago. Available at: https://chicagounbound.uchicago.edu/cgi/viewcontent.cgi?article=1323&context=law_and_economics (accessed 23 August 2018).

Toshiyuki, S. and Goto, M. (2010) Should the US Clean Air Act include CO_2 emission control?: examination by data envelopment analysis. *Energy Policy* 38(10), 5902-5911. DOI: 10.1016/j.enpol.2010.05.044.

UN (2015) *The Millennium Development Goals Report 2015.* Available at: http://www.un.org/millenniumgoals/2015_MDG_Report/pdf/MDG%202015%20rev%20%28July%201%29.pdf (accessed 23 August 2018).

UNEP (1992) *Rio Declaration on Environment and Development.* UN Environment Programme. A/CONF.151/26 (vol. I); 31 ILM 874. Available at: http://www.un.org/en/development/desa/population/migration/generalassembly/docs/globalcompact/A_CONF.151_26_Vol.I_Declaration.pdf (accessed 22 August 2018).

UNEP (2015) *The Emissions Gap Report 2015.* UN Environment Programme. Available at: http://uneplive.unep.org/media/docs/theme/13/EGR_2015_301115_lores.pdf (accessed 22 August 2018).

USEPA (1994) *Memorandum of Agreement between the California Environmental Protection Agency and the US Environmental Protection Agency.* US Environmental Protection Agency, Washington, DC.

USEPA (1995) *Inspection/Maintenance Flexibility Amendments; Final Rule. Federal Register, vol. 60.* US Environmental Protection Agency, Washington, DC, pp. 48029–48037.

USEPA (1996) *Inspection/Maintenance Flexibility Amendments (Ozone Transport Region); Final Rule. Federal Register, vol. 61.* US Environmental Protection Agency, Washington, DC, pp. 39032–39037.

USEPA (2000) *Additional Flexibility Amendments to Vehicle Inspection Maintenance Program Requirements; Amendment to the Final Rule. Federal Register, vol. 65.* US Environmental Protection Agency, Washington, DC, pp. 45526–45535.

USEPA (2001) *Amendments to Vehicle Inspection Maintenance Program Requirements; Final Rule. Federal Register, vol. 66.* US Environmental Protection Agency, Washington, DC, pp. 18156–18179.

Viriyo, A. (2012) Principle of sustainable development in international environmental law. *SSRN*. DOI: 10.2139/ssrn.2133771.

Wang, J. (2009a) *Emission Trading: Practice and Innovation – Proceedings of the International Symposium on Emissions Trading*. China Environment Press, Beijing, China.

Wang, J. (2009b) Thirty years' rule of environmental law in China: retrospect and reassessment. *Journal of China University Geosciences (Social Sciences Edition)*. Available at: http://en.cnki.com.cn/Article_ en/CJFDTOTAL-DDXS200905001.htm (accessed 23 August 2018).

Watson, R.T., Zinyowera, M.C. and Moss, R.H. (1996) *Climate Change 1995: The IPCC Second Assessment Report: Scientific-Technical Analyses of Impacts, Adaptations, and Mitigation of Climate Change*. Cambridge University Press, Cambridge.

World Bank (2010) *World Development Report 2010: Development and Climate Change*. World Bank, Washington, DC. Available at: https://openknowledge.worldbank.org/handle/10986/4387 (accessed 22 August 2018).

Young, O.R., Guttman, D., Qi, Y., Bachus, K., Belis, D. *et al.* (2015) Institutionalized governance processes: comparing environmental problem solving in China and the United States. *Global Environmental Change* 31, 163-173. DOI: 10.1016/j.gloenvcha.2015.01.010.

Yue, Y. (2015) Summary of environmental monitoring and enforcement for the first half year of 2015. *China Environmental News* 5 August, 82.

Zhang, S. (2004) Analysis of the marginalized reality of the environmental policy in China and discussion for future reforms. *China Population, Resources and Environment* 14, 14–18.

Zhang, S.Q. and Huang, D. (2011) Controlling fine particulate pollution and mitigating environmental health damage. *Environmental Protection* 16, 25–26.

Zhou, X. (2014) From the 'Law of Huang Zongxi' to the logic of the empire: the historical lead of the logic of Chinese State Governance. *Open Times* 4, 8.

Index

CABI – who we are and what we do

This book is published by **CABI**, an international not-for-profit organisation that improves people's lives worldwide by providing information and applying scientific expertise to solve problems in agriculture and the environment.

CABI is also a global publisher producing key scientific publications, including world renowned databases, as well as compendia, books, ebooks and full text electronic resources. We publish content in a wide range of subject areas including: agriculture and crop science / animal and veterinary sciences / ecology and conservation / environmental science / horticulture and plant sciences / human health, food science and nutrition / international development / leisure and tourism.

The profits from CABI's publishing activities enable us to work with farming communities around the world, supporting them as they battle with poor soil, invasive species and pests and diseases, to improve their livelihoods and help provide food for an ever growing population.

CABI is an international intergovernmental organisation, and we gratefully acknowledge the core financial support from our member countries (and lead agencies) including:

Ministry of Agriculture
People's Republic of China

Australian Government
Australian Centre for
International Agricultural Research

Agriculture and
Agri-Food Canada

Ministry of Foreign Affairs of the
Netherlands

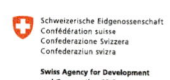

Schweizerische Eidgenossenschaft
Confédération suisse
Confederazione Svizzera
Confederaziun svizra

Swiss Agency for Development
and Cooperation SDC

Discover more

To read more about CABI's work, please visit: **www.cabi.org**

Browse our books at: **www.cabi.org/bookshop**,
or explore our online products at: **www.cabi.org/publishing-products**

Interested in writing for CABI? Find our author guidelines here:
www.cabi.org/publishing-products/information-for-authors/